An Introduction to Astrobiology

Compiled by a team of experts from The Open University in Milton Keynes, UK, this textbook has been designed for introductory university courses in astrobiology. It starts with a detailed examination of how life may have arisen on Earth and looks at fossil evidence of early life. The evidence for possible life on Mars is reviewed in detail and the potential for life on Europa and Titan is also examined. The possibility of life in exoplanetary systems is considered and the book concludes with a discussion of the search for extraterrestrial intelligence. Written in an accessible style that avoids complex mathematics, and illustrated in colour throughout, this book is suitable for self-study and will appeal to amateur enthusiasts as well as undergraduate students. It contains numerous helpful learning features such as boxed summaries, student exercises with full solutions, and a glossary of terms. The book is also supported by a website hosting further teaching materials:
http://publishing.cambridge.org/resources/0521546214

About the editors:

IAIN GILMOUR undertook PhD research in the Earth Sciences department at the University of Cambridge. He then spent several years in the US before returning to the UK to found the organic geochemistry laboratory in the Planetary and Space Sciences Research Institute at The Open University. His research interests are in astrobiology – primarily the origin of extraterrestrial organic matter, the Early Earth, large-scale planetary impacts and the biogeochemistry of stable isotopes. Dr Gilmour is now a Senior Lecturer and has authored several texts on the origin of life, in addition to winning a Europrix Multimedia Art Award for developing multimedia teaching materials.

MARK SEPHTON studied for a PhD in Organic Compounds in Meteorites at The Open University and subsequently spent two years investigating the Permian–Triassic extinction as a Research Fellow at the University of Utrecht. He returned to The Open University in 2000 to become a Lecturer in the Planetary and Space Sciences Research Institute and Department of Earth Sciences. Dr Sephton's research interests include extraterrestrial organic matter in meteorites and the causes and consequences of mass extinctions in the geological record. He teaches courses on the carbon-based record of life and environmental change.

Background image: a high-resolution image generated by the Mars Orbiter Laser Altimeter (MOLA), an instrument on board NASA's Mars Global Surveyor spacecraft, showing the topography of Mars. Low-altitude areas are shown in blue, highlands in yellow to red. The deep-blue region in the south is the giant Hellas impact basin, which is nearly 9 km deep and 2100 km across. The pale-blue area in the centre of the image is the Isidis Planitia basin. (NASA)

Thumbnail images: (from left to right) a false-colour image of the North Pole showing oceans with abundant plankton and land areas with significant vegetation in green; part of Europa; DNA; an image taken by the High Resolution Stereo Camera onboard ESA's Mars Express orbiter. The area is 100 km across and shows a channel (Reull Vallis) once formed by flowing water. ((from the left) first and second images NASA; third image © 1999 Photo Disc Inc; fourth image ESA)

An Introduction to Astrobiology

Edited by Iain Gilmour and Mark A. Sephton

Authors:

Andrew Conway

Iain Gilmour

Barrie W. Jones

David A. Rothery

Mark A. Sephton

John C. Zarnecki

PUBLISHED BY THE PRESS SYNDICATE OF THE UNIVERSITY OF CAMBRIDGE

The Pitt Building, Trumpington Street, Cambridge, United Kingdom

CAMBRIDGE UNIVERSITY PRESS

The Edinburgh Building, Cambridge, CB2 2RU, UK

40 West 20th Street, New York, NY 10011–4211, USA

477 Williamstown Road, Port Melbourne, VIC 3207, Australia

Ruiz de Alarcón 13, 28014 Madrid, Spain

Dock House, The Waterfront, Cape Town 8001, South Africa

http://www.cambridge.org

First published 2003

This co-published edition first published 2004

Edited, designed and typeset by The Open University.

Printed and bound in the United Kingdom by Bath Press, Blantyre Industrial Estate, Glasgow G72 0ND, UK

A catalogue record for this book is available from the British Library

ISBN 0 521 83736 7 hardback

ISBN 0 521 54621 4 paperback

This publication forms part of an Open University course S283 *Planetary Science and the Search for Life*. Details of this and other Open University courses can be obtained from the Course Information and Advice Centre, PO Box 724, The Open University, Milton Keynes MK7 6ZS, United Kingdom: tel. +44 (0)1908 653231, e-mail general-enquiries@open.ac.uk

Alternatively, you may visit the Open University website at http://www.open.ac.uk where you can learn more about the wide range of courses and packs offered at all levels by The Open University.

To purchase a selection of Open University course materials visit the webshop at www.ouw.co.uk, or contact Open University Worldwide, Michael Young Building, Walton Hall, Milton Keynes MK7 6AA, United Kingdom for a brochure. tel. +44 (0)1908 858785; fax +44 (0)1908 858787; e-mail ouwenq@open.ac.uk

1.1

CONTENTS

CHAPTER 1
ORIGIN OF LIFE

Shortly after the formation of the Earth some 4.6 Ga ago, our planet was a lifeless and inhospitable place. Yet if we examine rocks that were created about a billion years later, we can find evidence that by 3.5 Ga ago life had established a firm foothold on Earth. It is what happened in the intervening period that is the focus of this chapter. In other words, we will try to understand how life began.

During the course of this chapter you will examine how scientists have striven to define life and how unexpectedly difficult it is to perform this seemingly simple task. You will also examine the chemistry and function of entities that make up a living system. Then you will study the sites in the Universe and on the Earth where life's raw materials could have been formed before life had even begun. Finally, you will cover the mechanisms by which non-biological raw materials may have been combined into the first living organism.

1.1 What is life?

We will begin our attempts to define life in good company, as some of the world's most famous scientists have contributed to our current level of knowledge. Throughout history, questions have been raised about how and when life arose. Initially, many considered that life arose spontaneously and repeatedly on the Earth. These convictions were supported by what was thought to be the spontaneous generation of flies and maggots from rotting meat, lice from sweat, eels and fish from sea mud, and frogs and mice from moist soil. Occasionally, the idea of spontaneous generation was queried. For example in 1668, a Tuscan doctor called Francesco Redi (1627–1697) demonstrated that maggots were the larvae of flies and if the meat was kept in a sealed container, so that adult flies were excluded, no maggots appeared. However, when Dutch microscope maker Anthony van Leeuwenhoek (1632–1723) detected micro-organisms in 1676, spontaneous generation was the seemingly obvious explanation for such ubiquitous creatures. The matter was finally laid to rest in 1862 when, in an attempt to win a prize offered by the French Academy of Science, Louis Pasteur (1822–1895) (Figure 1.1) performed a convincing series of experiments. Pasteur showed that if a broth or solution was properly sterilized and excluded from contact with micro-organisms, it would remain sterile indefinitely.

Figure 1.1 Louis Pasteur (1822–1895), who disproved the idea that life could be generated spontaneously. (Robert Thom)

Pasteur had answered an important question by disproving spontaneous generation as the origin of life, but inevitably he had raised a new and more difficult question. If all life comes from existing life, where did the first life come from? Ironically, in demolishing the long-held idea that life arose spontaneously from inanimate matter, it was an inescapable and logical conclusion that the very first life may have done exactly that – arisen from non-living materials present in the Universe.

1.1.1 A definition of life

If we are to establish *when* and *how* life originated, we must first define exactly what life is.

Most biologists would identify two key features that indicate life:

- the capacity for self-replication, and
- the capacity to undergo Darwinian evolution.

Let us explore these criteria in slightly more detail. For an organism to self-replicate it must be able to produce copies of itself. For Darwinian evolution to occur, imperfections or mutations must occasionally arise during the copying process and these new mutations must be subjected to natural selection (Box 1.1). Nature favours particular characteristics under particular environmental conditions and those individuals best suited to the existing conditions are most likely to survive. For this process to bring about an evolutionary change any advantageous features brought about by mutation must be passed on to future generations.

BOX 1.1 NATURAL SELECTION AND DARWINIAN EVOLUTION

Figure 1.2 Charles Darwin (1809–1882), who established the theory of natural selection. (© The Natural History Museum, London)

After graduating from Cambridge with a degree in theology at the age of 22, Charles Darwin (1809–1882) set sail as a naturalist on the British Navy's HMS *Beagle* mapping expedition (1831–1836). His voyage brought him to the Galapagos Islands in the Eastern Pacific Ocean. Within the Galapagos archipelago he found a wide variety of plants and animals. It was there that he began to formulate ideas about the process of evolution. Darwin recognized that any population consists of individuals that are all slightly different from one another. Those individuals having a variation that gives them an advantage in staying alive long enough to successfully reproduce are the ones that pass on their traits more frequently to the next generation. Subsequently, their traits become more common and the population evolves. Darwin called this 'descent with modification'.

The Galapagos finches provide an excellent example of this process. For instance, among the birds that ended up in arid environments, the ones with beaks that were better suited for eating cactus seeds got more food than those birds with beaks that were less suitable. As a result of the additional food, they were in better condition to mate. In a very real sense, nature selected the best-adapted varieties to survive and to reproduce. Darwin called this process 'natural selection'.

Darwin did not believe that the environment was producing the variation within the finch populations. He correctly thought that the variation already existed and that nature just selected the best-adapted varieties – in our example the birds with the most suitable beak shape for eating cactus seeds as against less-favourable shapes. Darwin described this process as the 'survival of the fittest'. It was not until 1859 when Darwin was 50 years old that he finally published his theory of evolution in a book entitled *The Origin of Species by Means of Natural Selection*. Today, the concept of natural selection and its influence on successive generations is called **Darwinian evolution**.

From our short list of two characteristics a working definition of *life* can be created. Gerald Joyce of the National Aeronautics and Space Administration (NASA) proposed the following definition: 'a self-sustaining chemical system capable of undergoing Darwinian evolution'. However, any definition of life is likely to fail in certain circumstances. For example, the mule is the offspring of a donkey and a horse. A mule cannot breed and therefore is incapable of taking part in the processes of self-replication and Darwinian evolution, yet few would deny that it is alive. But, for the majority of cases, our definition of life will be a satisfactory one.

For life to be self-sustaining and capable of Darwinian evolution both energy and materials must be extracted from the surrounding environment to allow growth and replication. Furthermore, some sort of living apparatus must be present to govern and facilitate the chemistry of life. In the following sections we will examine just what kinds of chemical system characterize life and what types of energy might have been available to primitive life on the early Earth.

1.1.2 Why carbon?

There is only one element that can form **molecules** of sufficient size to perform some of the functions necessary for life as we know it. This element is carbon.

Carbon can form chemical bonds with many other atoms, allowing a great deal of chemical versatility. Commonly, organic compounds also contain the elements hydrogen, oxygen, nitrogen, sulfur and phosphorus. A range of metals such as iron, magnesium and zinc also bond with carbon.

Carbon can form compounds that readily dissolve in liquid water and, as you will see shortly, water is essential for life on Earth. Elements fundamental to the development of living organisms must be able to interact readily with one another, and that occurs most readily in the presence of water.

You will often encounter the term 'organic' in relation to how life may have originated. 'Organic' usually signifies the influence of biology, but to the chemist the term simply covers the chemistry of compounds based on carbon.

All currently known life utilizes carbon-based organic compounds.

The relative abundance of the more common elements in the Universe indicates that the Universe is well stocked with the elements needed to construct these organic compounds (Table 1.1). The four most common elements that are utilized by life on Earth (hydrogen, oxygen, carbon and nitrogen) are the most abundant non-noble gas elements in the Universe. Sulfur and phosphorus (not listed, but the 15th most abundant element in the Universe) are also important for life on Earth.

QUESTION 1.1

Examine Table 1.1 and, extrapolating from our discussion of carbon above, explain why the amounts of noble gas elements in living organisms are so small.

Table 1.1 The ten most abundant elements in the Universe, Earth and life (expressed as atoms of the element per 100 000 total atoms).

Order	Universe		Whole Earth		Earth's crust		Earth's ocean		Humans	
1	H	92 714	O	48 880	O	60 425	H	66 200	H	60 563
2	He	7 185	Fe	18 870	Si	20 475	O	33 100	O	25 670
3	O	50	Si	14 000	Al	6 251	Cl	340	C	10 680
4	Ne	20	Mg	12 500	H	2 882	Na	290	N	2 440
5	N	15	S	11 400	Na	2 155	Mg	34	Ca	230
6	C	8	Ni	1 400	Ca	1 878	S	17	P	130
7	Si	2.3	Al	1 300	Fe	1 858	Ca	6	S	130
8	Mg	2.1	Na	640	Mg	1 784	K	6	Na	75
9	Fe	1.4	Ca	460	K	1 374	C	1.4	K	37
10	S	0.9	P	140	Ti	191	Si	–	Cl	33

Noble gases are highly unreactive and, until recently, were known as the inert gases. They include helium (He), neon (Ne) and argon, (Ar). These gases do not usually bind chemically with any other elements to form compounds.

1.1.3 Why water?

Liquid water also appears to be an essential requirement for life. Living systems need a medium in which molecules can dissolve and chemical reactions can take place. Water has been called the universal solvent because it performs this function so well. Few other solvents can match the abilities of water to facilitate life. Water exists as a liquid in a temperature range that is not too cold to sustain biochemical reactions and not too hot to stop many organic bonds from forming. An occasionally proposed alternative, ammonia, would be liquid on other worlds much colder than ours, but at such low temperatures chemical reactions that could lead to life would operate sluggishly and living systems may struggle to become established.

1.2 The building blocks of life

We have discovered that life on Earth relies mainly on four elements, hydrogen, oxygen, carbon and nitrogen, with smaller amounts of two other elements, sulfur and phosphorus. Yet these six elements are found in a wide variety of organic combinations. Each combination has its own role in maintaining and perpetuating living systems. To fully appreciate how living systems operate we must begin to think of our elements in terms of the molecules in which they are contained.

1.2.1 Water

Before discussing the major classes of organic molecules found in living things we will pause and consider the role of water. The water molecule is the major component of living tissues, generally accounting for 70% of their mass.

- What clues are there in Table 1.1 to suggest there are relatively large quantities of water present in living systems?
- Hydrogen and oxygen, the elements that combine to form water, are the two most abundant elements in the human body.

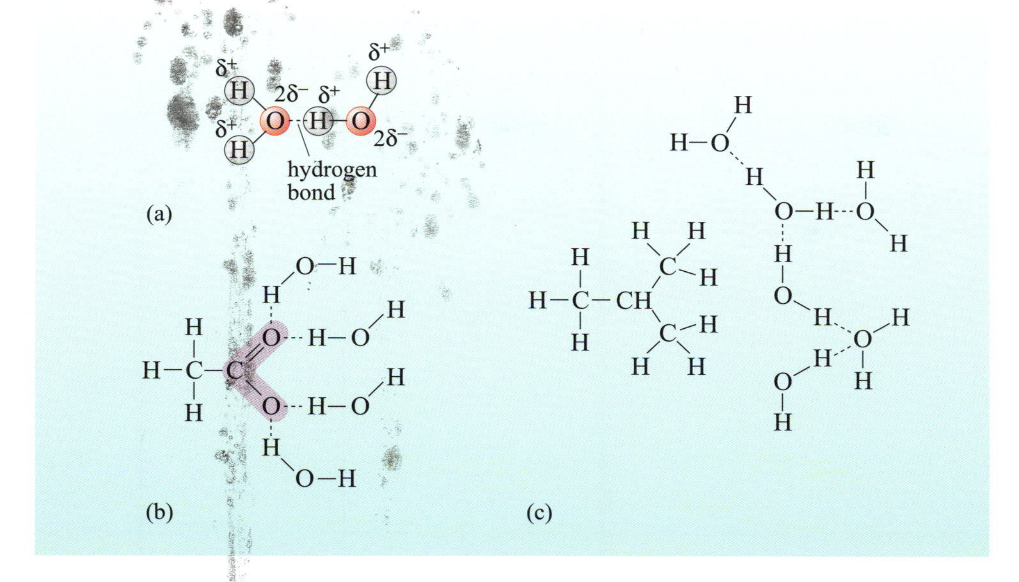

Figure 1.3 Water molecules and their interactions. (a) Water molecules carry a partial positive charge (δ^+) on their hydrogen atoms and a partial negative charge (δ^-) on their oxygen atoms that can interact to form a hydrogen bond. (b) Water molecules interact with polar organic molecules. (c) Apolar organic molecules do not interact with water molecules. (Zubay, 2000)

Because living systems contain so much water it is inevitable that the majority of other components will exist in an aqueous environment. Water is a **polar** solvent, in other words each side of the molecule carries a different electrical charge. In detail, the hydrogen atoms in water are positively charged while the oxygen atom is negatively charged. Because opposites attract, two water molecules can interact to form a hydrogen bond between the hydrogen of one water molecule and the oxygen of another (Figure 1.3a).

Water molecules not only interact with themselves but can also exert a significant influence on organic molecules. When considering organic solutions it is useful to remember the saying 'like dissolves like'. What this means is that polar solvents such as water will dissolve polar organic molecules (Figure 1.3b). Conversely, **apolar** (non-polar) organic molecules do not readily dissolve in water (Figure 1.3c). These characteristics allow us to define two classes of organic molecule:

- Those that are polar, have a high affinity for water, and are therefore soluble are termed hydrophilic (water lovers) molecules.
- Those that are apolar, have a low affinity for water, and are therefore relatively insoluble are called hydrophobic (water haters) molecules.

We shall see later that living systems exploit the hydrophobic and hydrophilic nature of different molecules to perform specific functions.

Table 1.2 lists the major constituents found in a bacterium. Notice the dominant constituent is water, but other chemicals are present.

■ Are the majority of organic molecules in a living system present as small or large structures?

❑ Most of the organic molecules are very large.

Table 1.2 Types and abundances of the molecules that make up a bacterium.

	Percent of total weight	Number of types of molecule
water	70	1
inorganic ions, e.g. Na^+, K^+ and Ca^{2+}	1	20
small organic molecules (< 1000 atomic units), e.g. fatty acids, sugars, amino acids, nucleotides	7	750
large organic molecules (> 100 000 atomic units), e.g. collections of lipids, carbohydrates, proteins, nucleic acids	22	5000

An atomic unit, or more correctly atomic mass unit, is defined as one-twelfth the mass of an atom of carbon and approximates to the mass of a single proton or neutron.

So, except for water, most of the molecules in a living system are large organic molecules or 'macromolecules'. These macromolecules can be subdivided into four different types: **lipids**, **carbohydrates**, **proteins** and **nucleic acids**.

These macromolecules are usually the products of combining many individual organic molecules called **monomers** (from the Greek for single-parts) to create **polymers** (from the Greek for many-parts). Each of these types of macromolecule has a specific function in living systems. We will now examine these different types of macromolecule and the roles that they perform.

1.2.2 Lipids (fats and oils)

Lipids are a diverse group of molecules that have one hydrophobic and one hydrophilic end (Figure 1.4). Overall they are poorly soluble in water, a feature which ensures that they are rarely found as individual molecules. Lipids arrange themselves into weakly bonded aggregates that can be considered macromolecules. Lipids are a convenient and compact way to store chemical energy and the weak bonding within their macromolecular structure results in a high degree of flexibility that is useful in membranes.

1.2.3 Carbohydrates

Carbohydrates are molecules that have many hydroxyl groups (–OH) attached, as shown in Figure 1.5. These hydroxyl groups are polar so carbohydrates are soluble in water. Sugars are common carbohydrates that form ring-like structures when dissolved in water (Figure 1.5). Sugars with five carbon atoms are called pentoses while those with six carbon atoms are called hexoses. Large carbohydrate structures are called polysaccharides and consist of sugar monomers connected together. This process whereby monomers are linked together to give a polymer is called polymerization. Polymerization occurs by reactions that involve the loss of water and result in a linear or branched network as in polysaccharides. Polysaccharides are useful energy stores and can also provide structural support for organisms.

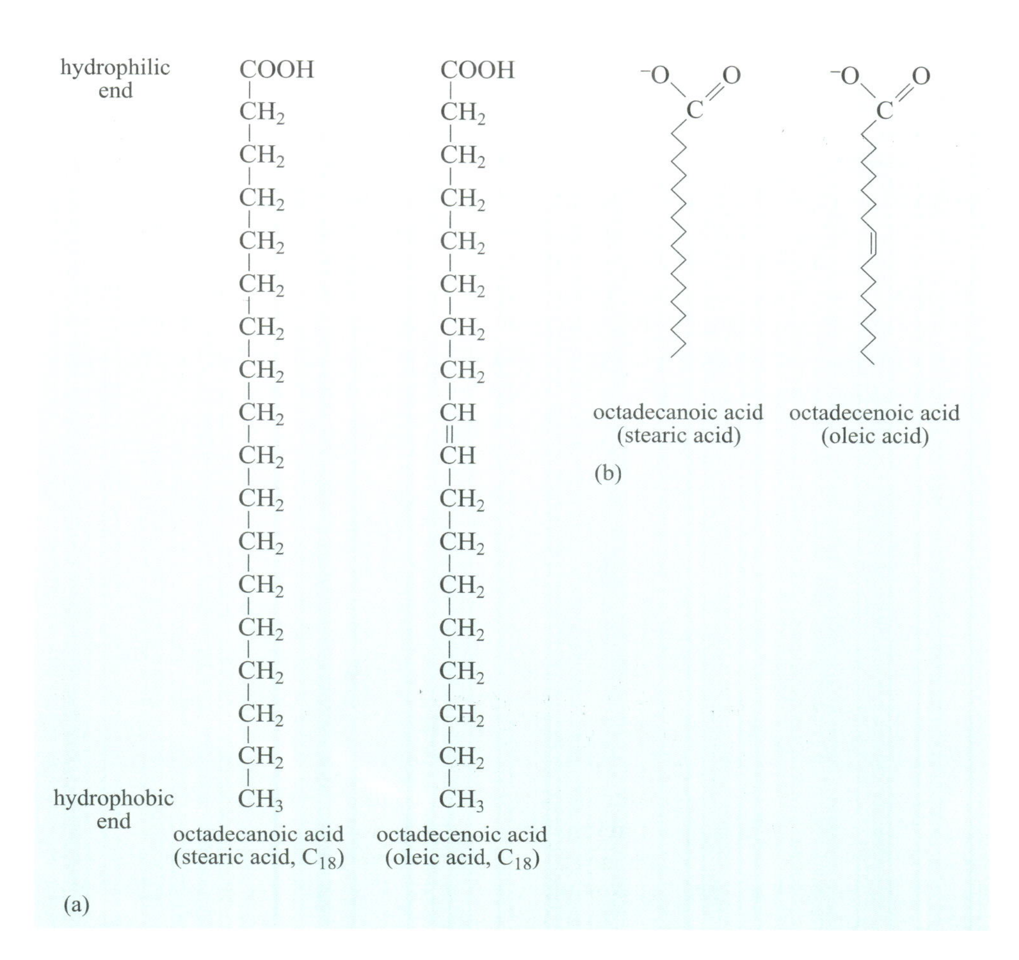

Figure 1.4 The structure of two common lipids containing eighteen carbon atoms, octadecanoic acid and octadecenoic acid: (a) detailed structures; (b) abbreviated structures. (Zubay, 2000)

Figure 1.5 (a) The structure of two common five- and six-carbon sugar molecules. (b) Sugar monomers polymerize by simple reactions that involve the loss of water to form (c) polysaccharides. (Zubay, 2000)

1.2.4 Proteins

Proteins (from the Greek *proteios* or primary) are the most complex macromolecules found in living systems. They consist of long 'trains' of amino acids linked together. As with polysaccharides, they are linked together by simple reactions that involve the loss of water (Figure 1.6). There are 20 different amino acids found in the proteins of living systems and it is the particular sequence of amino acids that gives a protein its function. Proteins are perhaps the most important of life's chemicals and have an enormous number of different roles. For example, they provide structure (e.g. in human fingernails and hair) and act as **catalysts** (e.g. aiding digestion in our stomachs). Proteins with catalytic properties are called **enzymes**.

A catalyst is a substance that increases the rate of a reaction but is not itself used up in the reaction.

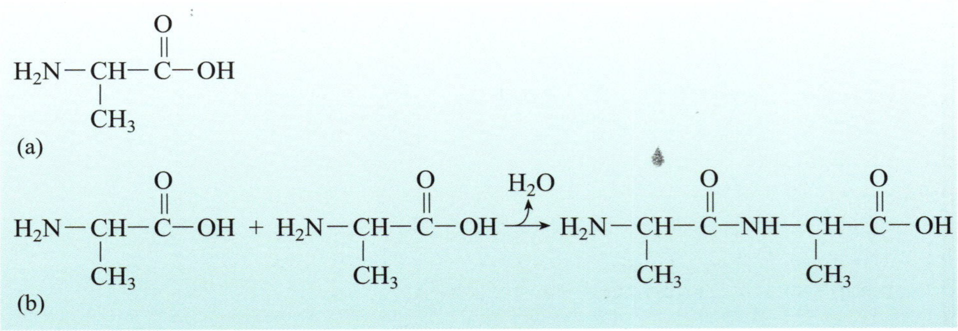

$$H_2N-CH(CH_3)-C(=O)-OH$$

(a)

$$H_2N-CH(CH_3)-C(=O)-OH + H_2N-CH(CH_3)-C(=O)-OH \xrightarrow{H_2O} H_2N-CH(CH_3)-C(=O)-NH-CH(CH_3)-C(=O)-OH$$

(b)

Figure 1.6 (a) An amino acid. (b) Amino acid monomers polymerize by simple reactions that involve the loss of water. Many reactions of this type will eventually produce proteins.

1.2.5 Nucleic acids

Nucleic acids are the largest macromolecules found. They exist as a collection of individual **nucleotides** (Figure 1.7a) linked together in long linear polymers (Figure 1.7b). As with sugars and amino acids, nucleotides can be linked together by simple reactions that involve the loss of water. Nucleotides contain:

- a five-carbon sugar molecule
- one or more phosphate groups
- a nitrogen-containing compound called a nitrogenous base.

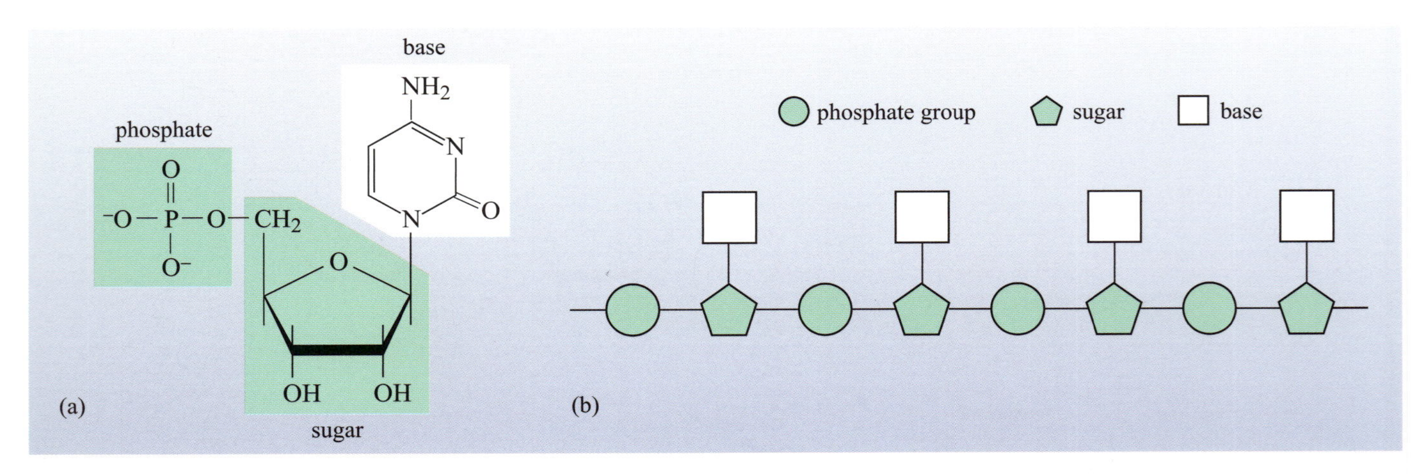

Figure 1.7 (a) The structure of a nucleotide consisting of a phosphate group, sugar molecule and nitrogenous base (cytosine in this instance). (b) Nucleotides polymerize by simple reactions that involve the loss of water to form nucleic acids. ((a) Zubay, 2000)

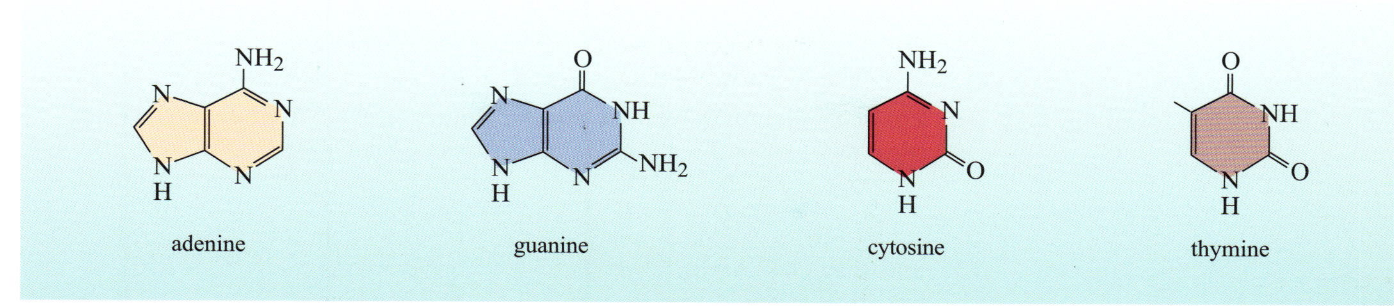

Figure 1.8 The bases found in DNA.

The most famous nucleic acid is deoxyribonucleic acid or **DNA**. Another important nucleic acid is ribonucleic acid or **RNA** (we will examine the role of RNA later in this section). Prior to 1953 it was known that DNA contained four different nucleotides, each possessing identical sugar and phosphate groups but different bases. These bases are adenine, guanine, cytosine and thymine, sometimes referred to in shorthand by the letters A, G, C and T (Figure 1.8). However, exactly how these components were arranged was unknown.

In 1953, James Watson and Francis Crick recognized that DNA consists of two long molecular strands that coil about each other to form a **double helix** (Figure 1.9). Bonds that resemble the steps of a spiral staircase connect the two helical strands. The steps consist of two nucleotides, with each nucleotide forming half of the step. The bases in the centre of the helix are joined by weak hydrogen bonds. The bases always match – adenine in one nucleotide is always paired with thymine in the other and, likewise, guanine is always linked to cytosine. Hence, the sequence of bases on one strand strictly determines the base sequence on the other.

DNA strand	DNA strand
A	T
T	A
G	C
C	G

The bases are attached to their helical strands by sugar groups, which in turn are connected together along the exterior of the helix by phosphate groups.

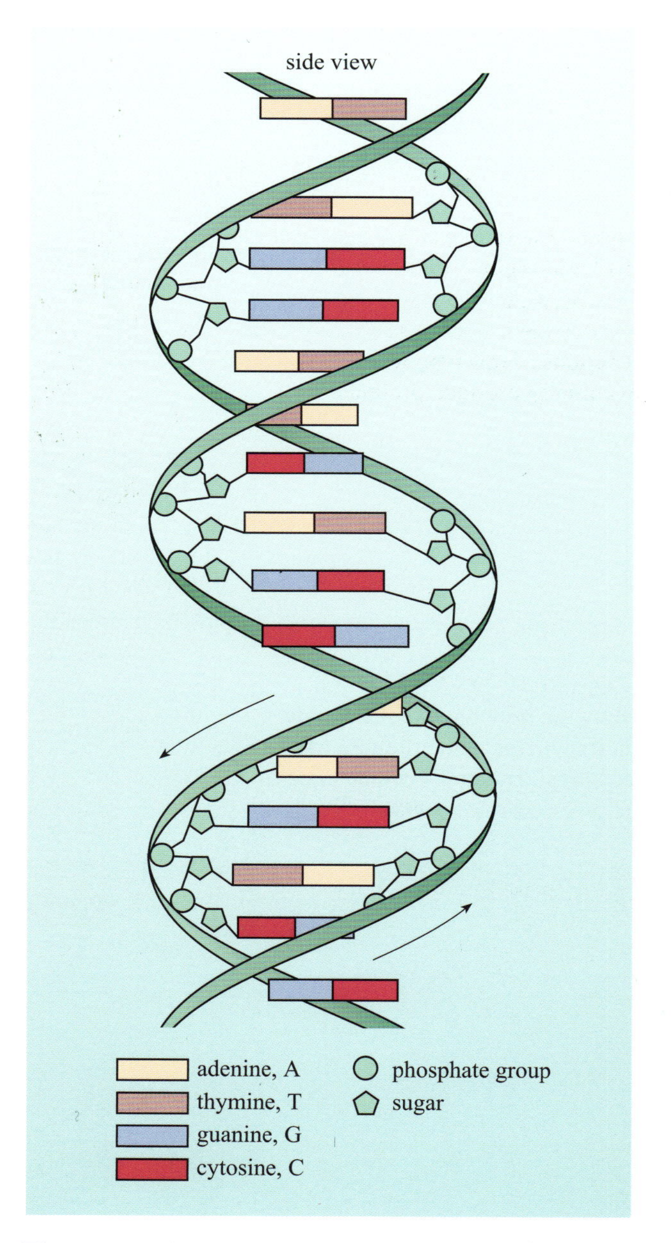

Figure 1.9 The DNA double helix. Note that the 'ribbons' are not real, but are there to illustrate the nature of the double helix.

■ Examine the DNA shorthand above and compare it with the idealized structure of nucleic acids in Figure 1.7. Which parts of the nucleic acid structure are omitted in the shorthand?

□ The sugar and phosphate backbone is not represented because the types of sugar and phosphate groups do not change within a particular nucleic acid.

■ Can you suggest how this complementary system allows DNA to pass on genetic information?

□ Since the sequence of bases on one strand of the helix determines the sequence on the other, 'unzipping' the double helix provides two templates that can be used to produce two new DNA molecules from the single parent.

Special protein enzymes separate the strands of the double helix. The single strands hook up with spare nucleotides present in the liquid surrounding the molecule. Each base in the unzipped strand latches on to its complementary base. The sugar and phosphate groups of the newly acquired nucleotides then join together into helical strands, so two identical double-helix molecules are formed, exactly like the original (Figure 1.10).

The discovery of DNA structure provided the basis for understanding one of the key characteristics of life: the mechanism that enables biological molecules to replicate themselves.

Different DNA sequences also account for the variations between individuals of the same species, and for the differences between species (Box 1.2).

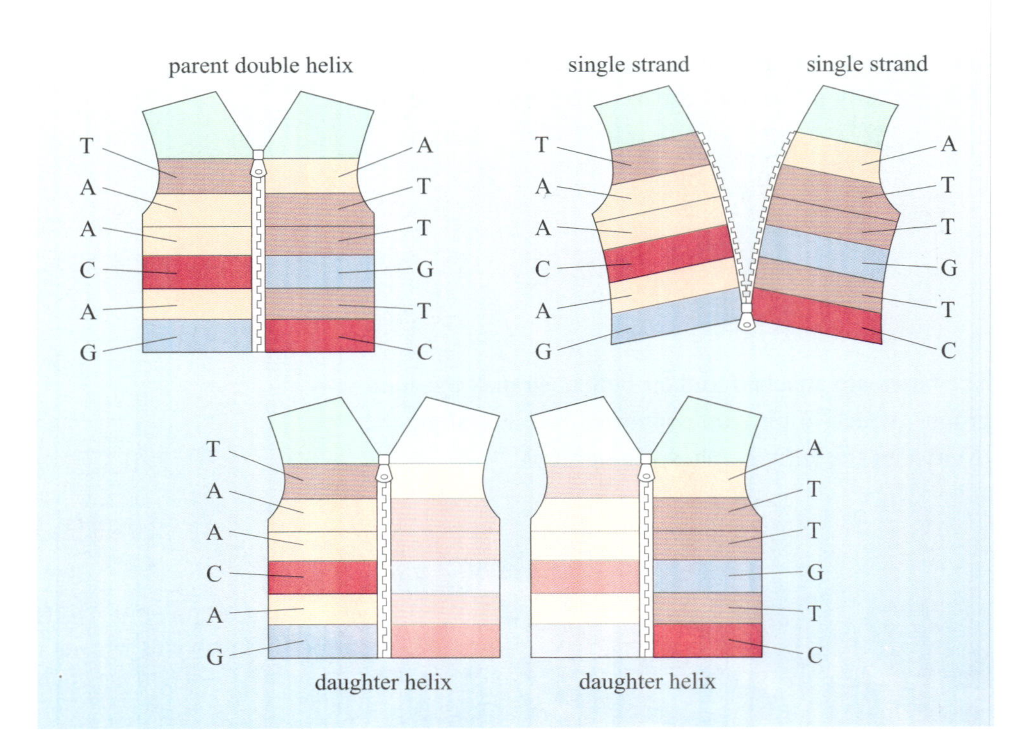

Figure 1.10 DNA replication showing how one parent double helix 'unzips' and produces two identical daughter double helices. (Lowestein *et al.*, 1998)

BOX 1.2 DNA HYBRIDIZATION – THE 'WHO'S WHO' OF GENETICS

We have talked about unzipping the DNA double helix to produce two identical DNA strands and it is reasonable to assume that single strands of DNA from the same species can be zipped back together. But what of single DNA strands from closely or distantly related species, can they be spliced together? The answer is yes – but only partly. Closely related species will have similar, but not identical, sequences of nucleotides and will splice relatively well, whilst distantly related species will have dissimilar sequences of nucleotides and will splice relatively poorly.

This concept has been exploited to produce a technique known as DNA hybridization. The DNA helix from one species is unzipped by heating and then combined with unzipped DNA from another species. When the mixture cools some of the different strands splice together. If the species are closely related the strands will almost match and the new double helix will be strongly bound together. If the species are not closely related the opposite is true – the new double helix will not be strongly bonded together. The strength of the bonding, and therefore the strength of the species relationship, is revealed when the new mixture is heated again – the weakly bound (less-related) helices unzip at lower temperatures.

DNA hybridization has revealed that some species that appear similar and were thought to be related, actually have quite different ancestries. For example, the belief that owls were akin to falcons and hawks (Figure 1.11) was mistaken as these birds are more closely related to nightjars. Similarly, starlings are not close relatives of crows but are related more to mockingbirds.

Figure 1.11 Appearances can be deceptive. The DNA of owls (centre) reveals that they are more closely related to nightjars (left) than falcons and hawks (right). (© Dan Sudia)

In addition to self-replication and passing on information from generation to generation, DNA also governs protein synthesis. DNA contains a set of 'instructions' called the **genetic code** which is expressed by the sequence of bases in the molecule. For example ATGC would be one part of a genetic code, while ATGG would be another. The genetic code directs the production of thousands of proteins needed for the structure and function of living systems. In the process of protein synthesis, the DNA message is first transcribed (copied) into an RNA message. RNA is very similar to DNA, but has slight differences. The sugar component in RNA is ribose rather than deoxyribose, and the base uracil is present instead of thymine (Figure 1.12). Hence, the four RNA bases are adenine, guanine, cytosine and uracil or in shorthand A, G, C and U. When bonding with DNA, uracil replaces thymine and forms a base pair with the adenine of DNA.

DNA strand	RNA strand
A	U
T	A
G	C
C	G

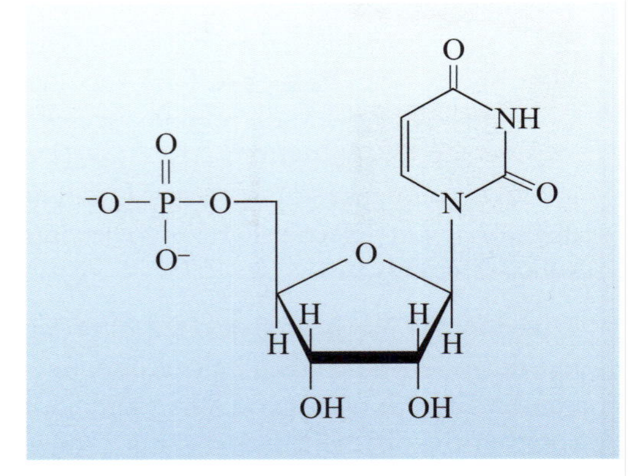

Figure 1.12 An RNA nucleotide containing the sugar and base that makes RNA distinct from DNA.

During transcription, the DNA makes messenger RNA (mRNA). To do this, the DNA helix first unzips as if to replicate itself. Next, instead of each of the nucleotides seeking out a matching DNA nucleotide to build a new DNA double helix, they seek out RNA nucleotides to produce a strand of mRNA. The mRNA strand is then released and the DNA helix zips itself together again. The released mRNA carries its own version of the DNA sequence into a region that contains free amino acids where molecular factories called ribosomes use the mRNA to combine amino acids into long protein chains.

1.2.6 The cell

Many different molecules must be in close association for living systems to operate. This is because in chemistry the rate of reaction generally increases with the concentration of the reactants. Yet what is there to stop molecules simply drifting off in solution and bringing a halt to the chemistry of life? The answer is the cell. In its simplest form, a cell is a small bag of molecules that is separated from the outside world (Figure 1.13). At the centre of the cell, strands of DNA are devoted to the storage and use of genetic information. The DNA is surrounded by the cytosol, which is a salt water solution containing enzymes and the ribosomes. The cell contents are surrounded by a soft membrane. This is called the cell membrane and consists of lipids and proteins. The cell membrane restricts the movement of molecules into and out of the cell and thereby protects the cell's contents. Finally, a tough cell wall consisting of carbohydrate molecules and short chains of amino acids provides the cell's rigidity. So cells provide an environment in which biochemical processes can occur and genetic information can be stored. Cells are the basic structural unit of all present-day organisms on the Earth, but vary in number, shape, size and function. For instance, bacteria are single-celled organisms whereas humans contain around 10^{12} cells.

Simple cells such as the one in Figure 1.13 can reproduce by splitting in two. This process begins when DNA is replicated and the two new DNA molecules attach themselves to different parts of the cell membrane. Next the cell begins to divide, separating the two DNA-containing regions. Finally, when cell division is complete, two identical daughter cells have been produced from the parent.

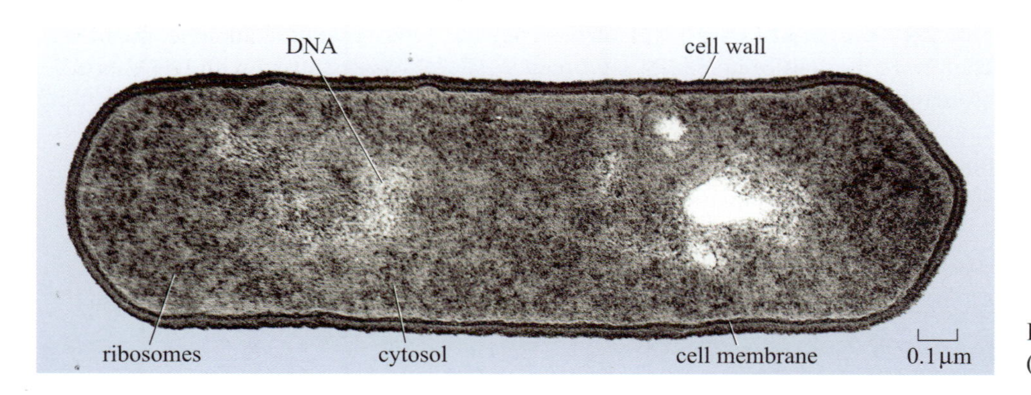

Figure 1.13 A simple cell. (Courtesy of I. D. J. Burdett)

1.3 How to study the origins and remains of life

Now that we have explored what we mean by life, what is necessary for life to exist and what living organisms are made of, we can turn to the detection of life. At this point, it is also appropriate to consider how we will perform our investigations into life's origins. As you will see, there is more than one approach to the problem.

1.3.1 Identifying past and present life

'Biological marker', or 'biomarker' for short, is a term initially used in petroleum exploration. Petroleum geochemists attempt to discover when and where the correct environments existed in the geological past to produce and accumulate fossil fuels. One of their most useful tools is the recognition of molecular fossils or biomarkers that are specific to particular organisms in organic-rich rocks. The value of a biomarker increases if the organism was restricted to a certain environment, thereby increasing its diagnostic value.

Recently, astrobiologists have adopted the term biomarker and its definition has been extended. Today, the word 'biomarker' is no longer used exclusively for organic material but is used for any evidence that indicates present or past life detected either in situ or remotely. In 1999, astrobiologists David Des Marais and Malcolm Walter listed the following categories of biomarker:

1 Cellular remains.
2 Textural fabrics in sediments that record the structure and/or function of biological communities.
3 Biologically produced (biogenic) organic matter.
4 Minerals whose deposition has been affected by biological processes.
5 Stable isotopic patterns that reflect biological activity.
6 Atmospheric constituents whose relative concentrations require a biological source.

A brief glance at these criteria will reveal the extreme subjectivity included in the definition. Establishing whether or not textural fabrics or organic matter in a sample is biogenic might be a very difficult process. For example, aromatic hydrocarbons are a class of organic molecule that can be generated by heat and pressure on the

Aromatic hydrocarbons are molecules built up of units containing six carbon atoms joined in a ring by alternating single and double bonds. If several carbon rings are present the term polyaromatic hydrocarbon, or PAH for short, is used.

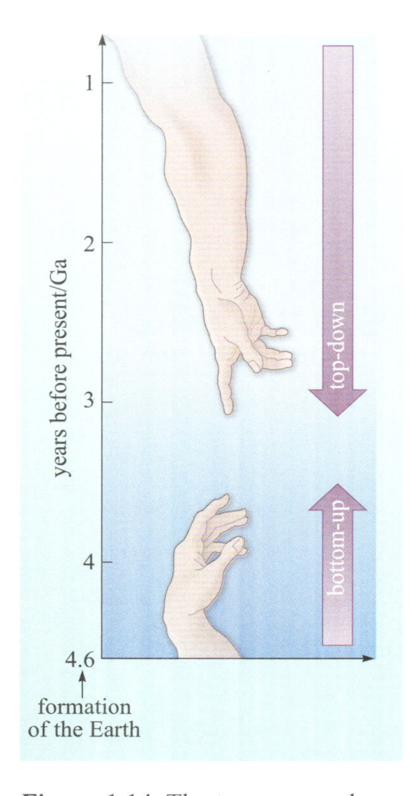

Figure 1.14 The two approaches to the study of life's origins. (Lahav, 1999)

biological remains of living organisms. They are a major component of coal. Coal comprises the fossil remains of land plants. The problem, however, is that where aromatic hydrocarbons are concerned, many roads lead to Rome. These molecules can just as easily be produced in internal combustion engines, garden barbecues or giant stars. It is easy to imagine the controversy that could arise from detecting such organic compounds and attributing their presence to once-living organisms.

1.3.2 Two approaches to the origin of life

Studies into life's origins can be categorized into two general types of enquiry:

- a 'bottom-up' strategy, and
- a 'top-down' strategy.

The bottom-up approach focuses on a collection of inanimate elements, molecules and minerals with known properties and attempts to figure out how they may have been combined in the past to create a living organism. By contrast, the top-down method looks at present-day biology and uses the information to extrapolate back towards the simplest living entities. In the following sections we will use both the bottom-up and top-down tactics to arrive as close as possible to an answer to the question of how life on Earth arose. Of course, there is no guarantee that we can take both approaches far enough so that the two arrive at a mutual destination (Figure 1.14).

1.4 Organic matter in the Universe

Let us begin our look at the origin of life using the bottom-up approach. Organic matter is a fundamental constituent of living systems and represents one of the inanimate substances that life must have been generated from on the early Earth. Furthermore, it is inevitable that the distribution of organic matter in the Universe will have a direct bearing on where life could originate. So, to put our own planet in context, we will now examine the major environments in which it is thought that organic matter is created.

The production of organic matter was occurring in our region of the Galaxy long before our Solar System formed. The circumstellar (surrounding a star) envelopes around carbon-rich red giant stars churn out large amounts of extraterrestrial organic molecules (Figure 1.15). It has been postulated that the chemical reactions that form organic molecules in these regions are similar to those observed in a simple candle flame on Earth and the dominant products are aromatic hydrocarbons. Figure 1.16 presents infrared spectra from a circumstellar envelope around IRAS21282+5050. The close match between the spectra of the starlight and two laboratory aromatic hydrocarbon standards suggests that these organic molecules are common constituents of that part of space.

The name IRAS21282+5050 denotes that the object was identified in the data returned by the InfraRed Astronomical Satellite (IRAS) which surveyed the sky in 1983. The numbers refer to the object's position on the sky.

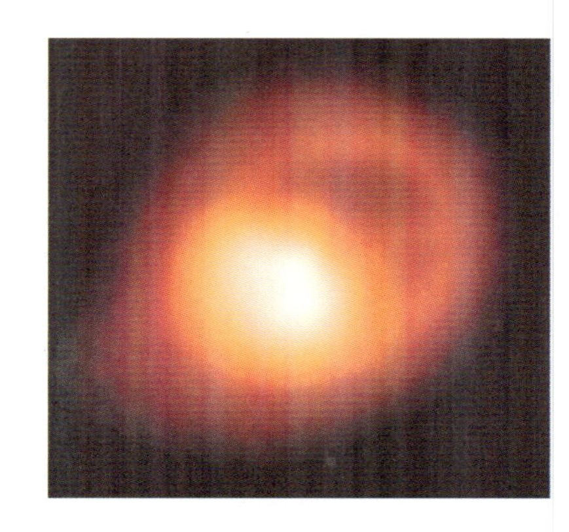

Figure 1.15 A circumstellar envelope in which large amounts of extraterrestrial organic matter are found. (© European Space Agency)

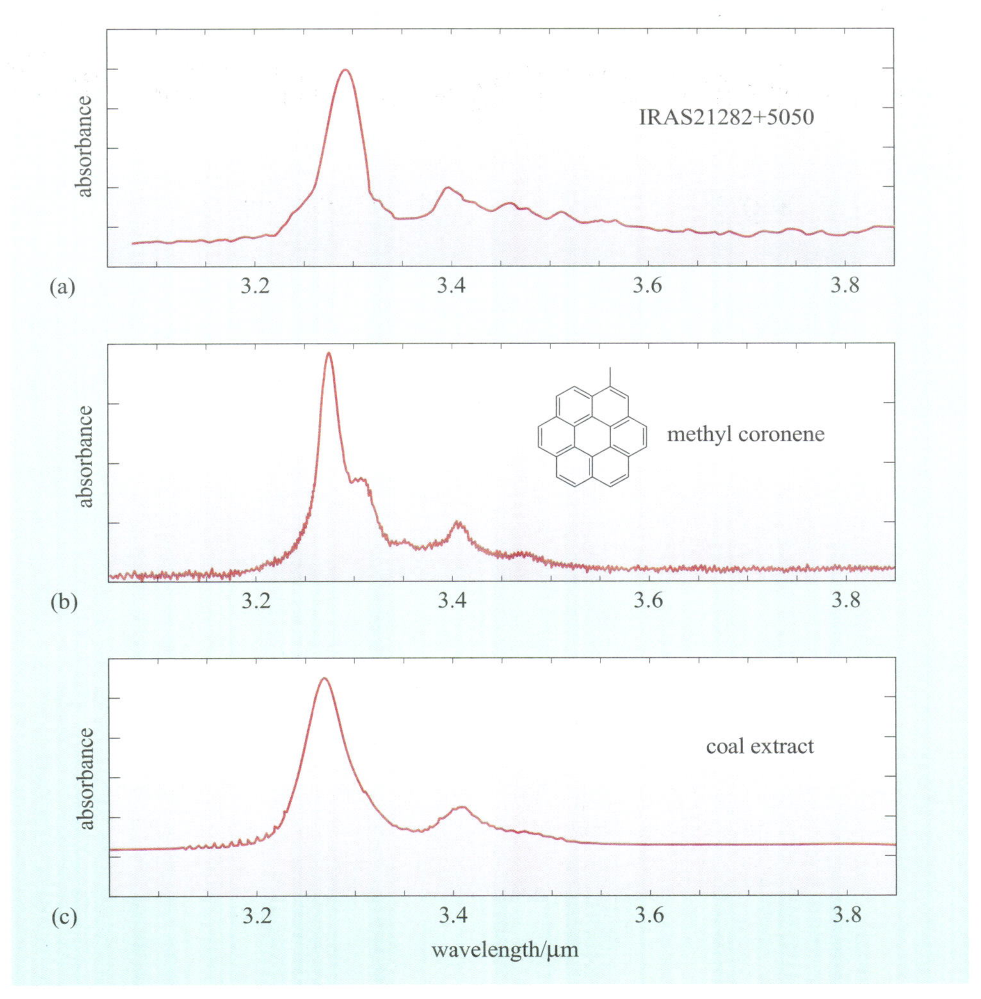

Figure 1.16 Partial infrared spectra of a circumstellar envelope from IRAS21282+5050 compared to a single aromatic hydrocarbon (methyl coronene) and an aromatic hydrocarbon mixture (coal extract). (Adapted from de Muizon *et al.*, 1986)

The star's wind expels these molecules into interstellar space. From there they can be caught up in other environments where organic matter may be created. Some of these environments are molecular clouds (Figure 1.17) that represent the coldest (10–20 K) and densest parts of the **interstellar medium** and play a key role in the evolution of the Galaxy. Every star and planetary system was formed inside a molecular cloud; the other types of interstellar clouds are too warm and diffuse to allow the generation of stars. Star formation occurs when deeply embedded clumps of interstellar gas and dust collapse under their own gravitational attraction.

Numerous different molecules have been identified in interstellar clouds and circumstellar envelopes (Table 1.3) and the list continues to expand. It may seem surprising that the largest quantity of organic molecules in our Galaxy is found not on the Earth but in giant molecular clouds.

Figure 1.17 Hubble Space Telescope images of molecular clouds in the Eagle Nebula. The gaseous clouds are several light years long. (NASA)

Table 1.3 The molecules detected in interstellar space and in circumstellar envelopes. Note: D is deuterium, a form of hydrogen.

hydrogen species					
H_2	HD	H_3^+	H_2D^+		
hydrogen and carbon compounds					
CH	CH^+	C_2	CH_2	C_2H	C_3
CH_3	C_2H_2	C_3H (lin)	C_3H (circ)	CH_4	C_3H_2 (circ)
H_2CCC (lin)	C_4H	C_5	C_2H_4	C_5H	H_2C_4 (lin)
CH_3C_2H	C_6H	H_2C_6	C_7H	CH_3C_4H	C_8H
hydrogen, carbon and oxygen compounds					
OH	CO	CO^+	H_2O	HCO	HCO^+
HOC^+	C_2O	CO_2	H_3O^+	$HOCO^+$	H_2CO
C_3O	CH_2CO	HCOOH	H_2COH^+	CH_3OH	HC_2CHO
C_5O	CH_3CHO	C_2H_4O (circ)	CH_3OCHO	CH_2OHCHO	CH_3COOH
CH_3OCH_3	CH_3CH_2OH	$(CH_3)_2CO$			
hydrogen, carbon and nitrogen compounds					
NH	CN	NH_2	HCN	HNC	N_2H^+
NH_3	$HCNH^+$	H_2CN	HCCN	C_3N	CH_2CN
CH_2NH	HC_2CN	HC_2NC	NH_2CN	C_3NH	CH_3CN
CH_3NC	HC_3NH^+	C_5N	CH_3NH_2	CH_2CHCN	HC_5N
CH_3C_3N	CH_3CH_2CN	HC_7N	CH_3C_5N	HC_9N	$HC_{11}N$
hydrogen, carbon (possibly), nitrogen and oxygen compounds					
NO	HNO	N_2O	HNCO	NH_2CHO	
other species					
SH	CS	SO	SO^+	NS	SiH
SiC	SiN	SiO	SiS	HCl	NaCl
AlCl	KCl	HF	AlF	CP	PN
H_2S	C_2S	SO_2	OCS	HCS^+	SiC_2(circ)
NaCN	MgCN	MgNC	H_2CS	HNCS	C_3S
$HSiC_2$	SiC_3	SiH_4	SiC_4	CH_3SH	C_5S

Note: (circ) denotes circular molecules and (lin) denotes linear molecules.

- Using Table 1.3, which is the smallest molecule and which is the largest molecule detected in interstellar and circumstellar environments?
- H_2 is the smallest molecule and $HC_{11}N$ is the largest molecule.

These various types of molecules are formed through a complicated network of chemical reactions inside the interstellar or circumstellar clouds where they are found. In dense molecular clouds the temperatures are so low that any gas hitting a dust grain of solid matter will immediately freeze out to form an icy mantle. Once the organic compounds are attached to a grain, chemical reactions are catalysed by the grain surface and any reaction products are processed further by ultraviolet and cosmic rays. The products of grain mantle chemistry may have an opportunity to take part further in organic chemistry when they are incorporated in the warm (200–400 K) and dense (>100 H_2 molecules/cm^{-3}) areas of gas located around recently formed stars. These areas have been shown to be the most chemically diverse regions in the interstellar medium and are called hot cores. Many of the species in Table 1.3 have only been seen in hot cores. Various interesting chemical compounds are produced when the icy grain mantles are heated in hot cores. When a star is formed nearby, its radiation evaporates the ice and the molecules return to the gas phase.

Once star formation is underway, a spinning disc of dust and gas called a **solar nebula** is produced. The solar nebula inherits a variety of organic molecules from its molecular cloud (Figure 1.18) although some organic matter may be synthesized anew. Processes similar to the gas phase reactions proposed for circumstellar shells and the grain catalysed reactions in molecular clouds may have operated in the nebula. Eventually, the solar nebula is replaced by a solar system containing stars and planets.

Figure 1.18 An artist's impression of the solar nebula – a rotating disk of dust and gas that gave rise to the Sun and planets. (NASA)

1.5 Synthesis of organic molecules on the early Earth

Following the formation of a solar nebula, the surfaces of newly formed planets and their atmospheres provide the next opportunity for the generation of organic molecules. It is the processes taking place on planets that we turn to next, by considering in detail the reactions that may have occurred following the birth of the planet we understand best, our own Earth.

1.5.1 Energy sources

It appears that much of the chemistry of the Universe is organic chemistry and there are a number of extraterrestrial environments that are capable of generating organic compounds that may be biologically useful. But these organic compounds are relatively simple and are a far cry from the highly organized organic systems found in living organisms. The laws of physics dictate that systems will inevitably trend towards disorder if left to their own devices. Energy is required to generate and sustain order. So to understand how life began on the early Earth we must appreciate what energy sources were available.

Energy may have had two roles in the origin of life: it could have fuelled reactions that synthesized organic matter on the early Earth and it was certainly utilized at some point to sustain primitive life.

Table 1.4 lists the main sources of energy available on the present-day Earth.

Table 1.4 Present-day sources of energy averaged over the Earth.

Source	Energy /J m^{-2} yr^{-1}
total radiation from the Sun	1 090 000.0
ultraviolet light	1 680.0
electric discharges (lightning)	1.68
cosmic rays	0.000 6
radioactivity (to 1 km depth)	0.33
volcanoes	0.05
shock waves (atmospheric entry)	0.46

- What is the largest source of energy on the Earth, and by what factor does it exceed the next largest?
- Sunlight is the largest energy source, by a factor of 1 090 000/1680 = 650.

The strength of the Sun on the early Earth is thought to have been up to 20–30% weaker than it is today, but it would still have been the major energy source available for the synthesis of organic matter. As we shall see later, electrical discharges have been used in many experiments that have attempted to recreate the synthesis of organic molecules on the early Earth. But given the small amounts of energy provided in this way, it seems unlikely that they were important sources of energy for organic synthesis. The decay of radioactive forms of uranium and potassium would have provided heat from the Earth's interior. Primordial heat would have been generated as the Earth's accretion released gravitational energy and volcanic activity would have led to the eruption of lavas at temperatures well over 1000 °C. The shock waves generated as meteors and **meteorites** passed through the atmosphere would have also contributed to the energy available to synthesize molecules. Yet all of these energy sources would have been small relative to that supplied by the Sun.

1.5.2 Miller's origin of life experiment

In the early 1950s, Stanley Miller was working for his PhD at the University of Chicago under the guidance of Harold C. Urey (1893–1981). In an attempt to recreate the types of chemical reactions that may have occurred on the early Earth, Miller used a flask of water to represent the primordial ocean. He then heated the flask. Water vapour circulated through the apparatus, setting his primitive hydrosphere in motion (Figure 1.19). Another flask, placed higher than the one containing water, represented the atmosphere and contained methane (CH_4), ammonia (NH_3) and hydrogen (H_2) all of which became mixed with the invading water vapour. Next, in a move reminiscent of a low-budget Frankenstein movie, he subjected the gases to a continuous electrical discharge that represented lightning. The electrical energy caused the gases to interact and the reaction products accumulated lower down in a water-filled trap. The Miller–Urey experiment was allowed to run for one week. When the reaction products were analysed it became clear that a number of organic compounds needed for life, notably amino acids, had been produced relatively simply and abiotically in a reducing atmosphere. The ease at which these compounds could be produced suggested that they should be abundant and widespread in the Universe.

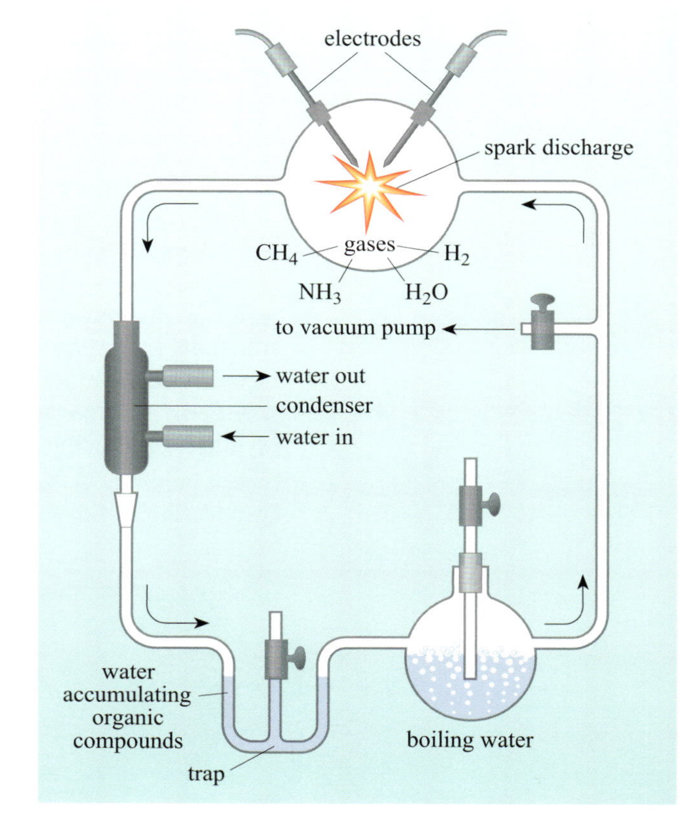

Figure 1.19 Miller and Urey's apparatus used in the abiotic synthesis of amino acids. The lower flask containing boiling water represents the primordial ocean. Water vapour enters the upper flask, representing the primordial atmosphere, and mixes with methane, hydrogen and ammonia. Electrical discharges cause the gases to combine into amino acids, which then accumulate in a water-filled trap.

1.5.3 The fall of the Murchison meteorite

On a Sunday morning in September 1969 the tranquillity of a small town called Murchison, near Melbourne, Victoria in Australia was shattered by a sonic boom that heralded the arrival of a rare type of carbon-rich meteorite. The Murchison meteorite was a **carbonaceous chondrite**, a piece of **asteroid** that had sat out in space, somewhere between Mars and Jupiter, and had remained unaltered since shortly after the birth of the Solar System (Figure 1.20). The first suggestions that the Murchison meteorite contained organic molecules came from the initial eyewitness reports that commented on solvent smells emanating from the stone. The early investigations of the organic inventory of the Murchison meteorite were performed in laboratories that had been preparing for the return of samples from the Apollo lunar missions. Several classes of organic compounds were quickly recognized, including amino acids. These discoveries indicated that the early Solar System must have been a place in which much organic chemistry was taking place. Table 1.5 compares the types and amounts of amino acids synthesized in the Miller–Urey experiment to those found in the Murchison meteorite.

- How good a match is the amino acid content of the Miller–Urey products and the Murchison meteorite?
- The match is remarkably good. Both the types and abundances of amino acids appear similar.

The discovery of similar organic compounds in the Murchison meteorite and Miller–Urey experiment supported the idea that the production of the basic organic building blocks of life is a widespread and probably common feature of the Universe. It appeared that simple organic molecules would have been available in reasonable concentrations in the early Solar System and on the early Earth. It was not difficult to imagine that, eventually, simple molecules would be polymerized into macromolecules that would achieve the properties of life.

The Miller–Urey experiments have been reproduced many times using different energy sources and slightly different starting mixtures. Each time, biologically useful small molecules have been created from mixtures of reduced gases. However, as you'll see in Section 2.4.4, recent models of atmospheric evolution indicate that the early Earth would not have had a methane- and ammonia-rich reducing atmosphere because these gases are easily destroyed by sunlight. It now seems that the more stable molecules carbon dioxide, nitrogen and water dominated the Earth's early atmosphere. Under these less reducing conditions Miller–Urey synthesis is much more difficult. So it appears that the environment of the early Earth might have been fit for sustaining life but less suitable for the in situ production of life's organic raw materials.

Figure 1.20 The Murchison carbonaceous chondrite, a type of meteorite that preserves organic matter from the early Solar System.

Table 1.5 Abundances of amino acids synthesized in the Miller–Urey experiment and those found in the Murchison meteorite. The number of dots represents relative abundance. Those amino acids used by life (i.e. in proteins) are indicated.

Amino acid	Abundance of amino acids		Found in proteins on Earth
	synthesized in the Miller–Urey experiment	Found in the Murchison meteorite	
glycine	●●●●	●●●●	yes
alanine	●●●●	●●●●	yes
α-amino-*N*-butyric acid	●●●	●●●●	no
α-aminoisobutyric acid	●●●●	●●	no
valine	●●●	●●	yes
norvaline	●●●	●●●	no
isovaline	●●	●●	no
proline	●●●	●	yes
pipecolic acid	●	●	no
aspartic acid	●●●	●●●	yes
glutamic acid	●●●	●●●	yes
β-alanine	●●	●●	no
β-amino-*N*-butyric acid	●●	●●	no
β-aminoisobutyric acid	●	●	no
γ-aminobutyric acid	●	●●	no
sarcosine	●●	●●●	no
N-ethylglycine	●●	●●	no
N-methylalanine	●●	●●	no

1.6 Delivery of extraterrestrial organic matter to the early Earth

1.6.1 We are stardust

Without a reducing atmosphere on the early Earth, Miller–Urey reactions could not have produced the large amounts of organic matter that provided the raw materials for life. In this scenario, the early Earth would have been bereft of its own organic matter. In fact, in the early Solar System the paucity of organic matter available to create primitive life was not a problem exclusive to the early Earth. Figure 1.21 shows a plot of carbon abundance in the Solar System. In general, carbon can be considered as an indicator of the amounts of abiotic organic matter. Biotic organic matter (life) is also indicated. Inwards of the **asteroid belt** the amount of organic matter declines abruptly. So it appears that during the time that life originated, the whole inner Solar System contained little organic matter.

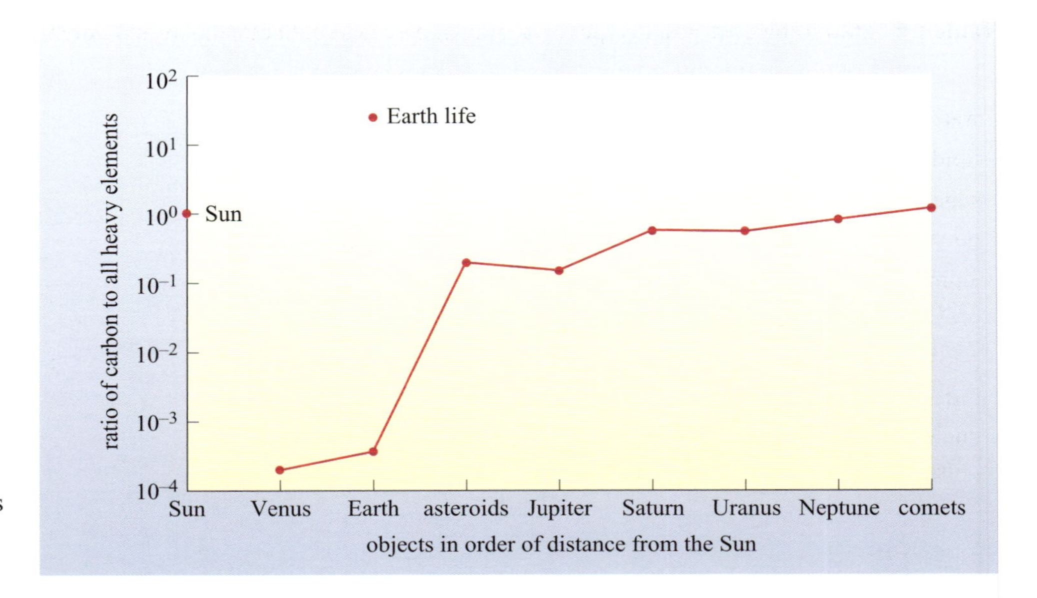

Figure 1.21 Ratio of carbon to heavy elements (all elements more massive than H and He) for various Solar System objects. The horizontal axis is not to scale.

We have mentioned that, in addition to organic matter, liquid water is a prerequisite for life. Liquid water has probably been stable on the Earth's surface since early in Earth's history and may have once been stable on the surface of Mars. Further out in the Solar System, liquid water may be maintained in subsurface environments of icy moons, but the zone in which liquid water has been stable on planetary surfaces over the last 4.6 Ga has not extended beyond about 1.7 AU (see Section 2.3.2).

This is a paradoxical situation as the two ingredients needed for the recipe of life (water and organic matter) appear, in a very general sense, to occupy different areas of our Solar System. In 1961 Juan Oró proposed a solution to the paradox of life's two key components occupying different parts of the Solar System.

■ Look again at Table 1.1. Is the elemental composition of life more similar to the Earth and its crust, or to the Cosmos as a whole?

❑ Life has elemental abundances that are more similar to the Cosmos than the Earth.

This relationship did not go unnoticed by Oró and other astrobiologists and led to the proposal that life could have been kick-started by organic matter delivered to the early Earth by extraterrestrial objects. In effect, meteorite and **comet** impacts would simply bring parts of the organic-rich region of the Solar System to those areas in which liquid water could survive. Table 1.6 compares the types of organic molecules found in living systems and those found in the abiotic mixtures in meteorites.

■ What is the main difference between the organic constituents of life and those of the Murchison meteorite?

❑ Meteorites contain simple organic molecules (monomers) whereas life also contains more complex polymerized versions of these molecules (polymers).

Table 1.6 The biological role and types of organic molecules (both monomers and polymers) found in life and in meteorites.

	Role	Life	Murchison meteorite
water	solvent	yes	yes
lipids (hydrocarbons and acids)	membranes, energy storage	yes	yes
sugars (monosaccharides)	support, energy storage	yes	yes
polysaccharides (polymerized sugars)		yes	no
amino acids	many (support, enzymes, etc.)	yes	yes
proteins (polymerized amino acids)		yes	no
phosphate	genetic information	yes	yes
nitrogenous bases		yes	yes
nucleic acids (polymerized sugars, phosphates and nitrogenous bases)		yes	no

QUESTION 1.2

Table 1.7 gives present-day mass ranges, estimated accretion rates and carbon contents of various extraterrestrial objects that fall to Earth today. The upper part of the table gives data for carbon in any form, i.e. both inorganic and organic forms. The lower part of the table gives data for carbon present only as organic matter. Using the data provided calculate the answers to the following questions.

(a) Complete the table by calculating the accretion rates of total carbon and organic carbon for each type of object listed.

(b) For total meteoritic matter, what object represents the greatest source of carbon per year? Is this carbon likely to arrive steadily over time?

(c) For organic matter, what object represents the greatest source of carbon per year? Is this carbon likely to arrive steadily over time?

(d) How does the total meteor carbon accretion rate compare with the organic carbon accretion rate for meteors? What does this tell you about the overall state of carbon in meteors?

(e) How much *total carbon* would be supplied by meteoritic matter in (i) ten years, (ii) a hundred years, and (iii) a hundred thousand years?

(f) How much *organic matter* would be supplied by meteoritic matter in (i) ten years, (ii) a hundred years, and (iii) a hundred thousand years?

Table 1.7 Accretion rates on Earth today.

Sources	Mass range /kg	Mass accretion rate (estimated) /10^6 kg yr^{-1}	Carbon %	Carbon accretion rate /10^6 kg yr^{-1}
meteoritic matter				
meteors (from comets)	10^{-17} to 10^{-1}	16.0	10.0	
meteorites	10^{-2} to 10^{5}	0.058	1.3	
crater-forming bodies	10^{5} to 10^{15}	62.0	4.2	
unmelted material contributing organic matter				
meteors (from comets)	10^{-15} to 10^{-9}	3.2	10.0	
meteorites, non-carbonaceous	10^{-2} to 10^{5}	2.9×10^{-3}	0.1	
meteorites, carbonaceous	10^{-2} to 10^{5}	1.9×10^{-4}	2.5	

QUESTION 1.3

The mass of carbon in the biosphere is 6.0×10^{14} kg. How long would it take for meteoritic materials to supply (a) a similar amount of carbon, and (b) a similar amount of organic carbon? Would the supply of extraterrestrial carbon and organic carbon to the early Earth have been greater or lesser than that of today?

The answer to Question 1.3 indicates that significant amounts of carbon arrive in the form of meteorites and meteors but only a fraction of this is organic matter. The largest outside contributors to the Earth's organic inventory are the meteors which rain down steadily on the planet but at present-day rates it would still take around a billion years to create the equivalent carbon content of Earth's current biosphere. In addition to any organic matter contained within meteorites and meteors, the shock waves generated from these objects travelling through the Earth's atmosphere may have forced gases to combine, producing organic matter. Furthermore, when large meteorites arrive at the Earth's surface they, and the rocks they hit, may be vaporized and organic matter generated as the gases recombine. On the early Earth between 4 Ga and 3.8 Ga ago all of the processes outlined above would have been more relevant, simply because the rate at which meteors and meteorites arrived at the planet's surface would have been much higher. Evidence for this comes from the Moon where craters indicate that the asteroid and comet impacts that scarred the lunar surface reveal a final flurry of devastation. It is reasonable to assume that the Earth experienced a similar punishment. This period is called the **late heavy bombardment** and the supply of extraterrestrial objects and their associated organic matter to the Earth's surface would have been much greater and the raw materials of life more abundant.

1.6.2 Chirality

The connection between extraterrestrial organic matter and terrestrial life is not proven, but the detailed chemical properties of amino acids found in meteorites and those found in life reveal a startling similarity. Molecules with identical chemical formulae can differ in how their atoms are arranged and these different structural arrangements are called **isomers**. Some isomers separate into left-handed and right-handed forms. This chemical property is **chirality** (pronounced 'ky–rality'). Life exhibits a preference when it comes to the use of chiral molecules.

What is chirality?

If you hold a plain ball up to a mirror, the mirror image of the ball will be identical to the ball itself. It is easy to imagine taking the reflected image of the ball and superimposing it on the real item (Figure 1.22a). However, mirror images are not always identical. For example, place your hands side-by-side in front of you with your palms facing up. You will see that your hands are mirror images of each other but they are not the same (Figure 1.22b). It is for this reason that we need right-hand and left-hand gloves. Try laying one of your hands on top of the other, again palms up. You will see that the thumbs and fingers do not lie in the same position on both hands. In other words they are non-superimposable. Any object whose mirror image is non-superimposable, such as a hand, will be chiral. In fact, the word *chiral* is Greek for 'hand-like'. Any object whose mirror image is superimposable, such as a ball, is termed 'achiral'.

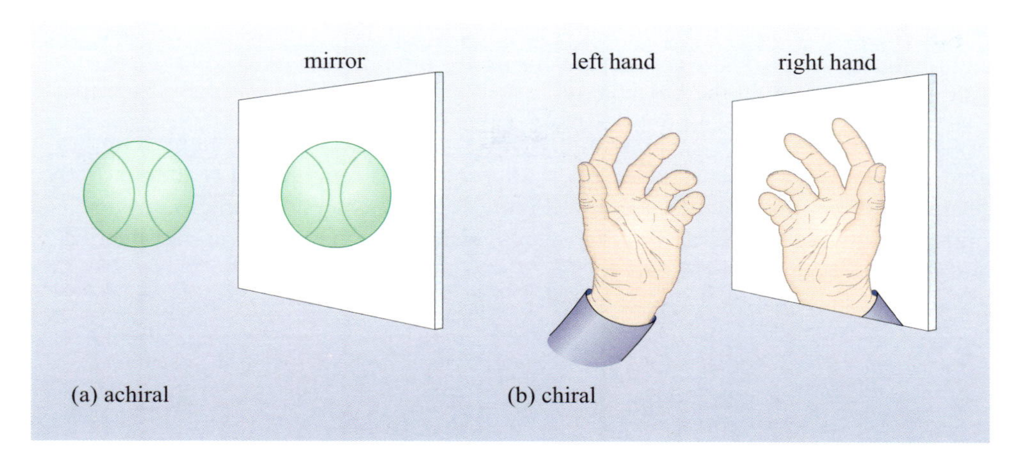

Figure 1.22 (a) The mirror image of a ball is superimposable on the original object and is therefore achiral. (b) The mirror image of a hand is not superimposable on the original and is therefore chiral.

- What other human appendages are non-superimposable or chiral?
- Feet are chiral objects. If you need convincing of this point, try to put your left shoe on your right foot. Ears are also chiral.

Molecules are equally capable of exhibiting achirality and chirality. When all the carbon atoms in a molecule have less than four different structures attached to them then the molecule will be superimposable on its mirror image and will be achiral. For example, the molecule in Figure 1.23a has a mirror image that can be superimposed on the original molecule. However, whenever a molecule has a carbon atom that has four different structures bonded to it, it will not be superimposable on its mirror image and will be chiral. For example, the mirror image of the molecule in Figure 1.23b cannot be superimposed on the original molecule.

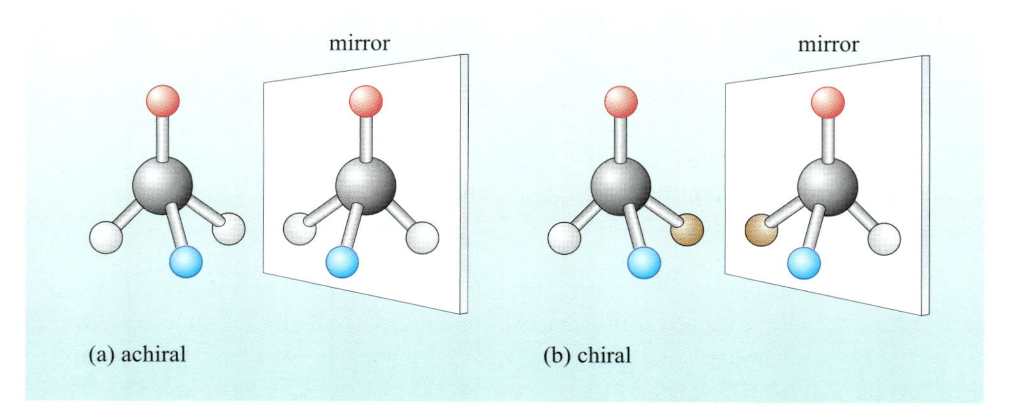

Figure 1.23 (a) An achiral molecule. Its mirror image can be superimposed on the original. (b) A chiral molecule. The mirror image cannot be superimposed on the original.

When discussing chiral molecules it is usual to talk in terms of left-handed and right-handed forms. The right-handed forms are often abbreviated to D (dextro) and the left-handed forms to L (levo).

We can recognize achiral and chiral molecules in amino acids. The simplest amino acid glycine (Figure 1.24a) has a superimposable mirror image and is achiral. More complex amino acids such as alanine (Figure 1.24b) have a non-superimposable mirror image and are chiral. In other words, alanine can be present in left-handed and right-handed forms (Figure 1.25).

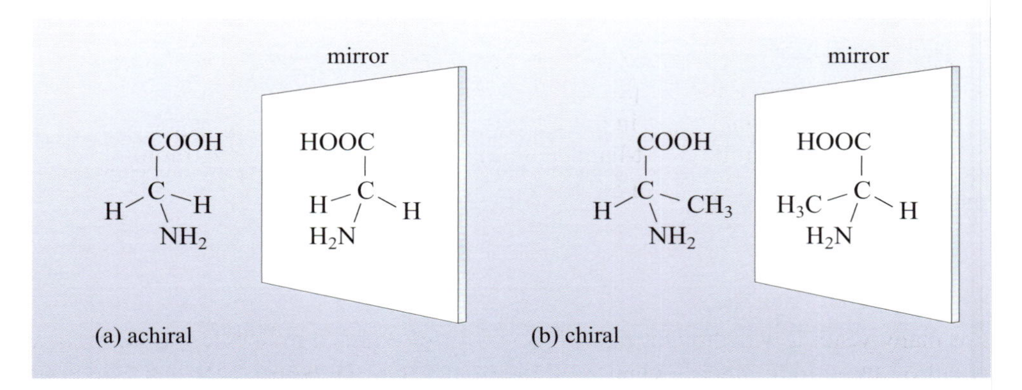

Figure 1.24 (a) The achiral amino acid glycine. Its mirror image can be superimposed on the original. (b) The chiral amino acid alanine. The mirror image cannot be superimposed on the original.

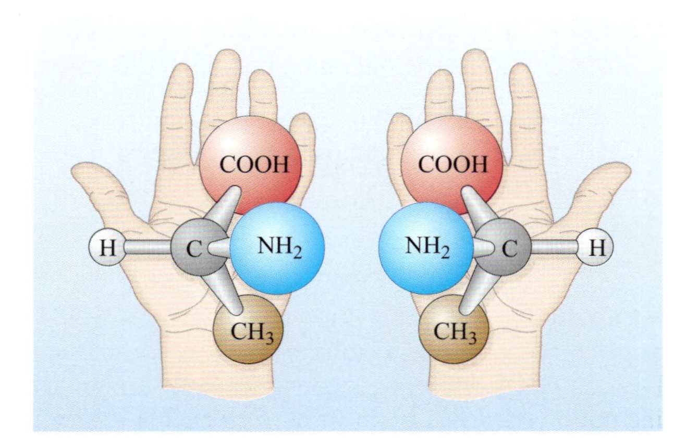

Figure 1.25 The chiral amino acid alanine can be present as left- and right-handed forms.

Chirality and life

In the absence of life, the chemical reactions that make amino acids generally create equal numbers of left- and right-handed forms and such assemblages are called **racemic** mixtures. Yet what is remarkable is that all life on Earth uses only the left-handed forms of amino acids for producing proteins. Proteins are found in every living thing and there can be no life without them. Proteins are constructed from a number of amino acids, and of the 20 amino acids used to make proteins, 19 can exist as left-handed and right-handed forms. The exception is the simple molecule glycine (Figure 1.24a) which has a central carbon atom with fewer than four different structures attached and therefore is achiral.

Mixing left- and right-handed forms of amino acids in proteins would produce structures that hindered the proteins from performing their biological functions. So at some point in the Earth's history, life must have begun with the left- rather than the right-handed amino acids. Once underway, biology was locked into this preference.

The origin of chirality

For many years it was thought that life's use of left-handed molecules was simply a result of an original random chance on the early Earth. However, recent research on organic-rich meteorites has suggested that this preference was inherited from life's starting materials that may have come from space. It has been discovered that an excess of the left-handed form of amino acids is present in meteorites, such as the Murchison meteorite, indicating that these amino acids existed in Solar System material before there was life on Earth.

One possibility for generating a left-handed excess in extraterrestrial organic molecules is a particular type of starlight called **ultraviolet circularly polarized light**, or UVCPL for short. With UVCPL, the electric field direction rotates along the beam. As rotation can occur in a left- or right-hand direction, UVCPL is a chiral phenomenon. Chiral substances have different absorption intensities for left and right UVCPL. Since photolysis (destruction by light) occurs only when light photons are absorbed, UVCPL light destroys one form of the molecules more readily than the other form.

Polarized light has light waves that have electromagnetic vibrations in only one direction. Ordinary light vibrates in all directions perpendicular to the direction of propagation.

Recent work has detected large amounts of UVCPL from the star-forming regions of Orion (Figure 1.26) which will be imposing a chiral preference on any abiotic molecules present there. As this part of the Orion nebula is a site for star formation, chiral amino acids may be available in the soon to be created star and planetary systems. It is tempting to imagine our own Solar System forming in such a region of circularly polarized light, ultimately leading to the excess of left-handed amino acids that we see in meteorites.

It appears therefore that life's left-handed preference may have originated by the action of UVCPL on chiral molecules. When the Solar System formed, some of this organic material arrived on the early Earth via impacts of comets, meteorites and dust particles. These molecules were then part of the **prebiotic** material available for the origin of life, and may have tipped the scales for life to develop with left-handed amino acids. Yet some scientists have suggested that extraterrestrial environments may not only have delivered organic matter to the early Earth but also viable micro-organisms themselves, a theory termed **panspermia** (Box 1.3).

Figure 1.26 Reflection nebulae in Orion, a good source of UVCPL. At the top of the picture is a loose grouping of bright stars. The fronds beneath these stars are the reflection nebulae. (© Anglo-Australian Observatory, photograph by David Malin)

BOX 1.3 PANSPERMIA

In 1908 a Swedish chemist, Svante Arrhenius (1859–1927), published a book which contained the proposal that life in the form of spores could survive in space and be spread from one solar system to another. Spores drifting in the upper atmosphere of a life-rich planet would be forced into interstellar space by the pressure of a nearby star's radiation. Eventually, some of the spores would fall upon another planet where they would flourish and the process would begin again. However, the theory soon attracted criticism because of the large doses of fatal radiation that would have been encountered during a lengthy trip through space.

William Thomson (Lord Kelvin) (1824–1907) proposed a variation on panspermia in which spores are carried through space by meteorites. Although it is unknown whether impact events could launch rocks between solar systems, transport to and from worlds within solar systems occurs frequently. Evidence for interplanetary transport is present in the form of meteorites on Earth that have come from the surfaces of Mars and the Moon. The most likely candidates for a source of organisms in the early Solar System are Mars and Venus. Both have hostile surface environments at the present day but over 3 Ga ago the situation was likely to have been very different and the exteriors of both worlds more hospitable to life.

In 1996 NASA scientists made the highly controversial announcement that a Martian meteorite contained fossil evidence of alien micro-organisms. The meteorite, Alan Hills (ALH) 84001, has a complex and fascinating history:

4.5 Ga	crystallized from magma on Mars
4.0 Ga	battered but not ejected by an asteroid impact
3.6–1.8 Ga	altered by water to produce carbonate minerals
16 Ma	blasted into space by an asteroid impact
1984	discovered in Antarctica
1996	NASA announced the discovery of Martian life

Although the claims of fossil life in (ALH) 84001 have now largely been discounted, the concept that micro-organisms could be transported between planets has begun to attract serious scientific attention.

1.6.3 Keeping your concentration – organic matter accumulation

The amount of organic material estimated to have fallen to Earth per hundred million years around the time of the origin of life is 10^{16}–10^{18} kg. If this material were spread across the surface of the Earth, it would form a layer ranging from 1.6 cm to 1.6 m in thickness. Although this amount of material would represent a significant source of organic carbon in the prebiotic environment if it all survived and accumulated, most of the cometary and meteoritic infall surviving atmospheric entry would presumably fall into oceans and be buried in sediments.

EXAMPLE

Calculate how much organic carbon would be delivered to the Earth each 100 Ma using the present-day rates in Table 1.7. Explain any differences between your calculated value and the value given for the early Earth.

SOLUTION

Present day rate of organic matter accumulation

$$= (0.32 \times 10^6) + (2.9 \times 10^{-6}) + (4.7 \times 10^{-6})$$

$$= 0.32 \times 10^6 \text{ kg yr}^{-1}.$$

Over 100 Ma the amount of organic matter delivered to Earth = 3.20×10^{13} kg.

These rates are less because the early Earth was in the middle of the late heavy bombardment.

As stated earlier, the rate of a chemical reaction increases with the concentration of reactants. So some form of concentration mechanism would have been necessary for the organic matter arriving in the form of meteorites, meteors and comets to take part in chemical reactions that may have led to the production of a living system. After all for two molecules to react, they have to come into contact. Several possible concentration mechanisms can be considered.

1. Marginal marine environments such as lagoons or tidal pools provide a means of concentrating dilute solutions. The solutions are temporarily cut off from the main ocean and evaporation causes the residual liquid to contain a higher proportion of organic molecules.
2. Freezing an aqueous solution also causes an increase in the concentration of any dissolved organic compounds because the water freezes first.
3. The surfaces of clays and other minerals provide sites for trapping organic matter. Clays in particular are useful minerals as they can accommodate organic molecules on and within their structure.

Concentration processes are important for providing enough localized raw materials for the creation of primitive living systems – in much the same way as it is impossible to build a house if the bricks and mortar are repeatedly scattered over a wide area. Once the raw materials required for the origin of life were in place, the serious work of construction could begin.

1.7 Achieving complexity

The mechanisms by which organic molecules could have been assembled into complex living organisms are very poorly understood. Yet it is interesting to examine how these fundamental steps may have occurred. In this section you will encounter several processes that could have aided the development of increasingly complex organic systems that, eventually, became the direct forerunner of life.

1.7.1 The ties that bind – creating polymers and macromolecules

The data in Table 1.2 indicate that macromolecules are important to life. Initially, this appears to be a cause for concern for our ability to understand the origin of life as great complexity appears necessary for living systems to function. It is appropriate therefore to consider just how difficult it is to generate macromolecules from simple organic molecules.

- ■ Carefully examine Figures 1.5 and 1.6 and state how the production of polymers from simple monomers occurs.
- ❑ The polymerization of monomers occurs by a reaction involving the loss of a water molecule.

For example, the –OH groups from two sugar monomers can combine to form a bond following the release of a water molecule (Figure 1.5). Similarly, the combination of the $-NH_2$ group and –COOH group of two amino acids also forms a bond following the release of water (Figure 1.6). Similar reactions also occur during the polymerization of nucleic acids.

- ■ Following the formation of a bond between sugar monomers in Figure 1.5, are all of the reactive –OH groups used up?
- ❑ No, the ends of the new larger molecule still contain reactive OH groups.

In a similar fashion, following the formation of a bond between two amino acids, the ends of the new larger molecule still contain reactive $-NH_2$ and –COOH groups. These features ensure that polymerization reactions can continue indefinitely to form larger and larger organic structures.

Making polymers in the presence of water is one of the most problematic processes in the origin of life. Water is a destructive compound that will break down polymers not build them up.

So large molecules can be generated by polymerization reactions that involve the loss of water. It is highly likely that most primitive polymerizations occurred in this way. However, the high level of order and complexity of some of the macromolecules used in living systems will require more sophisticated methods, and life makes good use of the catalytic properties of protein enzymes. Enzymes make the chemistry of life more efficient, but with water-loss polymerization reactions we have at least made a start towards organic complexity.

1.7.2 Formation of boundary layers

As we discussed in Section 1.2.6, all life today has cells, tiny packets of chemicals surrounded by membranous boundary layers. Hence, a question we musk ask is how did cellular life arise? There were no large molecules like nucleic acids and proteins available on the prebiotic Earth to control the assembly processes characteristic of life. So the first forms of life must have arisen through a self-assembly process.

Let us explore how such processes operate. We have talked about hydrophobic and hydrophilic compounds but certain kinds of organic compounds have both properties at either end of the molecule: a polar hydrophilic head and a hydrophobic tail. These compounds are said to be **amphiphiles**, which means they 'love both'. The polar heads carry a small electrical charge, which makes them soluble in polar solvents such as water. The uncharged tails are much less soluble in water.

If amphiphilic molecules are added to water they tend to sit at the surface with the hydrophilic heads in the water and the hydrophobic tails in the air. In this way they can create a single layer of molecules – a monolayer, as shown in Figure 1.27. The **monolayer** is just one molecule thick, so can be thought of as a two-dimensional surface, or membrane.

- What form do you think amphiphilic molecules will take when introduced within the water by shaking the mixture?
- They tend to gather together and form small spherical structures where the hydrophilic heads face the water while the hydrophobic heads are tucked inside, shielded from the water (Figure 1.28). These spherical structures are termed '**micells**'.

Larger collections of amphiphilic molecules can form a double-layer structure. This arrangement is called a **bilayer** (Figure 1.29a). Imagine two sheets of molecules sandwiched together so that all the hydrophilic heads are on the outside (in contact with the water) and the hydrophobic tails are inside (away from the water). Just

Figure 1.28 A lipid micell – spherical collections of lipids with hydrophilic heads on the exterior and hydrophobic tails in the interior.

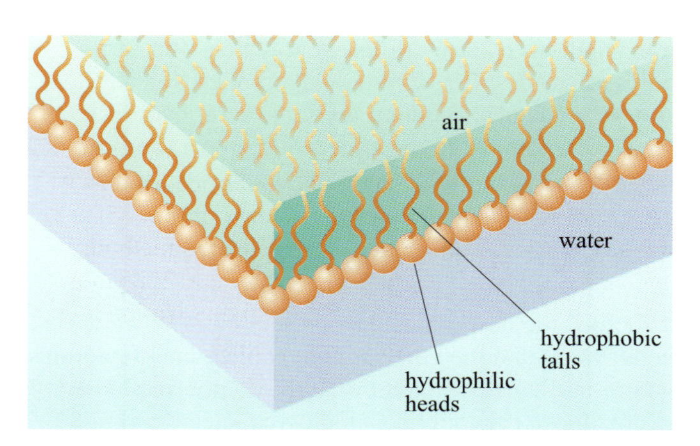

Figure 1.27 A lipid monolayer – a single layer of lipid molecules with hydrophilic ends in the water and hydrophobic ends in the air.

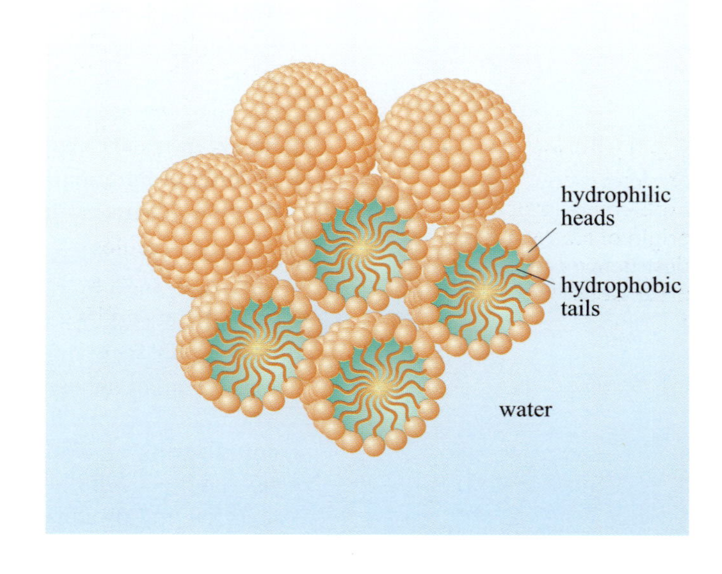

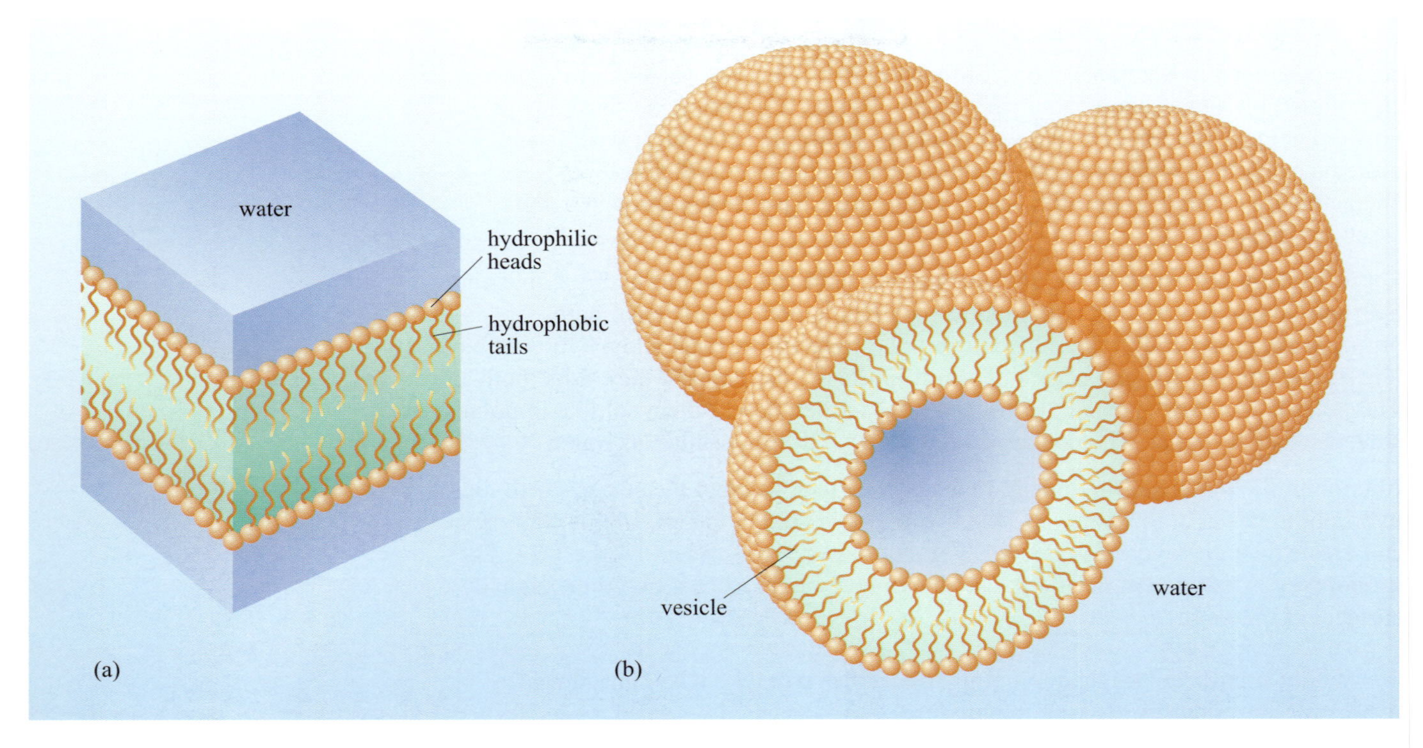

Figure 1.29 (a) A lipid bilayer and (b) bilayer vesicle.

like monolayers, bilayers are a form of membrane and while the spherical form of a single layer of molecules is called a micell the double layer equivalent is termed a '**bilayer vesicle**' (Figure 1.29b). Bilayers are particularly interesting because many organisms use this type of membrane to preserve the integrity of their cells. Monolayers, micells, bilayers and bilayer vesicles are structures that form spontaneously and perhaps provided the original membrane-bounded environment required for cellular life to begin.

The importance of membranes to primitive life has led to a number of experimental investigations aimed at determining how they could have formed under prebiotic conditions. In 1924, Alexander Oparin (1894–1980), a Russian biochemist, showed that proteins, when added to water, group together to form droplets (Figure 1.30a). These droplets are called **coacervates**, a name derived from the Latin for clustered or heaped. Coacervates form in solutions of many different polymers, including proteins, nucleic acids and polysaccharides. Coacervation is a property of physical chemistry related to the polarity of molecules and their ability to form hydrogen bonds in water. Many substances when added to the coacervate preparation can become preferentially incorporated into the droplets, providing a means by which prebiotic chemical factories could have been constructed.

In 1958, Sidney Fox heated dry mixtures of amino acids causing their polymerization by reactions involving the loss of water. The amino acid polymers resembled proteins and so Fox called the new molecules 'proteinoids'. When these proteinoids were dissolved in hot water and then the solution cooled, the proteinoids formed small spheres about 2 μm in diameter which Fox termed '**microspheres**'. The microspheres displayed a double wall resembling a biological membrane and could shrink or swell, depending on the salt concentration of the water. If left for several weeks the microspheres absorbed more proteinoid material from the solution and produced buds which occasionally separated to form second generation microspheres (Figure 1.30b).

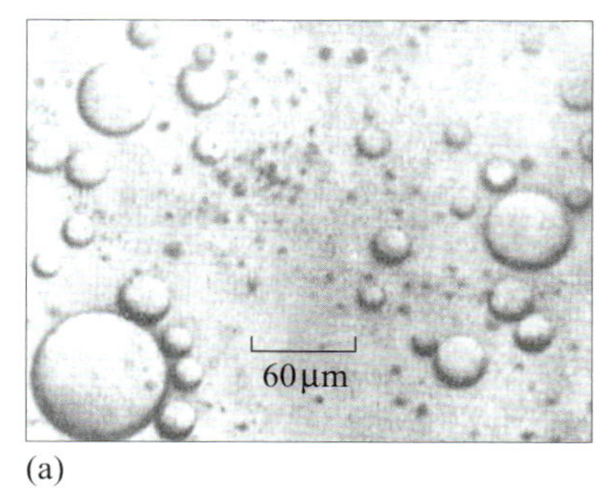

(a)

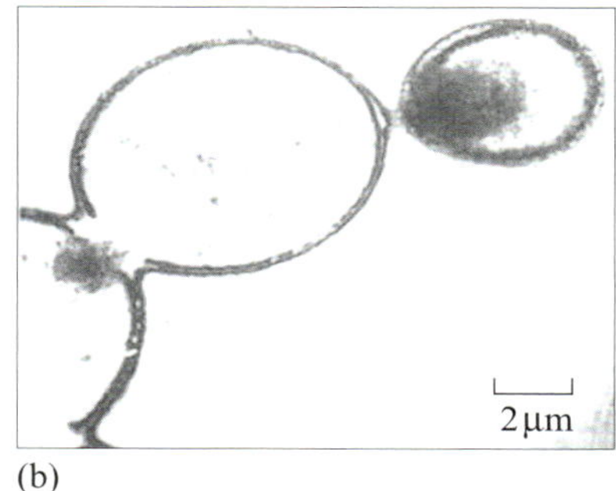

(b)

Figure 1.30 (a) Coacervates and (b) proteinoid microspheres. ((a) sourced from www.angelfire.com; (b) sourced from University of Hamburg website)

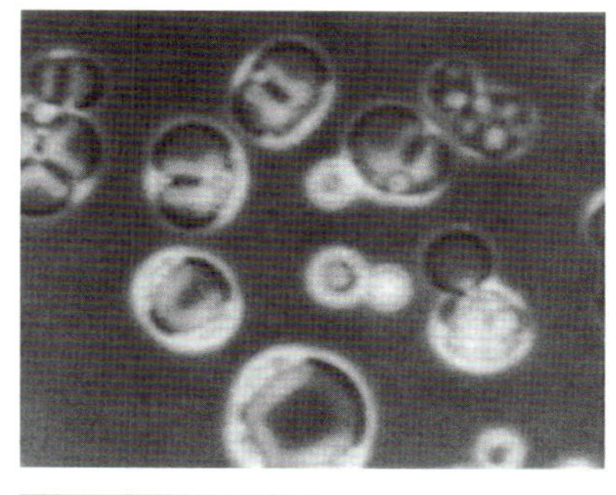

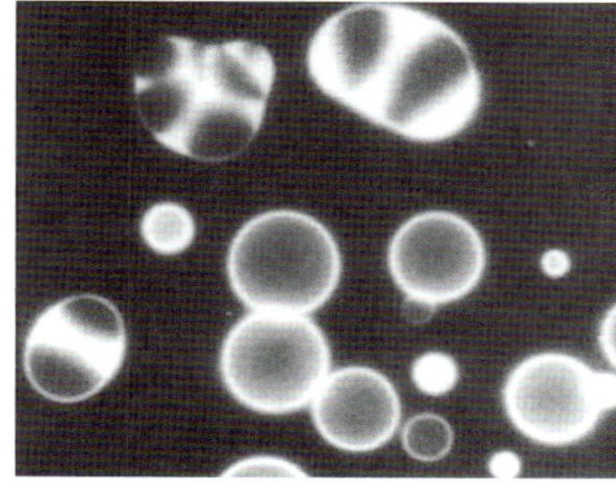

Figure 1.31 Bilayers generated from the Murchison meteorite organic matter. (Dr. David Deamer)

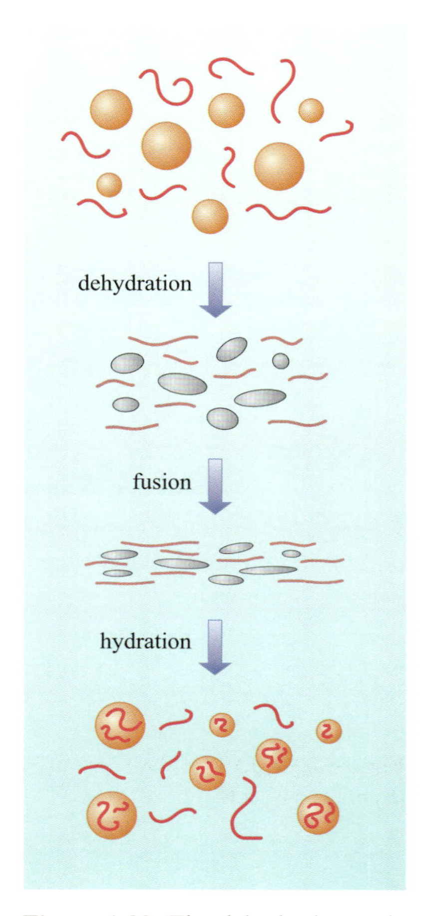

Figure 1.32 The dehydration and incorporation of molecules, and rehydration of membranes.

In 1985 David Deamer considered that amphiphilic molecules might have been delivered to the early Earth through extraterrestrial infall. He extracted organic matter from the Murchison meteorite and added it to water to explore the ability of meteoritic organic matter to form boundary layers. The Murchison molecules formed membrane-bound bubbles (Figure 1.31) providing strong evidence that, on the early Earth, mixtures of abiotic organic compounds could have helped to form membranes for primitive cellular life.

You can imagine how these membranes could have been mixed with a collection of molecules and then dehydrated on the early Earth. When hydrated again small chemical factories may have developed into primitive cells (Figure 1.32).

1.7.3 The role of minerals

The development of the first living system must have involved a sequence of chemical transformations which achieved a greater level of structure and complexity than the available starting materials. Many believe that minerals served a number of critical functions in this process. We can identify four key roles that minerals could have played in the origin of life: protection, support, selection and catalysis. Let's look at each of these in turn.

Minerals could have acted as hosts for assembling chemical systems thereby *protecting* them from dispersal and destruction. For example volcanic rock contains many small air pockets created by expanding gases while the rock was still molten (Figure 1.33) and some common minerals develop microscopic pits following weathering. Tiny compartments such as these could have housed small chemical mixtures which may have taken the first steps towards organized life.

Figure 1.33 Volcanic rock containing small air pockets that were created by expanding gases while the rock was molten.

Figure 1.34 Iron sulfide, which may have acted as a template, catalyst and energy source for the production of biological molecules.

The surfaces of minerals could have acted as a *support* structure for molecules to accumulate and interact. An effective way to assemble molecules from a dilute solution is to concentrate them on a flat surface. Experiments have been performed where solutions containing amino acids are evaporated in a vessel containing clays. The amino acids concentrate on the clay surfaces and then polymerize into short protein-like chains.

Minerals may have aided the *selectivity* of certain biologically useful molecules. Many have crystal faces that are mirror images of each other. Minerals such as calcite bond strongly to amino acids and when calcite crystals are submerged in racemic solutions of amino acids the left- and right-handed forms of the amino acids bond to different crystal faces. It is plausible that under the right conditions this selection and concentration process allowed protein-like molecules to form which were exclusively right or left handed. At some point natural selection chose the left-handed molecules and all subsequent life inherited this trait.

Minerals can act as *catalysts*. One of the elements that is required to generate biologically useful materials is nitrogen. However, although nitrogen is abundant in the Earth's atmosphere it is present as unreactive nitrogen gas. Primitive organisms must have found a way of converting nitrogen gas to a form assimilable by life. In industrial processes nitrogen and hydrogen are passed over metallic surfaces to generate ammonia. If similar reactions took place on the early Earth, the ammonia would have been a valuable source of nitrogen for biological reactions. This may have occurred in **hydrothermal vents** where nitrogen and hydrogen are passed over iron oxide surfaces.

Perhaps all of these functions operated simultaneously on the same mineral. For example, in 1988 Gunter Wächtershäuser, a German patent lawyer, suggested that iron and nickel sulfides could have served as a template, catalyst and energy source for the production of biological molecules (Figure 1.34). He took his proposal further by suggesting that the first living things may have been coatings stuck to the surfaces of these crystals. If the mineral acted as a catalyst it is possible that primitive metabolism proceeded in the absence of enzymes.

1.8 From chemical to biological systems

1.8.1 The RNA world

Recall from Section 1.2 that life as we know it uses DNA as a store of genetic information and RNA as the messenger that carries the information out into the cell. Importantly, nucleic acids have the genetic information necessary to reproduce themselves but need proteins to catalyse the reaction. Conversely, proteins can catalyse reactions but cannot reproduce without the information supplied by nucleic acids. This poses us with a dilemma in our bottom-up study of the origin of life – which one of these three key molecules (DNA, RNA and protein enzymes) could have existed without the other two? This has been called the 'chicken and egg' paradox.

Figure 1.35 The 'chicken and egg' paradox. Which came first, proteins which catalyse the production of nucleic acids, or nucleic acids that contain the genetic information needed to produce proteins?

In the mid-1980s it was discovered that RNA, unlike DNA, can perform some of the enzymatic functions needed for replication. The evidence came in a discovery made independently by the US biochemists Sidney Altman and Thomas Cech, which led to them being jointly awarded the Nobel Prize for Chemistry in 1989. RNA molecules that have catalytic properties similar to enzymes are called 'ribozymes'. In principle, because RNA molecules could store genetic information and act as catalysts, they would make proteins unnecessary for simple life. So a simpler 'RNA world' may have preceded the DNA-plus-protein world of today. RNA might have been able to replicate and evolve without specialized proteins – there are several observations that support this proposition:

- The nucleotides in RNA are more readily synthesized than the nucleotides in DNA.
- It is easy to imagine that DNA evolved from RNA and then, on account of its greater stability, DNA took over the RNA role.
- RNA is likely to have evolved before proteins because no plausible scenario can be envisaged where proteins can replicate in the absence of RNA.

Eventually, the evolving RNA organisms began transcribing DNA, which was a much more efficient replicator. The ability for RNA to create DNA is vividly illustrated by retroviruses (Box 1.4). Natural selection then saw to it that the more proficient DNA-plus-protein world outcompeted its parent RNA.

BOX 1.4 RNA AND RETROVIRUSES

A virus is a parasite without a cellular home. Viruses are fragments of nucleic acid within protein coats and are small (10–200 nm). When without a host, virus particles do not carry out the functions of living cells, such as **respiration** and growth. Yet once within a host cell they steal the cell's chemical energy and hijack its ability to synthesize protein and nucleic acids in order to replicate themselves.

Some viruses do not kill the host cells but persist within them in one form or another. Cancer-causing viruses and the human immunodeficiency virus (HIV) are of this type and evolve at about a million times the rate of nuclear DNA. These are 'retroviruses'. They reverse the normal cellular process of transcribing DNA into RNA: they multiply by transcribing RNA into DNA, which then takes over the cellular machinery to make more viral RNA.

The ability of RNA to transcribe DNA is a valuable piece of supporting evidence for the proposal of a past RNA world. It is tempting to consider retroviruses as the legacy of our darkest ancestors that can still wreak havoc in our modern complex biosphere.

1.8.2 Primitive biochemistries

In Section 1.5.1 we discussed the possible sources of energy available for life. These energy sources are captured by life and then utilized via metabolic processes. The most important source of energy on the Earth today is sunlight and this energy is captured by a process called **photosynthesis**. Photosynthesis is the production of carbohydrates from water and carbon dioxide. A photosynthetic reaction with a small energy barrier is one based on sulfur.

$$nCO_2 + nH_2S \longrightarrow (CH_2O)_n + nH_2O + nS \qquad (1.1)$$

Yet the photosynthesis reaction used most commonly today by plants is based on water.

$$nCO_2 + nH_2O + \text{energy} \longrightarrow (CH_2O)_n + nO_2 \qquad (1.2)$$

When organisms are in environments where sunlight is unavailable, alternative mechanisms must be employed to generate organic compounds. For example, in 1977 scientists in the submersible *ALVIN* were studying a mid-ocean ridge near the Galapagos Islands in the Pacific Ocean. They discovered underwater volcanoes with deep-sea hydrothermal vents (Figure 1.36) that were populated with a range of organisms. Seawater circulating through new, hot ocean crust at mid-ocean ridges is heated, and this hot seawater dissolves and exchanges chemicals with the rock. In some places along the ridges, this mineral-rich hot water vents back into the sea at temperatures of up to about 400 °C and only the great pressures generated by the column of seawater above stops the water from boiling. These sources of mineral-rich hot seawater support communities of organisms in which life depends not on light energy, but on chemical energy. The synthesis of organic matter in this way is called **chemosynthesis**. Today we know that hydrothermal systems are not restricted to mid-ocean ridges. They also occur deep in the Earth's crust where water and heat are present. Recently, it has been discovered that these hydrothermal regions also host simple chemosynthetic life forms. These ecosystems are now termed the 'deep hot biosphere' and it appears that the amount of organic matter deep in the Earth may actually rival that at the surface. You will encounter some of these unusual subsurface dwellers again in Section 2.5. Organisms that utilize photosynthesis and chemosynthesis to generate organic compounds from energy and simple inorganic substances are termed **autotrophs** (from the Greek *auto* meaning 'self' and *troph* meaning 'feed').

The carbohydrates generated by the energy-capturing processes of photosynthesis and chemosynthesis are used to generate energy-rich phosphate bonds. This stored chemical energy can then be tapped by the organism when needed. **Fermentation** and respiration are the two most common forms of metabolism on the Earth today. In fermentation, the carbohydrate glucose ($C_6H_{12}O_6$) is transformed into both CO_2 and ethanol (CH_3CH_2OH), or lactic acid ($C_3H_6O_3$) for a net gain of two energy-rich phosphate bonds. In respiration, the free oxygen in the Earth's atmosphere is used

Figure 1.36 A deep-sea hydrothermal (hot water) vent. (D. Thomson/GeoScience Features)

to extract more energy from the glucose molecules than is obtained by fermentation. The glucose is transformed into CO_2 and water with a net gain of 36 energy-rich phosphate bonds – a significant improvement.

$$(CH_2O)_n + nO_2 \longrightarrow nCO_2 + nH_2O + \text{energy} \qquad (1.3)$$

The metabolic mechanisms outlined above allow organisms to capture and store energy for use in the complex chemical reactions that keep biochemical systems operating. But what of organisms that seek to obtain their sustenance by consuming autotrophs and taking advantage of the hard work they have performed? These organisms are termed **heterotrophs** (from the Greek *hetero* meaning 'different').

1.9 The top-down approach – molecular phylogeny

Up to now, we have focused on the bottom-up approach to understanding the origin of life, that is attempting to build life from scratch. Now we turn to the top-down approach. We will try to extrapolate as far back as we can go to the origin of life by using information contained in the Earth's extant biology. Life on Earth has a history that extends back over almost 4 Ga and it has long been believed that this evolutionary history started with a single and simple common ancestor. Expressed another way, all life on Earth is related. Darwin subscribed to this idea and, in 1857, expressed the view that a time would come 'when we shall have very fairly true genealogical trees of each great kingdom of nature'.

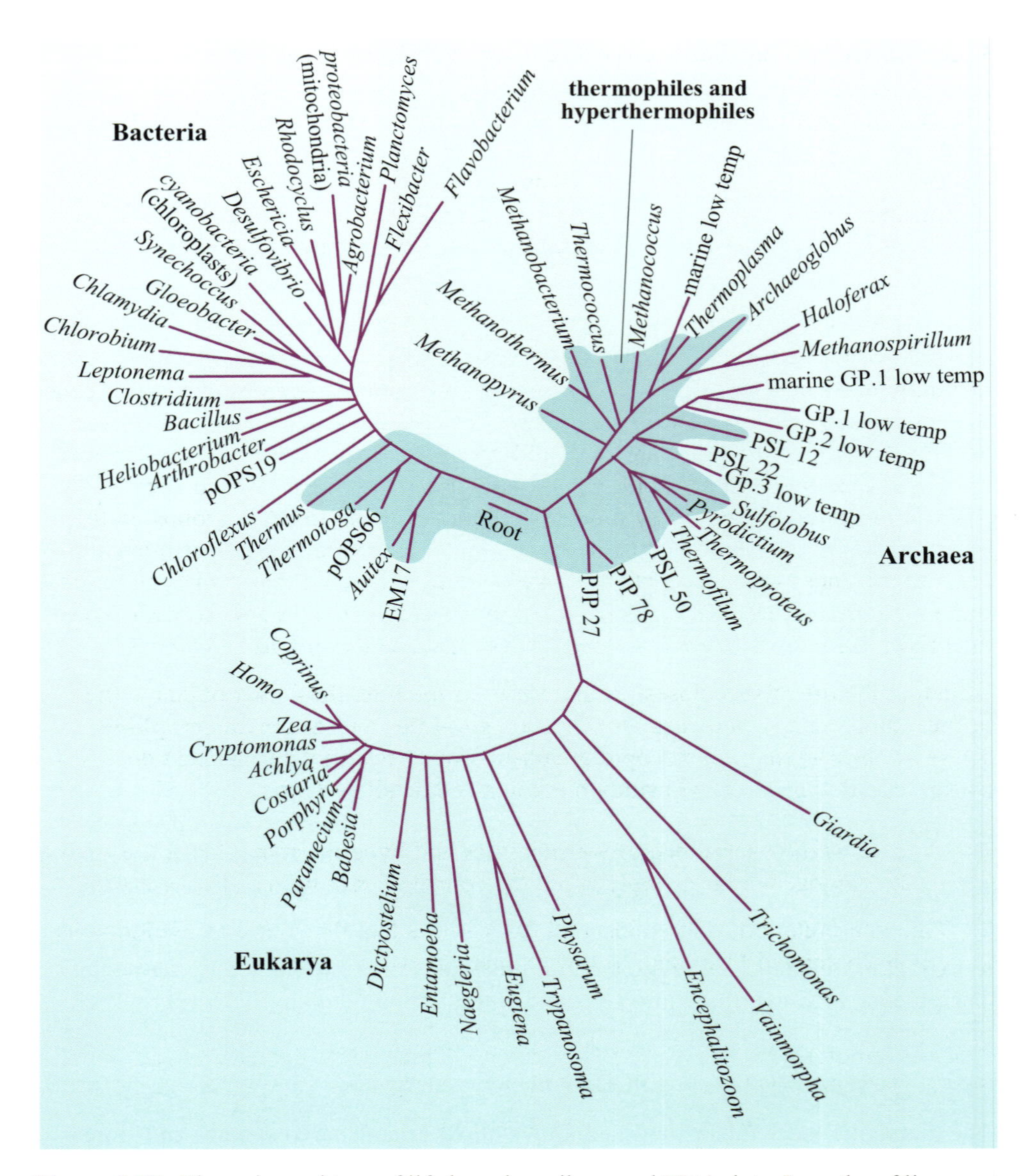

Figure 1.37 The universal tree of life based on ribosomal RNA data. Lengths of lines separating pairs of organisms correspond to the genetic differences between them. Life is divided into three domains (Bacteria, Archaea and Eukarya) as a result of constructing this tree. Deep, short branches indicated in the centre of the tree are populated by thermophilic and hyperthermophilic organisms.

Today Darwin's dream has been realized with the advent of genetically based **phylogenetic trees**. Life, it seems, does not reject what evolution has created, but simply builds on what has gone before. The biological record of this continuous addition and modification is present in genetic material, namely the sequence of nucleotides in RNA and DNA. So the basic method involved in constructing a phylogenetic tree is to examine similar molecules in different creatures and, if parts are found to be alike, then those parts must have been inherited by organisms from a common ancestor.

One of the more useful trees is built on the genetic information obtained from small sub-units of ribosomal RNA and comparisons between different organisms reveal a hierarchy of evolutionary innovation (Figure 1.37). The longer a branch, the greater the difference in ribosomal RNA sequences between the organisms at the start and the end. Three clear domains are evident: Bacteria, Archaea and Eukarya. Within the three domains, the branches with names associated with them refer to species, or groups of species, more closely related to each other than to other groups on the tree. Following these branches back to the points where they join other branches leads to the ancestral species of the named groups. Branches that lead back to the same main branch represent species that share a common ancestor. When studying the origin of life it is the *root* of this phylogenetic tree that interests us most.

Note that the organisms closest to the centre of the tree, those that populate the deepest and shortest branches, are the **thermophiles** and **hyperthermophiles**. These are heat-loving microscopic organisms found near hot springs and deep-sea hydrothermal vents. You will study these in more detail in Section 2.5.

- What does the occurrence of thermophiles and hyperthermophiles at the centre of the phylogenetic tree imply about the course of evolution of life on Earth?

- One interpretation of the ribosomal RNA tree is that the course of evolution has generally moved from high to low temperatures.

Another important feature of the ribosomal RNA tree is that the majority of the deepest branching organisms do not use light as an energy source. This suggests that photosynthesis may be a later development than processes utilizing geochemical energy sources. So can we use the tree to draw conclusions about the nature of the earliest life on Earth?

The phylogenetic tree seems to be telling us that our **last common ancestor** may have been similar to heat-loving chemosynthetic organisms that populate hydrothermal vents today.

Yet the term 'last common ancestor' highlights the fact that this was not necessarily the first organism on Earth.

It is possible that life began in a low-temperature environment then through adaptation colonized deep-sea hydrothermal systems. Once the hydrothermal environments were colonized, an ocean-sterilizing impact could have wiped out all but the heat-loving vent life which was then free to evolve and repopulate the now vacant cooler environments.

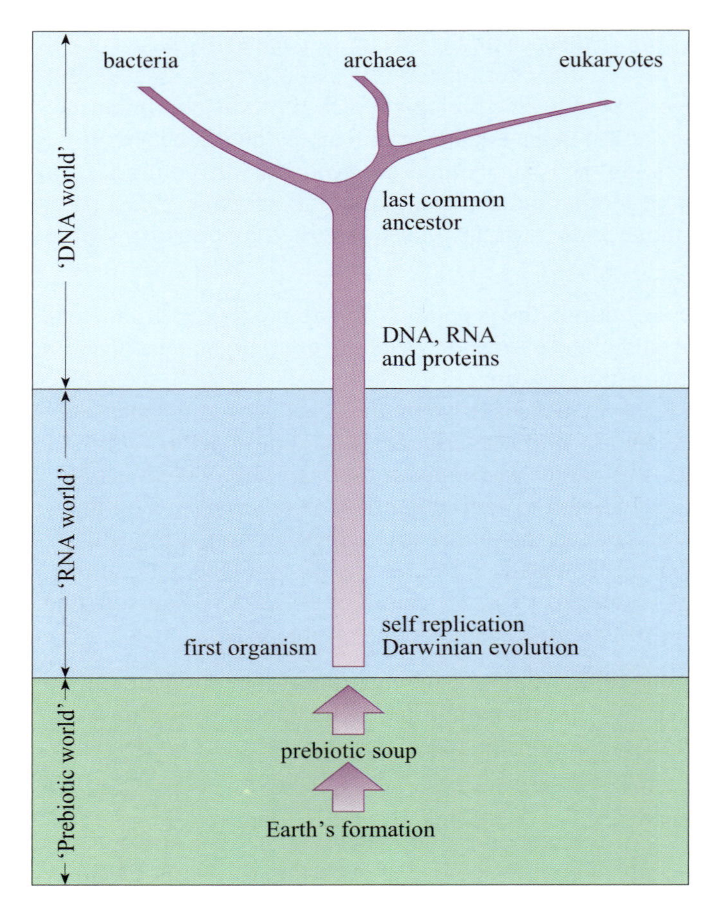

Figure 1.38 A synthesis of information from our bottom-up and top-down approaches to the origin of life. (Lahav, 1999)

1.10 A synthesis on the origins of life

In this chapter we have examined what life is and what it requires to exist. We have also attempted to construct plausible scenarios for the origin of life on Earth. But just how close have we come to understanding how life on Earth actually started? Figure 1.38 is a synthesis of the various lines of evidence we have put carefully into place. You will see that information from both the bottom-up and top-down approaches have been incorporated into the scheme. Our summary begins at the bottom, with the Earth's formation and the generation of a prebiotic soup, perhaps aided by the influx of organic matter from extraterrestrial objects. At some point in time, the prebiotic world produces the first living organism, possibly using minerals as catalysts or templates, and we are now in a biotic world in which self-replication and Darwinian evolution can take place. RNA probably played a significant role in primitive organisms, acting as both the store of genetic information and the catalyst for replication. Once DNA supplanted RNA as the main harbinger of genetic information the RNA world came to an end. Now, in the DNA world, the labour of life is shared between three molecules – DNA, RNA and proteins. No record of the first living organism or the first DNA-using organism remains so we must now take a giant leap to meet the top-down record provided by molecular phylogeny. The last common ancestor of life on Earth was probably a heat-loving organism similar to those found today at deep-sea hydrothermal vents.

1.11 Summary of Chapter 1

- Life can be described as a system that has the capacity to undergo self-replication and evolution. However, any definition of life is likely to be inadequate in specific circumstances.
- All known life is based on water and carbon. The constituent elements of life, the so-called biogenic elements, are abundant in the Universe.
- Life utilizes the ability of carbon to bond with many other elements and itself to create a variety of organic compounds that have specific biological functions. The main molecules of life are large macromolecules and include lipids, carbohydrates, proteins and nucleic acids.
- There are a number of extraterrestrial environments in which organic matter is produced without the aid of biological processes. These environments include the shells of carbon stars, molecular clouds and the solar nebulae.
- The organic products of extraterrestrial environments would have rained down on the Earth close to the time of the origin of life, adding to any organic matter synthesized on the Earth itself. Common characteristics, such as a left-handed preference in amino acid structures, appear to suggest a link between terrestrial biotic and extraterrestrial abiotic organic matter.
- Molecular complexity may have initially been achieved by simple chemical reactions although later, for greater efficiency, protein enzymes would have catalysed the reactions. Chemical reactions would also have been promoted by certain concentration mechanisms and the encapsulation of molecules within a membranous boundary layer.
- Early organisms are unlikely to have used the DNA, RNA and protein-based biochemistry common on the Earth today. It is more likely that RNA performed the functions of genetic information store and catalyst.
- The capture of energy by life involves the production of carbohydrates by autotrophic mechanisms. These carbohydrates are transformed into energy-rich phosphate bonds by metabolic processes such as fermentation and respiration.
- Molecular phylogeny indicates that the last common ancestor to all life on Earth was a heat-loving organism similar to those organisms that populate today's deep-sea hydrothermal vents.

CHAPTER 2
A HABITABLE WORLD

2.1 Introduction

In the previous chapter, you examined current theories as to how life on Earth might have originated from simple biogenic precursors. This raises the obvious question: why Earth? What, if anything, was special about conditions on the early Earth that enabled life to originate and evolve on this planet? Is an Earth-like planet essential for life to evolve elsewhere in the Universe? In this chapter you will examine what it is that makes the Earth a habitable planet, what conditions were like on the early Earth, and whether the life that has evolved on Earth can provide information as to the likelihood of life arising elsewhere in our own Solar System or beyond.

The Earth of today (Figure 2.1) is a very different place from the Earth that existed some 4.5 Ga ago shortly after the Solar System formed. Examining our planet from space, there are several pointers to the presence of conditions favourable to life and even the existence of life itself on its surface. One clue is the presence of liquid water, although this alone is not sufficient. However, a lot can be learned from a planet's colour, or more accurately the regions of the electromagnetic spectrum that the Earth reflects back into space. The Earth fluoresces in ultraviolet light, indicating that it has an atmosphere that contains almost 20% oxygen. This is demonstrated in Figure 2.2, a false colour image that illustrates how oxygen atoms in the Earth's atmosphere fluoresce as they absorb the Sun's ultraviolet radiation. Equally significant is the fact that green light (associated with complex organic molecules formed largely of carbon, in combination with lesser amounts of hydrogen, nitrogen, sulfur and other elements) is reflected from large areas of both land and sea (Figure 2.3).

Figure 2.1 From 4 million miles away on 16 December 1992, NASA's Galileo spacecraft, on its way to Jupiter, took this picture of the Earth–Moon system. The bright, sunlit half of the Earth contrasts strongly with the darker subdued colours of the Moon. (NASA)

The atmosphere is not a stable mixture of chemicals. Unless the carbon-rich compounds are continuously regenerated, nearly all of the organic material reflecting green light would have decomposed in the oxygen-rich atmosphere within a few hundred years.

Figure 2.2 This false colour image shows how the Earth glows in ultraviolet (UV) light. Reacting to, and absorbing the Sun's ultraviolet radiation, oxygen atoms fluoresce, appearing in Figure 2.2 as a cloak of gold closest to the Earth's surface, where the oxygen gas lies heaviest. As it thins with altitude, oxygen is coloured green, red, and finally blue. This image was taken with the Far UV Camera/Spectrograph deployed and left on the Moon by the crew of Apollo 16. (© NASA George Curruthers)

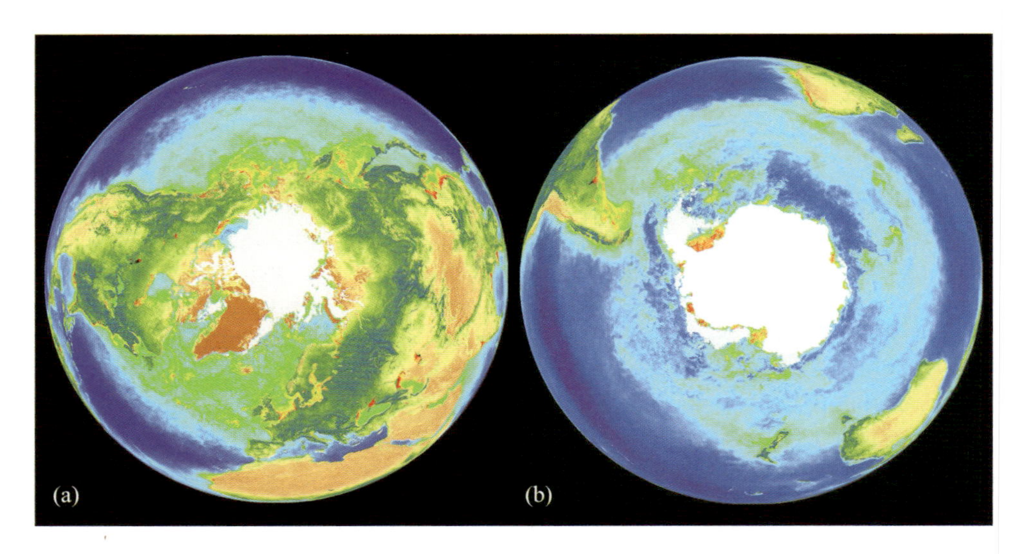

Figure 2.3 Oceans and life. This false colour image compiled from the SeaWiFS (Sea-viewing Wide Field-of-view Sensor) instrument on board the orbiting SeaStar satellite records reflectance spectra from the Earth's oceans. In these North (a) and South (b) Pole projections the colour depends on how sunlight is reflected by free-floating phytoplankton – photosynthesizing organisms that contain chlorophyll. Chlorophyll absorbs blue and red light and reflects green so that ocean areas with abundant plankton are shown in green as are land areas with significant vegetation. (NASA)

This would remove some of the oxygen from the atmosphere while the rest would have been used up in chemical weathering reactions (oxidation) involving the rocks of the Earth's surface. It seems, therefore, that some process operating on the Earth's surface continually regenerates complex carbon-rich compounds and maintains the oxygen content of the atmosphere.

Of course, we know that there is life on Earth: the carbon-rich compounds make up the plants and animals and these in turn use oxygen for respiration. So can we use the Earth as a model for our search for life elsewhere in the Solar System and beyond?

2.2 Defining a habitable planet

Life on earth has managed to survive a number of major climate changes such as large-scale glaciations. However, on occasions, dramatic changes in the Earth's environment have caused mass extinction events, wiping out large numbers of species.

Of necessity we will adopt the Earth as the reference of a habitable terrestrial planet since it is the only example we have. At least within our own Solar System, it appears that the Earth's habitability may be near optimal, especially for complex life. We also need to be clear what we mean by habitable. The conditions needed to sustain Earth-like animal life that uses oxygen are quite different from the much broader range of conditions that can support microbial life. For the former a habitable terrestrial planet would require an ocean and some dry land, moderately high O_2 (and low CO_2) abundance, and a reasonably stable climate. The moderately high O_2 and low CO_2 is a requirement for large mobile life on physiological grounds and also for the production of an ozone layer to provide shielding from the effects of harmful ultraviolet radiation. The Earth's oceans effectively regulate the planet's temperature on a global scale via the operation of a water cycle, which interacts with processes such as plate tectonics, and chemical weathering on land. Earth's long-term climate stability results from many astrophysical and geophysical constraints, including stellar evolution, comet and asteroid impact rate, the presence of a large natural satellite, and a long-term planetary heat source to drive plate tectonics. However, larger plants and animals have only been around on Earth for the last 500 Ma. For over 3 Ga the Earth has supported microbial life which can survive and evolve under much more extreme conditions, a topic you will examine in greater detail in Section 2.5. Even these simple forms of life share some common requirements: the presence of liquid water and long-term environmental stability (i.e. environmental conditions that have never been so extreme as to extinguish all life).

2.3 Habitable zones

2.3.1 Water and light

We have identified two properties that sharply differentiate the Earth from other planets in our Solar System and that enable it to support the abundant life on its surface. These are the liquid water that covers much of its surface and the planetary environment that maintains it. However, liquid water is rare in the Solar System; as you will see in Chapter 3 there is evidence, shown in Figure 2.4, that it once existed on Mars and you will explore the possibility that it may exist below the surface of Jupiter's satellite Europa in Section 4.1. For now, however, we will concern ourselves with the Earth and what makes it such a habitable planet.

Pure water exists as a liquid between 273 K and 373 K unless the pressure is too low, in which case the water sublimes to water vapour. The presence of liquid water on a planetary surface could therefore be used as a simple requirement for us to consider the planet as being habitable. As you will see, this single factor is unlikely to be either necessary or sufficient. However, it provides a useful guide for the conditions needed to support Earth-type life on terrestrial planets (or sufficiently large moons) in orbit around a star.

A **circumstellar habitable zone** is defined as encompassing the range of distances from a star for which liquid water can exist on a planetary surface.

The primary consideration in determining a planet's habitability is therefore temperature.

- What determines the average temperature of the atmosphere and surface of a terrestrial planet?
- The balance between incoming solar radiation and thermal emission from the planet.

The amount of sunlight received by a planet is determined by its distance from the star that it orbits and the amount of energy emitted by that star: its **luminosity**. The luminosity of a star represents the total output of radiant energy per second. In comparison with the Sun, therefore, a star 100 times as luminous would emit 100 times as much energy per second as the Sun does.

- Relative to incoming energy, how much energy must a planet radiate back into space in order to remain in equilibrium with its surroundings?
- For a planet not to get any hotter or cooler it must radiate the same amount of energy that it absorbs.

If we therefore assume that a planet undergoes no net heating or cooling in the short term, it is possible to estimate the temperature necessary for a planet to re-radiate all of the energy absorbed by the atmosphere and the surface (Box 2.1).

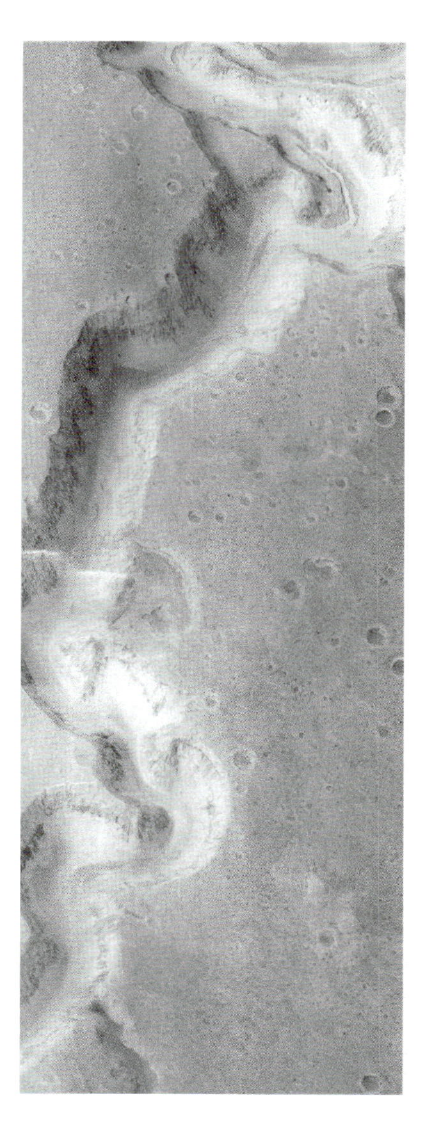

Figure 2.4 This Mars Global Surveyor image shows a portion of the meandering canyons of the Nanedi Valles system on Mars. The valley is about 2.5 km wide. The floor of the valley in the upper-right corner of the image exhibits a small channel, 200 m in width, which is covered by dunes and debris elsewhere on the valley floor. The presence of this channel is interpreted as indicating that the valley might have been carved by water that flowed through this system for an extended period of time. (NASA)

BOX 2.1 DETERMINING A PLANET'S EFFECTIVE TEMPERATURE

The temperature of the surface and atmosphere of a planet is determined by the balance between the energy that is absorbed and the energy that is emitted.

If a planet has no significant source of internal heat then the source of most of the energy reaching the atmosphere and the surface is the Sun. The solar flux density at the top of the Earth's atmosphere is about 1.38×10^3 W m^{-2}. Some of this energy is reflected back to space, the atmosphere absorbs some, and the rest reaches the surface, where it is either reflected or absorbed. The absorbed radiation heats the surface, which then re-radiates this energy, mainly in the infrared region.

On the assumption that a planet undergoes no net heating or cooling in the short term, it is possible to estimate the temperature necessary for a planet to re-radiate all of the energy absorbed by the atmosphere and the surface. This temperature, called the effective temperature, T_e, is defined as follows:

$$T_e^4 = \frac{L}{4\pi R^2 \times 5.67 \times 10^{-8}} \qquad (2.1)$$

where L is the *total* power radiated by the planet in watts, R is the radius of the planet in metres (its surface area is $4\pi R^2$, and radiation is emitted from the whole surface) And 5.67×10^{-8} is a constant which has the units Wm^{-2} K^{-4}.

This equation was originally derived for thermal sources or black bodies, but it now serves to define effective temperature, regardless of the form of the spectrum of the emitted radiation.

■ Look at Equation 2.1. How will the effective temperature of a planet vary as a function of the luminosity of its star?

□ The effective temperature in K is raised to the power four, it will therefore vary as the fourth root of the luminosity of the planet's star.

This is illustrated in Figure 2.5 which shows how the effective temperature (T_e) of an Earth-sized black body would vary if it were in orbit around stars of differing luminosity.

The balance between the radiation absorbed and emitted by a planet determines the temperature of its surface and atmosphere. Thus, in order to estimate T_e, the power lost by radiation must be equated with that absorbed from solar radiation. Since a star's radiation arrives from one direction, a planet is heated over only half of its surface at any time. The planet therefore casts a disc-shaped shadow of area πR^2, where R is the radius of the planet. The power absorbed depends on this area, and also on the solar flux density at the distance of the planet from the Sun.

■ Is all of the solar radiation that reaches a planet absorbed?

□ No, a fraction of solar radiation is also reflected.

The total fraction of solar radiation that is reflected by a planet is called the **albedo**, a. The total fraction absorbed is simply $(1 - a)$. By equating the power, L, radiated by the Earth with the solar power absorbed, which can be readily estimated independently, an effective temperature of 255 K can be estimated from Equation 2.1.

Another way of looking at the effect of stellar luminosity on a planet's effective temperature is to consider the distance of a planet from its star. For example, if we replaced our Sun with a star of greater luminosity, how far from that star would

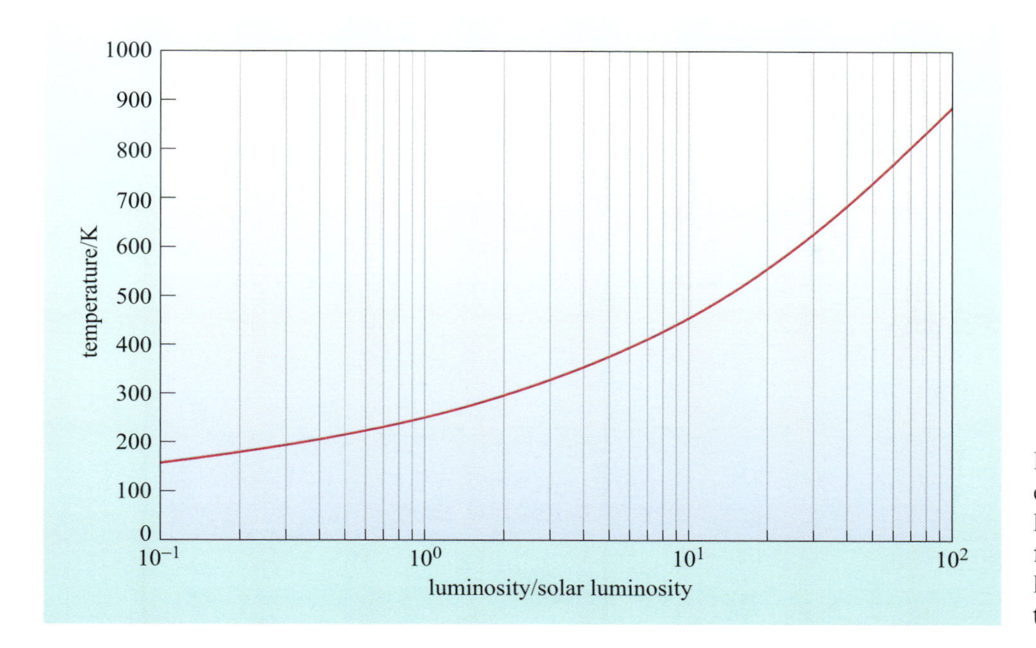

Figure 2.5 The variation in effective temperature (T_e) of an Earth-sized black body at 1 AU from stars of different luminosities. Luminosity is expressed relative to that of the Sun.

Earth have to be to maintain its current effective temperature? Since energy is conserved, electromagnetic radiation is not diminished as it travels through space. Consider a light source suspended in space that emits a flash of light that consists of a specific amount of energy. That energy will travel outwards in all directions from the light source, like a rapidly expanding sphere. At any particular moment the total energy in the expanding sphere is exactly the same as the energy initially emitted by the light source.

- What will happen to the surface area of the sphere as its radius increases?
- As the sphere expands its surface area increases.

Thus the initial energy is spread over a larger and larger area of space so that the amount of energy in a square metre of the surface of the sphere decreases. Since the surface area of the sphere is related to its distance from the light source then it follows that the further a planet is from its star, the less energy it will receive and the lower its effective temperatures will be. Thus the amount of energy (E_{in}) received by a planet, referred to as its solar flux density, is defined by:

$$E_{in} = \frac{\text{luminosity}}{4\pi R^2} \quad (2.2)$$

where R is the distance of the planet from the star. The distance scales with the square root of the star's luminosity, assuming that the effective temperature is to remain the same, and is illustrated in Figure 2.6. So if we were to replace the Sun by a star 10 times as luminous, the Earth would have to increase the radius of its orbit by the square root of 10, that is 3.16, if we wanted to maintain its present effective temperature at 255 K. This would correspond to an orbit in the middle of the asteroid belt between the orbits of Mars and Jupiter.

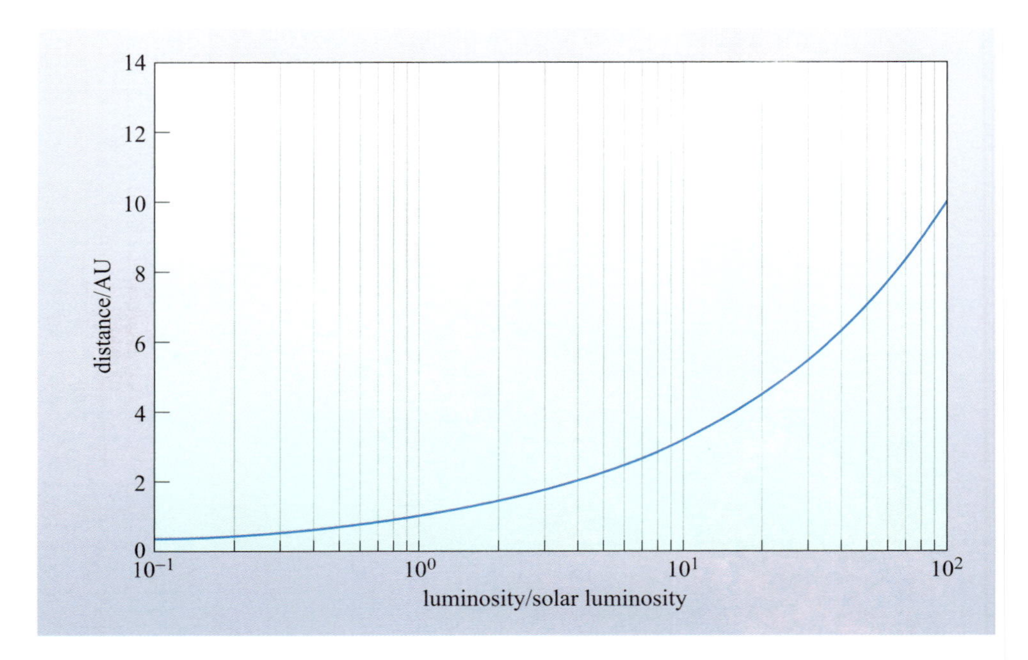

Figure 2.6 The distance from a star that is required to keep an Earth-sized body with an effective temperature of 255 K with increasing stellar luminosity.

- ■ The mean surface temperature of the Earth today is 288 K, some 33 K higher than its effective temperature. Why is this?
- ❑ The effective temperature calculation in Equation 2.1 does not take account of the trapping of heat energy by the Earth's atmosphere, the so-called greenhouse effect, which will raise the surface temperature as will internal heat from the Earth's interior.

We will look at the consequences of albedo and the greenhouse effect on the extent of circumstellar habitable zones in Section 2.3.2.

If we were to replace our Sun with a star ten times as luminous as our Sun, the effective temperature on the Earth would increase as the fourth root of 10, that is 1.78. So the effective temperature on our planet would become $(255 \times 1.78) = 453$ K, which is about 180 °C.

QUESTION 2.1

What would the effective temperature of the Earth be, if the Sun were to be replaced by a star 10 000 times as luminous as the Sun?

QUESTION 2.2

Where would the Earth have to orbit to have today's temperatures, if the Sun were to be replaced by a star 10 000 times as luminous?

2.3.2 The Sun's habitable zone

In fact, the Sun's luminosity has not remained constant throughout the history of the Solar System. In common with all **main sequence stars**, it has slowly increased from a level around 4 Ga ago estimated to be around 70% of its present value.

- What effect will a slowly increasing luminosity have on a star's circumstellar habitable zone?
- The habitable zone will migrate away from the star.

One consequence of the increase in the Sun's luminosity is that there has been a region around the Sun that has remained habitable throughout the history of the Solar System (Figure 2.7).

This region, in which a planet may reside and maintain liquid water throughout most of a star's life, is called the **continuous habitable zone.**

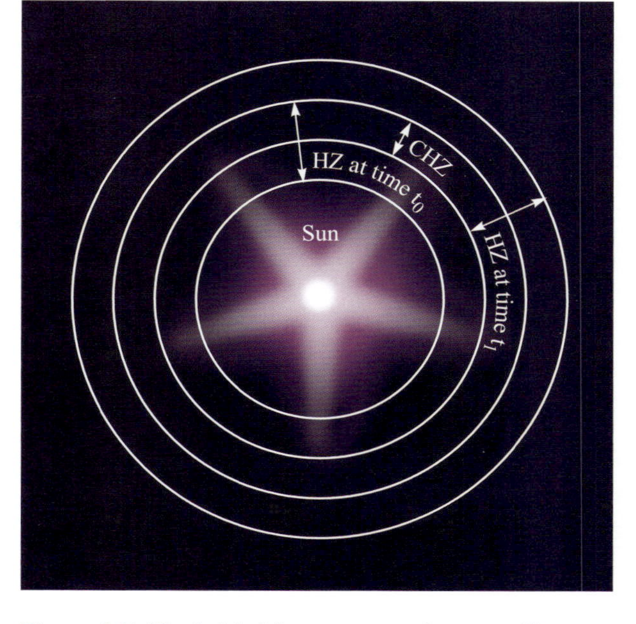

Figure 2.7 The habitable zone around a star will move outwards as the star's luminosity increases. The region that remains continuously habitable is the continuous habitable zone, CHZ.

We could use the relationships given in Equations 2.1 and 2.2 to determine the inner and outer radii of the Sun's habitable zone by equating the total power radiated by a planet (L in Equation 2.1, which is proportional to its effective temperature raised to the fourth power, T_e^4) to the solar flux density (proportional to the star's luminosity/R^2, from Equation 2.2). By using lower and upper temperatures of 273 K and 373 K for the freezing point and boiling point of water it would be possible to determine the inner and outer radii of the habitable zone. In essence, this is what was done when the concept of a habitable zone was first proposed in the late 1950s.

However, we've already seen that this is a simplistic approach since we obtain an effective temperature for the Earth of 255 K, well below the freezing point of water. It is important to note that the Earth is *not* an ideal black body. The majority of energy from the Sun is incident on the Earth at visible wavelengths, to which the Earth's atmosphere is transparent. However, the Earth radiates this heat away at infrared wavelengths. Since our atmosphere is not completely transparent at these wavelengths, because the so-called greenhouse gases (carbon dioxide, methane, water vapour and nitrous oxide) and the chlorofluorocarbons absorb infrared light, the planet is therefore forced to warm up in order to remain at equilibrium.

- What will be the consequences of variations in planetary albedo and the greenhouse effect on the extent of the habitable zone around a star?
- By reflecting a portion of stellar luminosity back into space, higher planetary albedos would move the inner edge of the habitable zone towards the star. On the other hand, the greenhouse effect, by raising a planet's temperature, would extend the outer edge of the habitable zone away from the star.

In the early 1990s, James Kasting and his co-workers proposed a means of determining the Sun's habitable zone that took into account the effects of albedo and greenhouse gases. They used a climate model to estimate the width of the habitable zone around our Sun and around other main sequence stars. They employed the basic premise that they were dealing with Earth-like planets with $CO_2/H_2O/N_2$ atmospheres and that habitability required the presence of liquid water on the planet's surface. The inner edge of the habitable zone was determined in this model by loss of water via its breakdown to oxygen and hydrogen by photolysis and the loss of hydrogen to space. With the water gone, then life as we understand it would not be possible. Kasting's climate model gives an estimate for the inner edge of both the habitable zone and continuous habitable zone in our own Solar System of 0.95 AU.

The distance at which CO_2 and other greenhouse gases can no longer compensate for the lower solar flux determines the outer edge of the habitable zone. Kasting's model indicated that the width of the habitable zone is greatly extended by the existence of a natural *carbon dioxide thermostat* that tends to regulate the temperatures of Earth-like planets, keeping them from getting too hot or too cold for liquid water to exist. He suggested that atmospheric CO_2 levels would tend to rise as a planet's surface becomes colder. The reason is that removal of CO_2 by silicate weathering, followed by carbonate deposition (Box 2.2), should slow down as the climate cools, and would cease almost entirely if the planet were to globally freeze. On planets such as Earth that have abundant carbon (in carbonate rocks) and some mechanism for recycling this carbon, for example plate tectonics, volcanism should provide a more-or-less continuous input of CO_2 into the atmosphere.

BOX 2.2 SOURCES AND SINKS OF CARBON DIOXIDE

Major CO_2 sources

On planets like Earth that have a means of recycling carbon, decarbonation (thermally decomposing carbon-containing rock and releasing CO_2) and volcanic outgassing are major sources of CO_2. Oceanic sea floor is continuously created at mid-ocean ridges and destroyed in subduction zones, where oceanic plates are pushed into the Earth's mantle. Here, calcium carbonate ($CaCO_3$) and silicate minerals are heated to high temperatures and pressures and chemically react with each other to produce CO_2:

$$\text{silicate minerals} + CaCO_3 \rightarrow \text{'new' silicate minerals} + CO_2 \qquad (2.3)$$

The CO_2 and other volatiles produced by decarbonation escape and are ultimately emitted to the atmosphere by volcanoes, hot springs etc.

■ Can you think of another major reservoir of carbon on Earth?

❑ Organic carbon.

On Earth, deeply buried organic sediments can, over millions of years, become uplifted to the surface during episodes of mountain building. Once exposed, the organic carbon in these sediments can be oxidized, returning CO_2 to the atmosphere. This provides a major additional source of CO_2. The organic carbon on Earth is overwhelmingly the result of biological processes; its recycling is an example of how the presence of life itself has a role to play in modifying planetary environments.

Major CO_2 sinks

Carbon dioxide in the atmosphere dissolves in rainwater to produce a weak acid called carbonic acid (H_2CO_3). In contact with rock at the surface, the carbonic acid can remove ions such as calcium and sodium from parent minerals – this process is called chemical weathering and it results in the production of bicarbonate ions (HCO_3^-). If the rock is a silicate, the reaction can be simplified as:

$$\underset{\text{(carbonic acid)}}{CaSiO_3 + 2H_2CO_3} + H_2O \rightarrow Ca^{2+} + 2HCO_3^- + \underset{\text{(silicic acid)}}{H_4SiO_4} \qquad (2.4)$$

In water, the calcium ions released by weathering can recombine with bicarbonate ions to form calcium carbonate as a solid deposit:

$$Ca^{2+} + 2HCO_3^- \rightarrow CaCO_3 + CO_2 + H_2O \qquad (2.5)$$

Note that for every two atoms of carbon removed from the atmosphere as CO_2 dissolved in rainwater, one atom of carbon is precipitated as calcium carbonate and the other is returned to the atmosphere as CO_2 gas.

- ■ Can you think of another major sink of CO_2 on Earth?
- ❑ Photosynthesis (Equation 1.2) removes CO_2 from the atmosphere (or CO_2 dissolved in water) and converts it to organic matter, which can then be buried to form fossil organic matter.

Kasting's climate model gave two estimates for the outer edge of the Sun's habitable zone. The first, based on the point at which CO_2 would start to condense from the atmosphere, gave a limit to the outer edge of the Sun's habitable zone of 1.37 AU. The second estimate considered the point at which a maximum greenhouse effect would operate, i.e. the point where there would be enough CO_2 and H_2O in a planet's atmosphere to raise temperatures to 273 K, and gave a limit to the outer edge of the habitable zone of 1.67 AU.

- ■ Where do the orbits of Venus and Mars fall in relation to the present habitable zone?
- ❑ Venus, with a mean distance from the Sun of 0.72 AU, is well inward of the inner boundary of the habitable zone for our Solar System. Yet the runaway greenhouse effect on that planet has resulted in surface temperatures some 500 K higher than its effective temperature. Mars, with a mean distance from the Sun of 1.52 AU, falls within the 'maximum greenhouse' limit of the habitable zone, but remains outside the first CO_2 condensation limit.

It is evident from the preceding discussion that trying to establish firm limits to the extent of a habitable zone, even in our own Solar System, requires a considerable understanding of processes on the planets concerned if the models are to be accurate. Indeed, Kasting's model gave an estimate for the width of the 4.6-Ga continuously habitable zone as 0.95 AU to 1.15 AU.

- Does Kasting's model for the extent of the continuous habitable zone fit with our knowledge of early Mars?
- We've already seen in Figure 2.4 that there is evidence of channels in the old cratered terrain that indicate that flowing liquid was widespread on Mars during the first 1–2 Ga of the Solar System.

Early Mars remains a real puzzle. Although it lies beyond the continuous habitable zone in Kasting's model, its surface was once carved by streams of some flowing liquid. Whether this implies that the early Martian climate was warm, or whether it was kept warm by geothermal heat, is still debated. If the climate was indeed warm, then the models are overlooking a key element of the climate system. Two additional warming mechanisms have been suggested:

1 The presence of additional greenhouse gases, especially CH_4. It is estimated that 0.1–1% methane may have been sufficient to supply the additional greenhouse warming.

For this process to work the cloud cover would have to be nearly complete.

2 The presence of CO_2 ice clouds analogous to cirrus clouds on Earth, which can create a substantial greenhouse effect. Such clouds primarily scatter outgoing infrared radiation and their net effect is to warm since they scatter more efficiently at infrared wavelengths than at solar wavelengths. We'll return to the environment of early Mars in the next chapter.

QUESTION 2.3

Why is Mars presently too cold to sustain life? (Hint: Consider the role that carbon sources and sinks play in regulating the Earth's climate.)

Plate tectonics or volcanism is required to recycle carbon.

Your answer to Question 2.3 should suggest a possible additional requirement for a planet to remain habitable, namely that it is large enough to maintain active plate tectonics or at least some form of volcanism throughout its lifetime. Where this cut-off lies is uncertain, but it is somewhere between 1 and 0.1 times the mass of the Earth as Mars is about one-tenth of the Earth's mass. We'll look at the role of plate tectonics on maintaining Earth's habitability in more detail in Section 2.4.2.

Are there circumstances in which liquid water may exist beyond the strict definition of a star's circumstellar habitable zone? So far, we've been concerned almost entirely with temperatures related to solar radiation, and the subsequent possibility of liquid water, on a planet's surface. However, tidal heating of the satellites around giant planets, such as Jupiter's satellite Europa, has raised the possibility of liquid water existing below the surface of this ice-covered satellite. The interest in Europa comes from information and images acquired by the NASA Galileo spacecraft, which revealed its surface to be one of the brightest in the Solar System as a result of the satellite having a water-ice crust, 150 km in thickness. As you'll examine in detail in Chapter 4, evidence for cryovolcanism on Europa's surface has led to the suggestion of a liquid or semi-liquid water layer beneath a crust of ice.

2.3.3 Habitable zones elsewhere in the Universe

We'll examine the possibility of habitable planets around other stars in more detail in Chapter 8. Here, we'll consider some of the characteristics of a star that will determine the extent of its habitable zone. We've already seen how changes in our Sun's luminosity throughout its history lead to the concept of a continuous habitable zone. However, the mass of a star will also determine both the size and the duration of a circumstellar habitable zone. The types of stars that can support Earth-type life on planets may be limited to those of lower masses since only these stars have long enough lives as stable luminous stars for planets to form and complex life to evolve. Although all main sequence stars generate luminous energy by converting hydrogen into helium through thermonuclear fusion, stars more massive than 1.5 times that of the Sun age too quickly to support the development of complex Earth-type life. Stars with less than half the Sun's mass are likely to tidally lock planets orbiting close enough to have liquid water on their surface so that one side is perpetually dark. However, atmospheric winds may smooth out temperature differences preventing the destruction of a life-sustaining atmosphere through condensation on the cold, perpetually dark side of the planet. The extent of the habitable zone around stars of different masses is summarized in Figure 2.8.

Tidal locking is the synchronous rotation of the star and planet.

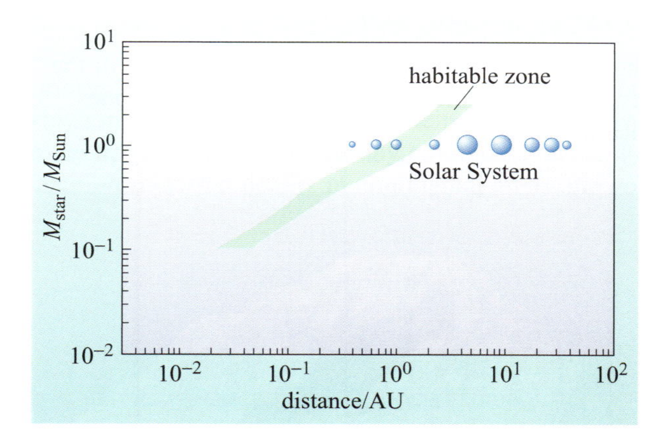

Figure 2.8 The extent of the continuously habitable zone around main sequence stars is bounded by the range of distances from a star for which liquid water would exist and by the mass of the star. Stars more massive than about 1.5 times the mass of the Sun evolve too quickly so planets would not have enough time to form complex life. Stars less than about 0.1 times the mass of the Sun will tidally lock planets close enough to have liquid water as well as subjecting them to stellar flares. (Adapted from Kasting *et al.*, 1993)

About half the stars in our Galaxy are in **binary systems**. In such systems, two kinds of stable planetary orbits exist: those in which the planet orbits both stars, called close binaries, and those in which the planet orbits one star well separated from its companion, called wide binaries. In each case only certain stable orbits exist: a planet must not be located too far away from either one star or too close to two 'home' stars or its orbit will be unstable. If that distance exceeds about one-fifth of the closest approach of the other star, then the gravitational pull of that second star can disrupt the orbit of the planet. In these cases the habitable zone would have to be calculated on an individual basis.

Scientists have recently proposed a theory that argues that certain regions of a Galaxy are more amenable to the development of complex life that others. In effect, they suggest that there are **galactic habitable zones**. Our own Milky Way Galaxy is unusual in that it is one of the more massive galaxies in the nearby Universe, and our own Sun has a relatively high concentration of elements heavier than helium, He. Based on studies of extrasolar planets, astronomers have noted that stars with higher concentrations of elements heavier than He are more likely to have planets

We've already made one important inference from the Isua rocks that helps explain this: the evidence that geological cycles were in operation 4 Ga ago. This implies that the Earth's internal structure was very similar to what we believe it to be today: the Earth had acquired its layered structure, as summarized in Figure 2.10, by 4 Ga ago.

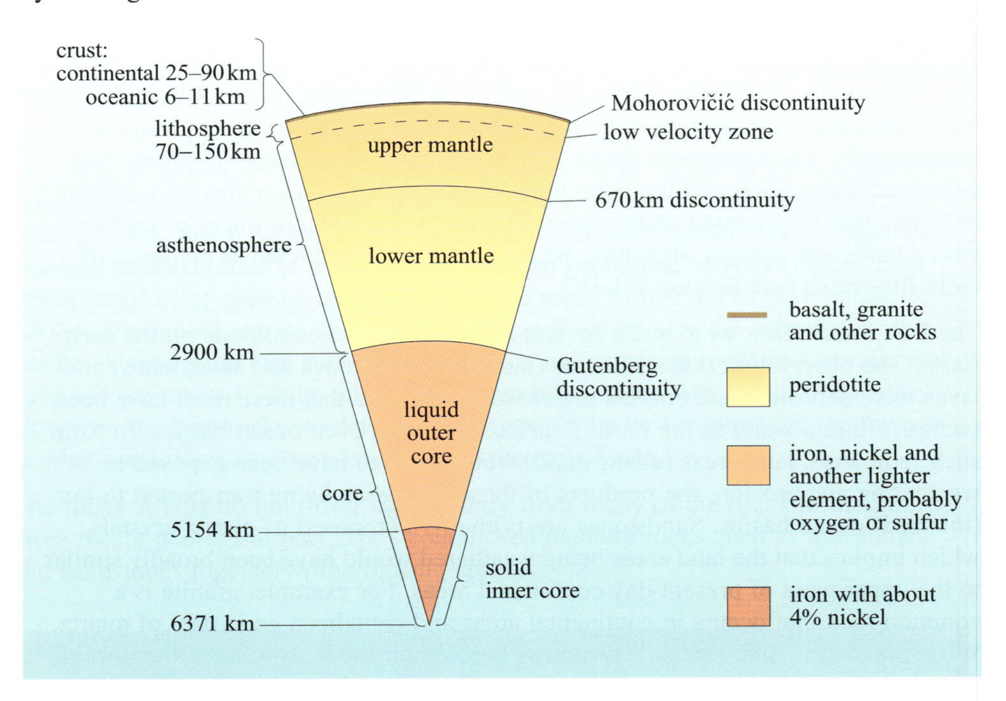

Figure 2.10 A schematic slice through the Earth, showing the major compositional features of the principal layers.

2.4.2 Plate tectonics on the early Earth?

Plate tectonic activity appears to be essential for maintaining the habitability of the Earth, since it continually recycles oceanic crust back into the mantle at subduction zones (Figure 2.11), and continually regenerates it at ocean ridges by the solidification of newly generated magma, some of which is erupted as lava on the sea-bed, to form the upper layers of the ocean crust. Without such a process, CO_2 and other atmospheric constituents would not be recycled back to the atmosphere. On planets like Earth that have abundant carbon (in carbonate rocks) and some mechanism, like plate tectonics, for recycling this carbon, volcanism should provide a more-or-less continuous input of CO_2 into the atmosphere.

- ■ Why is CO_2 such an important gas?
- ❑ As you saw in Section 2.3, CO_2 is a greenhouse gas that raises the temperature of the Earth's surface and atmosphere by retaining some of the energy received from the Sun.

The three processes that transfer internal heat to the Earth's surface and drive plate tectonic activity on the Earth are conduction, convection and advection. We know that modern internal Earth processes are mainly driven by heat that originates in approximately equal measures from two sources:

1 Heat energy from the radioactive decay of unstable isotopes, notably those of potassium, uranium and thorium.

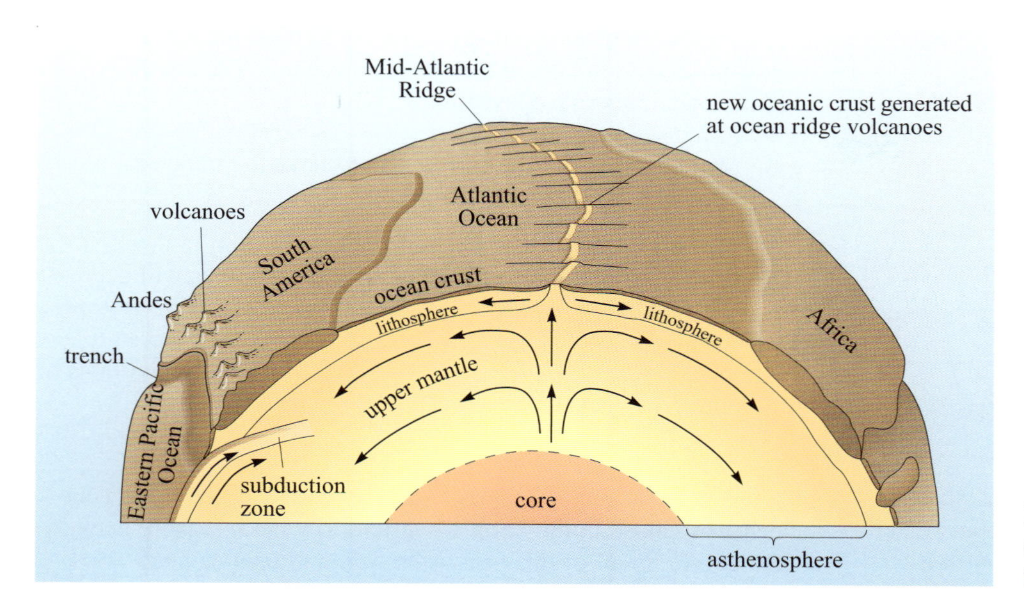

Figure 2.11 Schematic cross-section of the present-day plate tectonic cycle.

2 The remaining primordial heat.

This internal heat has somehow to be lost to space by thermal radiation from the surface. This can only happen if the heat is transferred upwards to the surface by conduction, convection or advection. However, silicate rocks are poor heat conductors so that convection and advection are the main processes by which the Earth transfers heat from its interior to its surface.

The rate at which a planet's internal heat is lost is therefore critical to the continuation of its tectonic activity.

- ■ What will determine the *rate* at which a planet loses its internal heat?
- □ Key factors are its size (larger bodies lose heat more slowly and will therefore remain active longer) and its composition. These factors determine the amount of heat available and the body's ability to convect.

While volcanism played a major role in the early history of Mars, the Moon, and probably Mercury, their small sizes relative to Earth resulted in the loss of internal heat at a much faster rate.

Tectonic activity on a planet will also be determined by its composition since rocks of different composition have different physical properties, which when heated will influence the ability of a body to convect. Under high pressures and temperatures, most solids can behave as highly viscous fluids, given enough time. When rocks are heated, they expand and their density decreases, making them more buoyant relative to cooler and denser ones. Thus, while hot rocks slowly rise in some regions, cooler rocks slowly descend in others, in a system of convecting cells similar to those shown in Figure 2.12 (but the convecting cells are considerably less uniform).

Figure 2.12 Highly schematic representation of simple convection cells, represented by arrowed flow lines. Hotter material is transferred upwards from below and cooler material is transferred downwards from above.

A planet's chemical composition will determine the amounts of heat-producing radioactive elements present and affect the likelihood of internal convection.

- ■ Would we expect the early Earth to have the same rate of convection as it has at the present day?
- ❑ No, since one of the heat sources comes from radioactive decay then the total amount of unstable isotopes will decrease with time. There would have been more radioactive isotopes decaying to produce heat in the early Earth than there are now, so there would have been more heat to lose and rates of convection would have been greater.

It has been estimated that there was about five times more heat being produced in the Earth's interior 4 Ga ago that there is today. However, not all of this heat came from radioactive heat-producing elements. In addition to primordial heat from accretion, a substantial amount of gravitational energy was released as heat when huge amounts of iron and nickel sank to the middle of the Earth, forming the core, probably about 4.5 Ga ago.

A greater degree of internal heating is one possible solution to keeping the Earth's surface temperature above 273 K in response to the reduced level of solar luminosity in the early Solar System. Plate tectonics has certainly played a role in keeping the Earth habitable, as you saw in Section 2.3.2, through the recycling of carbon. However, some scientists believe that there have been times when the mechanisms that maintain the constancy of the habitability of the Earth may have faltered with periods of dramatic climate change in which ice entombed the whole planet (Box 2.3).

BOX 2.3 SNOWBALL EARTH

We have emphasized the importance of the Earth's surface temperatures and how the cycling of carbon has helped maintain liquid water on the Earth's surface. However, what happens if that cycle is disrupted? Scientists have been aware for some time that the geological record suggests that the Earth experienced periods of dramatic climate change, as many as four times between 750 Ma and 580 Ma years ago, that resulted in periods of global glaciations: a hypothesis that has been given the name **Snowball Earth**.

Icehouse and greenhouse

The evidence for climate change comes from thick layers of sedimentary rocks deposited 750–580 Ma ago. But these sedimentary rocks are apparently full of contradictions. Let's take, for example, glacial deposits that were laid down at sea-level near the Earth's Equator at the time. Today, a glacier would have to be at an altitude of more than 5 000 m to survive in the tropics. Interspersed with the glacial deposits are layers of iron-rich rocks that should only have formed if the Earth's atmosphere and oceans contained very little oxygen – however, the atmosphere at the time would have had a composition not too different from that of today. Equally puzzling was the observation that immediately overlying the glacial deposits were layers of carbonates typically found in tropical environments today. If glaciers had extended all the way to the Earth's Equator, in effect covering the planet in ice, how did it manage to warm up again so rapidly?

These contradictions started to make sense when scientists began to contemplate that the Earth may have experienced periods of severe climate change. The Snowball Earth hypothesis considers an Earth that was globally frozen for periods of 10 Ma or more. Heat escaping from the Earth's interior prevents the oceans from freezing to the bottom. However, surface temperatures drop to around 223 K and ice forms to a thickness of a kilometre or more. Under such conditions, all but a small fraction of the Earth's primitive organisms become extinct.

- What effect would an ice-covered Earth have on the major sinks for CO_2 from the Earth's atmosphere?
- Recall from Box 2.2, that CO_2 in the atmosphere is removed through silicate weathering followed by carbonate depositions. An ice-covered Earth would halt this process.

However, volcanic activity would not stop on an ice-covered Earth so that volcanic outgassing of CO_2 would continue. The Snowball Earth hypothesis suggests that it would accumulate to 350 times present-day levels of CO_2 creating severe greenhouse conditions that would warm the planet and melt the ice in perhaps as little as a few hundred years. Organisms that survived the icehouse must now endure a hothouse.

The Snowball Earth hypothesis also explains the occurrence of what are normally very rare iron-rich layers between the glacial deposits. These layers are analogous to the iron formations found much earlier in the Earth's history (see Section 2.4.4) when the oceans and atmosphere contained very little oxygen, and iron could readily dissolve. However, given several Ma of ice cover the oceans would be deprived of oxygen, so that dissolved iron expelled from seafloor hot springs could accumulate in the water. Once a CO_2-induced greenhouse effect began melting the ice, oxygen would again mix with the seawater and force the iron to precipitate out.

An explosion of life

Did the recovery of the Earth's climate following these huge glaciations 750–580 Ma ago pave the way for the explosion of complex multicellular animal life that happened shortly thereafter? Eukaryotes, cells with a membrane-bound nucleus and from which all plants and animals descended, emerged almost 1.8 Ga ago. However, the most complex organisms that had evolved when the first of these large glaciations occurred were filamentous algae and simple unicellular protozoa. It has always puzzled scientists why it took so long for these primitive organisms to diversify into the more complex organisms that suddenly appear in the fossil record at around 670 Ma (Figure 2.21).

A series of such global freezing events followed by equally unpleasant greenhouse conditions would certainly have had a dramatic effect on the evolution of life on Earth, effectively filtering out earlier forms of eukaryotes. All of the eukaryotes around today would thus derive from the survivors of a Snowball Earth. Some measure of the impact these conditions would have had on the evolution of eukaryotes may be evident from the phylogenetic tree (Figure 1.37). This depicts the phylogeny of the eukaryotes as a delayed radiation at the end of a long, unbranched stem. The lack of any earlier branching may indicate that any pre-existing eukaryotic ancestors were, in effect, pruned by the Snowball Earth episodes. Those that survived such global glaciations may have done so by taking refuge near the surface of the ice where photosynthesis could be maintained or on the sea floor near energy-rich hydrothermal vents.

2.4.3 The Earth's early hydrosphere

So far we have emphasized the importance of liquid water both to the origin of life on Earth and to the habitability of an Earth-like planet. But where did the Earth get its water from? The distribution of water in the inner Solar System is poorly understood, but may be roughly in scale with the size of the body. For those bodies from which we have samples, Earth contains the most water, followed by Mars, while other bodies such as the Moon and some asteroids, from which we have meteorite samples, are relatively dry. We have no known samples of Venus or Mercury, but the direct detection of water in the Venusian atmosphere suggests that Venus did contain water when it first formed.

Until recently, there have been competing views as to the origin of the Earth's water. One view held that the Earth accreted as a dry body and its water was subsequently added through cometary impact. The competing viewpoint held that the Earth inherited its water from water-bearing minerals in the un-degassed interiors of planetary embryos. However, evidence from hydrogen isotopes (Box 2.4) suggests that comets are unlikely to be the source of the Earth's water.

The ratio of the two stable isotopes of hydrogen in comets is not the same as the ratio for terrestrial ocean water, implying that the Earth did not obtain the bulk of its water by cometary impact after the end of accretion.

BOX 2.4 OTHER SOURCES OF THE EARTH'S WATER?

Water from comets?

Given that comets contain significant quantities of water, the idea that they provided the Earth's water may seem feasible. However, there is evidence from the isotopes of hydrogen that can be used as an argument against this viewpoint. Table 2.1 lists the ratio of the two stable isotopes of hydrogen (^1H, hydrogen; ^2H, deuterium – abbreviated to D) in comets 1P/Halley and Hyakutake and of the Earth's oceans (the overwhelming majority of water in Earth exists in its oceans). The ratio of the isotopes ^2H to ^1H is usually referred to as the deuterium/hydrogen ratio (abbreviated to D/H).

Table 2.1 D/H ratio ($^2H/^1H$) of water in comets 1P/Halley, Hyakutake and the Bulk Earth.

Body	D/H isotopic ratio
Bulk Earth	1.5×10^{-4}
Comet 1P/Halley	3.16×10^{-4}
Comet Hyakutake	2.82×10^{-4}

These data indicate that comets contain roughly twice as much deuterium relative to ^1H as the Earth's oceans. Thus, if the two comets listed in Table 2.1 are broadly representative of all comets, which is plausible but as yet not testable, then most terrestrial water must have come from sources other than comets.

Water from the solar nebula?

It also seems unlikely that the early Earth scavenged volatiles such as H_2O, CO_2 and N_2 directly from the solar nebula. This is because the relative concentrations of other volatiles, notably the rare gases Ne, Ar, Kr, and Xe, were much higher in the solar nebula than in the present atmospheres of the terrestrial planets. The evidence for this is illustrated in Figure 2.13, which shows the noble gas abundances in the atmospheres of the terrestrial planets relative to the solar composition (which represents that of the primordial solar nebula). It would be difficult for the terrestrial planets to preferentially lose rare gases but retain other volatiles.

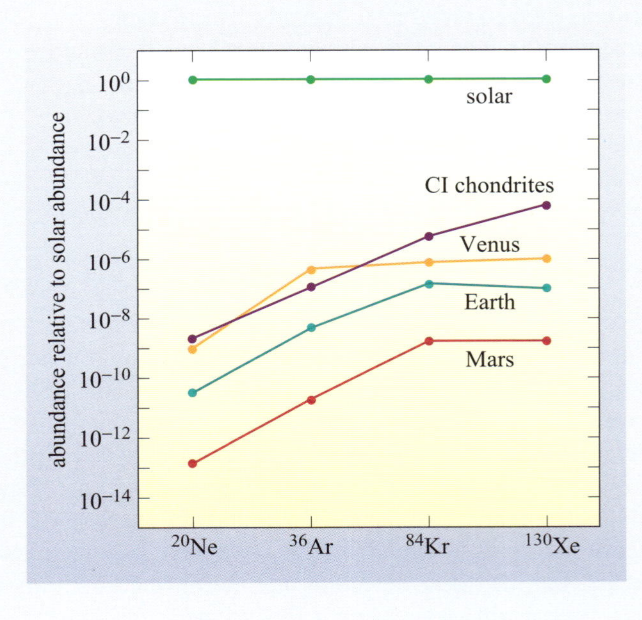

Figure 2.13 Noble gas abundances in the atmospheres of Mars, Earth and Venus relative to solar compositions. The data are plotted as the concentration of the rare gas relative to silicon divided by the corresponding solar ratio. (Adapted from Porcelli and Pepin, 2000)

At present, the most plausible model for the origin of volatile materials on the early Earth is from water-bearing grains that became incorporated in planetesimals and eventually planetary embryos. This model is not without its problems – one uncertainty is whether water-bearing planetesimals could have formed at 1 AU or whether they could only have formed at distant parts of the solar nebula, for example at the asteroid belt.

- Can you suggest how water could have been incorporated in material that condensed from the solar nebula at around 1 AU?
- It could have been incorporated in hydrated minerals, which condense at higher temperatures than the temperature at which water condenses to ice.

Evidence for the role of hydrated minerals comes from their presence in meteorites, for example in the carbonaceous meteorites such as Murchison that you met in Section 1.5.3. Current ideas for the origin of the Earth's hydrosphere are summarized in Figure 2.14. Large amounts of volatiles became incorporated into

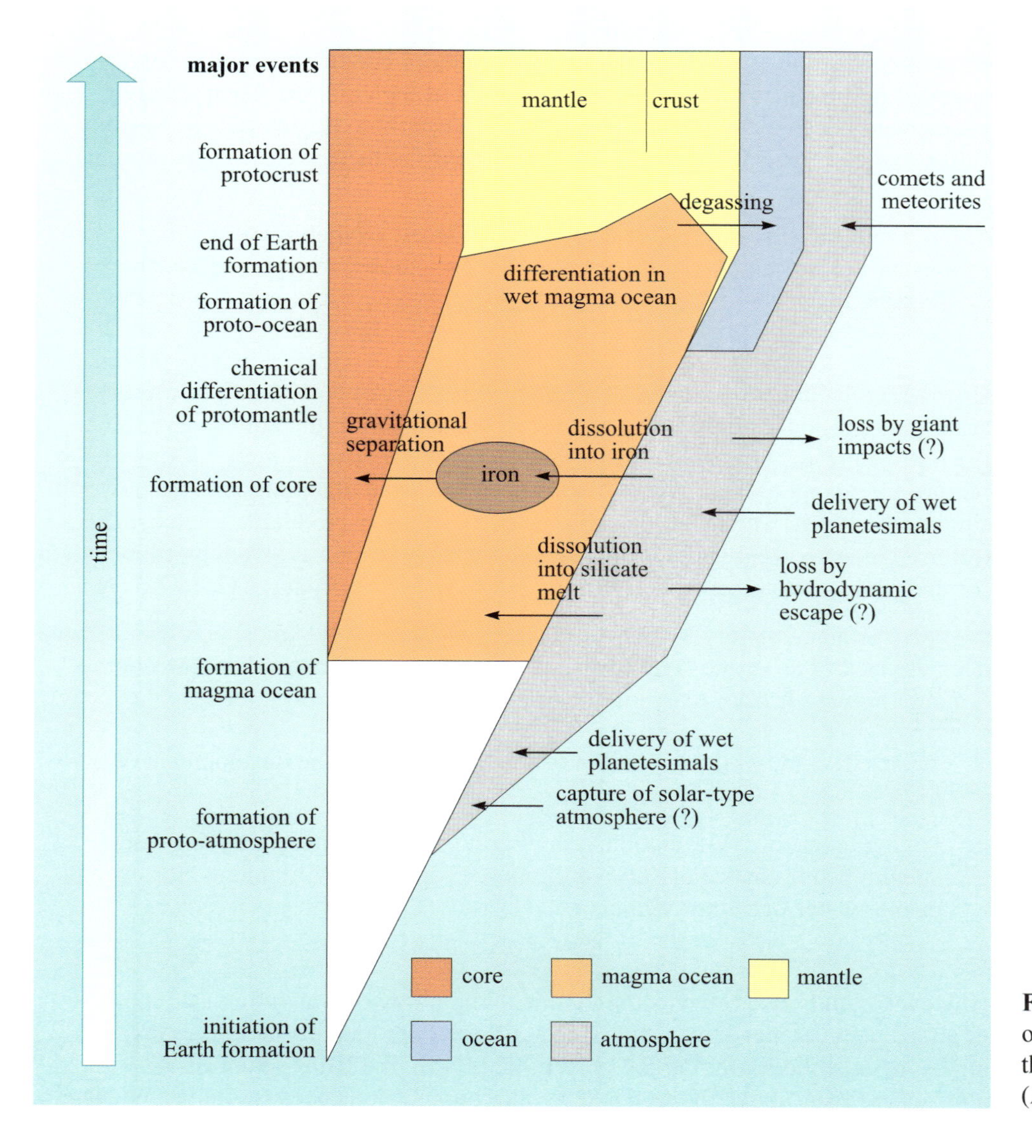

Figure 2.14 Schematic diagram of the behaviour of water during the early evolution of Earth. (Adapted from Abe *et al*., 2000)

the Earth as it formed. However, it seems likely that the heat generated by accretion, as well as during the formation of the Earth's core and the Moon, may have driven off much of these original volatile materials, while some volatile materials may have been incorporated into the Earth's core. The volatile materials that remained within the Earth have been outgassing from its interior ever since. Outgassing rates must have been much greater on the early Earth than they are now because rates of internal convection were greater. Most of the Earth's atmosphere and hydrosphere had probably originated from within the Earth by around 4 Ga ago, after formation of the core and the 'birth' of the Moon.

2.4.4 The Earth's early atmosphere

There is considerable debate over the oxidation state of the Earth' early atmosphere. What evidence we do have comes from the geological record. For example, the occurrence of oxidized sulfur compounds as sulfate (SO_4) deposits in rocks 3.5 Ga old from Australia implies that non-reducing conditions existed in some places, although many of the Earth's oldest rocks contain large quantities of the more reduced sulfur compound, pyrite (FeS). Other lines of evidence, such as the composition of gases emitted from modern volcanoes, our knowledge of the present composition of the atmosphere, and observations of conditions on other planets, have led to the conclusion that, after formation of the core, the Earth's early atmosphere became dominated by nitrogen (N_2), carbon dioxide (CO_2), possibly sulfur oxides (especially SO_2), and water vapour (as there were oceans present).

For about the first billion years of Earth's history, oxygen (O_2) was only present in trace amounts as a result of the breakdown of water vapour by UV radiation.

However, this inorganic mechanism of releasing oxygen compounds into the atmosphere produces only tiny amounts of free oxygen in the atmosphere. Since the Earth's present atmosphere is oxygen-rich and because all higher forms of life require free oxygen, there is an obvious need for some other, much more powerful source of oxygen. The most plausible source is oxygen-producing photosynthesis (referred to as oxygenic photosynthesis), a process that first evolved in cyanobacteria (or their immediate precursors) during the Archaean Era (from 3.8 Ga to 2.5 Ga ago).

One consequence of very low O_2 levels in the Earth's early atmosphere would have been the lack of an ozone layer to absorb incoming ultraviolet radiation from the Sun and prevent it from reaching the Earth's surface (as it does today).

■ What effect might the lack of an ozone layer have on the development of early life?

❑ Higher levels of UV radiation at the Earth's surface suggest that life would stand a better chance of survival in more protected environments, for example deeper water or within sediment.

However, clouds also reflect sunlight, including ultraviolet radiation, and the lack of an ozone layer may have resulted in a different thermal profile in the early atmosphere of the Earth. On the present-day Earth, the temperature of the atmosphere rises above the tropopause (at around 20 km) because the ozone layer absorbs ultraviolet light and gains energy. This energy is emitted as heat, effectively providing an invisible 'lid' for convection cells in the atmosphere. Without an ozone layer, atmospheric temperature could have continued to decrease to much greater altitudes on the early Earth. This would have allowed atmospheric convection cells to extend to much greater altitudes and, consequently, cloud formation may have extended to considerably greater heights than it does today. It is therefore possible that some shielding from ultraviolet light was provided by clouds, although we have no way of knowing how extensive any cloud cover would have been.

In Section 2.4.1 you saw that there were certainly bodies of water (possibly oceans) at least by 3.8 Ga ago, and also some land areas to provide sediments. Land formed from granitic material (like modern continental crust) must have been more limited in extent or we would have expected to find more examples of ancient granitic rocks on Earth. There may also have been small islands produced by basaltic volcanoes, which would not be preserved but would have been subducted and recycled back into the mantle.

Volcanic activity would have resulted in one particular environment wherever there were bodies of surface water. That is hydrothermal circulation, resulting from the circulation of seawater through cracks and fissures in hot basaltic rocks where lavas are being erupted to form new ocean crust or build volcanoes on the ocean floor. As you saw in Section 1.7.3, hydrothermal environments may have played a significant role in the origin of life on Earth.

The geological record also provides possible evidence for the changing nature of the Earth's early atmosphere as a result of the appearance of life. We'll examine evidence from carbon isotopes as to when oxygenic photosynthesis originated in the next section. However, one line of evidence for the appearance of free oxygen during the Archaean Era comes from **banded iron formations (BIFs)**. BIFs are amongst the oldest rocks on Earth. They occur throughout the world and are vast in extent. The example shown in Figure 2.15 from the Hamersley Ranges in Western Australia occupies a basin more than 300 km in diameter.

Figure 2.15 Mount Tom Price iron ore mine, Hamersley Ranges, Western Australia. (Hamersley Iron Pty Ltd.)

Banded iron formations are exactly what their name implies: they are rock formations that are characterized by finely banded dark brown, iron-rich layers alternating with lighter-coloured iron-poor layers, the layers range in thickness from less than a millimetre to about a centimetre. The iron-rich bands contain the highly insoluble iron oxides: haematite (Fe_2O_3), limonite ($Fe_2O_3.3H_2O$) and magnetite (Fe_3O_4). Chert, a rock composed of precipitated silica, occupies the iron-poor bands. Individual bands, often only a few millimetres thick, can extend for tens of kilometres. How BIFs formed is not entirely clear; we have no modern analogues to guide us. However, the involvement of iron oxides suggests that the process that led to the formation of BIFs must have affected the oxidation state of iron and hence BIFs could contain information about the oxidation state of the Earth's ocean and atmosphere at the time they formed.

QUESTION 2.4

The total mass of BIF deposits older than 2.5 Ga is estimated at 3.3×10^{16} kg. If a typical BIF consists of 30% haematite (Fe_2O_3), how much oxygen would have been incorporated in BIFs prior to 2.5 Ga ago?

Your answer to Question 2.4 shows that whatever the chemistry occurring in the formation of BIFs, large amounts of oxygen were incorporated into BIFs very early in the Earth's history. One interpretation is that this suggests oxygen was available in the shallow seas where most BIFs were formed.

Most theories for the origin of BIFs involve a significant role for hydrothermal activity on the early Earth. The seawater flowing through these hydrothermal systems would have dissolved iron-containing minerals so that iron in a reduced form was subsequently injected into the deep ocean through hydrothermal vents. It is generally accepted that the deep ocean on the early Earth was extremely oxygen deficient or anoxic so that iron escaped oxidation and precipitation at the vents themselves but was deposited in much shallower, more oxygenated water.

How did the iron get from the deep ocean to shallow water, crossing large expanses of oceans in so doing? One idea is that the iron was actually consumed by bacteria which flourished near the vents and that these bacteria then drifted away in vast colonies into shallow water where they died, depositing a thin film of organic-rich material. After a while the organic material would have been recycled, leaving the iron behind in its highly insoluble oxide form.

■ Hydrothermal vents provide a satisfactory source for the large amounts of iron involved in the formation of BIFs, but where might the large amounts of oxygen come from?

❑ Since oxygenic photosynthesis would have been a powerful generator of free oxygen, a strongly oxidizing local environment would result wherever it occurred.

2.4.5 Palaeontological and geochemical evidence for early life on Earth

The only direct evidence we have for life on the early Earth comes from palaeontological and carbon isotopic data obtained from the preserved geological rock record. These data suggest that life *may* have already been established by 3.5 Ga ago, possibly 3.8 Ga ago, which in turn suggests that life may actually have originated by 4 Ga ago. However, as you will see, these data are by no means unchallenged and are the subject of intense scientific debate.

If life was present on Earth 3.8 Ga ago, it is generally assumed that it would have arisen after the formation of the Earth's crust and oceans and after the end of the late heavy bombardment, i.e. some 4 Ga ago.

Stromatolites

In shallow coastal waters, characteristic mound-shaped structures called **stromatolites** (Figure 2.16) are built up by the accumulation of sediments that consist of thin gelatinous mats alternating with thin layers of calcium carbonate. The organisms that form these mats, and precipitate the calcium carbonate, include the simple photosynthesizing nitrogen-fixing cyanobacteria or blue–green algae. In cross-section, stromatolites have a layered structure like a stack of pancakes. Identical fossil structures occur in a variety of rocks that were produced in the first 2 Ga of the Earth's history. The oldest putative stromatolites so far reported come from the 3.46 Ga-old Apex cherts of the Warrawoona Group in Western Australia (Figure 2.17). The organisms that formed these stromatolites may also have been cyanobacteria, but this is not absolutely certain because some modern photosynthetic but not oxygen-producing bacteria form rather similar structures. This is an important point with far-reaching implications for the environment of the early Earth. Cyanobacteria (but not some other forms of photosynthetic bacteria) carry out oxygenic photosynthesis and release oxygen into the environment. The evidence that these 3.46 Ga-old rocks might contain the Earth's oldest fossils comes from structures resembling remarkably well-preserved bacterial and cyanobacterial microfossils that were first reported by the University of Los Angeles geologist Bill Schopf in the early 1980s. Schopf believes that the structures he observes are microfossils that contain evidence of *cells* and in some cases there are filaments and spherical structures that look remarkably like modern cyanobacteria (Figure 2.18). However, this interpretation has recently been challenged (Box 2.5).

Structures similar to stromatolites can also be produced by non-biological processes so fossil stromatolite occurrences can be controversial.

Figure 2.16 Modern stromatolites in Shark Bay, Western Australia. (Andrew A. Knoll)

Figure 2.17 Cross-section showing the layered structure of a stromatolite, 3.46 Ga old, from the Warrawoona Group, Western Australia. (Professor J. W. Schopf)

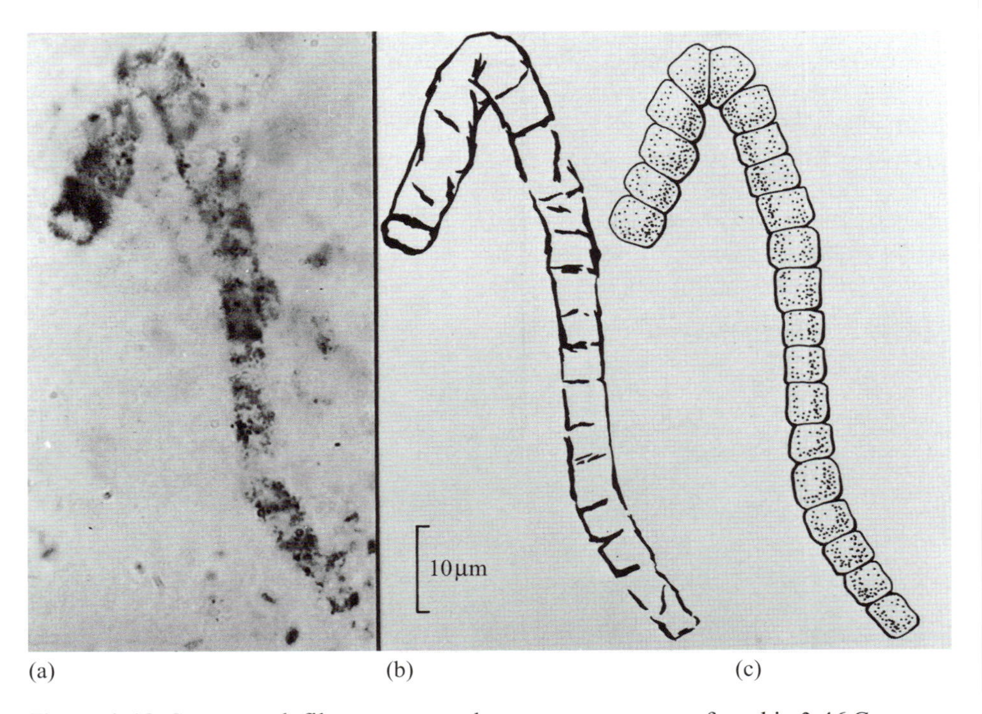

Figure 2.18 Segmented, filamentous, carbonaceous structures found in 3.46 Ga old fossil stromatolites resembling modern cyanobacteria: (a) photomicrograph, (b) line drawing, (c) reconstruction. (Commonwealth Palaeontological Collections of the Australian Geological Survey)

BOX 2.5 FOSSILS OR ARTEFACTS?

The interpretation that structures preserved in the 3.46 Ga old Apex cherts of the Warrawoona Group in Western Australia represent preserved microfossils of cyanobacteria was challenged in 2002 by a team of scientists led by Martin Brasier from the University of Oxford. Brasier and his colleagues believe the structures to be artefacts formed from carbon deposited by hydrothermal fluids passing through the rocks. They argue that the carbonaceous, filamentous structures are not consistent with those that might be expected from cyanobacteria, being much larger and containing branches not found in cyanobacteria. They also question whether the carbon found in the Apex cherts is of biological origin. If it is of biological origin, the estimated hydrothermal temperatures of between 250 °C and 350 °C would require extremely heat-tolerant organisms and they suggest that the carbon in these rocks may be produced by the non-biological catalytic synthesis of organic material.

Although biochemical remains are rare in rocks preserved from the early Earth, a portion of the organic matter they once contained is often preserved as a macromolecular material called *kerogen*, a word derived from the Greek *keros* meaning wax and used to describe the organic matter present in sediments. Most ancient sediments have been heated up by deep burial since they were deposited more than 3 Ga ago, and the kerogen in them is often highly altered. This makes much of the structural chemical information it once contained indecipherable. However, it is possible to use the abundance of the carbon isotopes in ancient kerogen as a tool for unravelling the biochemistry of the ancient Earth. Box 2.6 summarizes how biological processes will determine the relative ratios of ^{13}C to ^{12}C, expressed as $\delta^{13}C$ values, in living organisms. Since the carbon from those organisms is preserved when they die and can then become incorporated as kerogen in rocks, we can use the $\delta^{13}C$ values of kerogen to draw inferences about ancient biological processes on Earth and, should measurements become available, on other planets. The important point to take from Box 2.6 is that if a sample of organic matter is enriched in ^{12}C and depleted in ^{13}C relative to the inorganic carbonate standard, *the value of* $\delta^{13}C$ *will be negative*.

BOX 2.6 CARBON ISOTOPES AS INDICATORS OF BIOLOGICAL PROCESSES

Carbon isotopes

Carbon consists of a number of isotopes of which two, ^{12}C and ^{13}C, are stable. ^{12}C contains 6 neutrons and 6 protons; ^{13}C contains 7 neutrons and 6 protons. You may have come across the technique of carbon dating that uses one of the radioactive isotopes of carbon, ^{14}C, to determine the age of archaeological remains (^{14}C is an unstable isotope). Since stable isotopes do not undergo radioactive decay, they cannot be used to date rocks or other materials. However, physical processes do affect the ratio of the stable isotopes of an element. For example, in chemical reactions it takes less energy to break the bond $^{12}C—^{12}C$ than $^{12}C—^{13}C$ and, similarly, it takes less energy to make a bond between two ^{12}C atoms than between a ^{12}C atom and a ^{13}C atom.

- ■ If a simple chemical reaction involves the making of carbon bonds, will the products of the reaction contain more or less ^{12}C than the starting materials?
- ❑ Since it takes less energy to make a $^{12}C—^{12}C$ bond than a $^{12}C—^{13}C$ bond, the reaction will preferentially incorporate more ^{12}C into the products. The products will contain more ^{12}C and we would refer to them as being ^{12}C-enriched.

In any chemical reaction, molecules bearing the lighter isotope, ^{12}C, will, in general, react slightly more readily than those with the heavy isotope, ^{13}C.

The reason why bonds between elements containing the lighter isotope react more readily than those with the heavier one is due to differences in their physical properties (e.g. density, vapour pressure, boiling point and melting point) due to the greater vibrational energy of the lighter isotope, although the chemical properties of the isotopes of an element are the same.

Since biology involves a wide range of physical and chemical processes, it should come as no surprise that biological processes affect the isotopes of carbon. We refer to the separation of isotopes of an element during naturally occurring processes as a result of the mass differences between their nuclei as **isotope fractionation**. You should note, however, that most natural processes are not capable of completely separating the isotopes of an element, rather they tend to concentrate one isotope in preference to another.

On Earth, the natural abundance of ^{13}C is roughly one-ninetieth that of ^{12}C, that is ^{12}C is considerably more abundant. It is the ratio between these two isotopes that we are interested in and, by convention, we refer to the ratio of the minor isotope to the major isotope, i.e. $^{13}C/^{12}C$. Thus a typical carbonate rock on Earth might have a $^{13}C/^{12}C$ ratio of 0.01123722, while carbon from a living organism that has had its ratio of ^{13}C to ^{12}C affected by biological processes might have a $^{13}C/^{12}C$ ratio of 0.0109563. Differences in the natural abundance of carbon stable isotopes only begin to express themselves in three figures after the decimal point, i.e. a few parts per thousand variation. It is obviously not very practical to have to refer to the ratio between ^{13}C and ^{12}C using such small numbers. Furthermore, determining the absolute ratio of two isotopes is analytically very difficult so scientists adopt a standard and measure the ratio of $^{13}C/^{12}C$ relative to that standard. Enrichment or depletion in ^{13}C is then expressed in terms of a $\delta^{13}C$ value (δ is the lower-case Greek letter 'delta'). The ratio of $^{13}C/^{12}C$ for the sample being investigated is compared to that of a carbonate standard:

$$\frac{^{13}C/^{12}C\ \text{ratio of sample}}{^{13}C/^{12}C\ \text{ratio of standard}}$$

One is subtracted from this value and the whole is multiplied by 1000 to give a $\delta^{13}C$ value in terms of parts per thousand (‰):

$$\delta^{13}C = \left[\frac{^{13}C/^{12}C\ \text{sample}}{^{13}C/^{12}C\ \text{standard}} - 1\right] \times 1000 \qquad (2.6)$$

For carbon, the standard used is a Cretaceous fossil belemnite known as the Pee Dee Belemnite (abbreviated to PDB) after the rock formation from where it was recovered. It has a $^{13}C/^{12}C$ ratio of 0.01123722.

■ Using Equation 2.6, calculate the $\delta^{13}C$ value of a sample of organic matter with a $^{13}C/^{12}C$ ratio of 0.0109563.

❑ Substituting the values in Equation 2.6 we get:

$$\left[\frac{0.0109563}{0.01123722} - 1\right] \times 1000 = -25.0\text{‰}$$

One of the major carbon stable isotope fractionations that is the result of a biological process is that due to autotrophic photosynthesis. This is a complex process, but we can think of it as occurring in two stages: in the first stage carbon dioxide from the atmosphere is 'imported' into the cell; in the second stage, an enzyme known as ribulose bisphosphate carboxylase (abbreviated to **Rubisco**) is involved in forming carbohydrates $(CH_2O)_n$ from the imported CO_2. Both stages favour the incorporation of ^{12}C over ^{13}C so that the carbohydrates will have more negative $\delta^{13}C$ values than the CO_2.

■ Using Equation 2.6, will the value of $\delta^{13}C$ be negative, positive or zero when the ratio $^{13}C/^{12}C$ is (a) equal for standard and sample? (b) Greater in the sample? (c) Lower in the sample?

❑ Because one is subtracted from the ratio of ratios, for (a) the value will be zero; for (b) it will be positive and for (c) it will be negative.

The fractionation of the carbon stable isotopes produced by photosynthesis can be used as a biomarker for past biological processes since the organic matter produced by living organisms can become incorporated in rocks as organic carbon (abbreviated to C_{org}). However, it is important to realize that we also need an isotopic value for the source carbon used in photosynthesis, i.e. atmospheric CO_2 or CO_2 dissolved in ocean water. Direct measurements of the $\delta^{13}C$ values of past atmospheric CO_2 or oceanic dissolved CO_2 are rarely possible. Instead, scientists use the $\delta^{13}C$ value of carbon from carbonate rocks (abbreviated to C_{carb}) which represents oxidized carbon and so reflects the $\delta^{13}C$ values of the atmosphere at the time of carbonate rock formation. C_{carb} $\delta^{13}C$ values have remained around 0‰ throughout Earth history (Figure 2.20), from which scientists infer that the $\delta^{13}C$ value of atmospheric CO_2 has remained roughly constant over the last 3.8 Ga.

Biological processes result in large isotopic fractionations, i.e. they will preferentially use one isotope of carbon over another. Indeed, biological processes are the most important cause of variations in the isotopic composition of carbon on Earth. For the most part, the largest fractionation in carbon isotopes occurs during the initial production of organic matter by autotrophs. In general autotrophic organisms preferentially use the isotope ^{12}C over ^{13}C when they fix carbon, for example from atmospheric CO_2, to produce organic matter.

- If we were to analyse the carbon isotopic composition of a blade of grass, would it contain more ^{12}C or less ^{12}C that the carbon dioxide it fixes from the atmosphere?

- The grass, in common with all plants, will preferentially use ^{12}C over ^{13}C, so it will contain more ^{12}C than atmospheric CO_2. The grass will therefore have a more negative $\delta^{13}C$ value than atmospheric CO_2.

Figure 2.19 shows the range of $\delta^{13}C$ values for living (extant) autotrophs, present-day marine bicarbonate (in solution) and atmospheric CO_2. There is quite a range of values among autotrophs, reflecting their different styles of carbon fixing-reactions, but the $\delta^{13}C$ values are mostly between about –10‰ and –40‰.

Atmospheric CO_2 and marine bicarbonate are the principal carbon sources used in photosynthesis.

C3, C4 and CAM plants are groups of plants that fractionate carbon isotopes to differing extents.

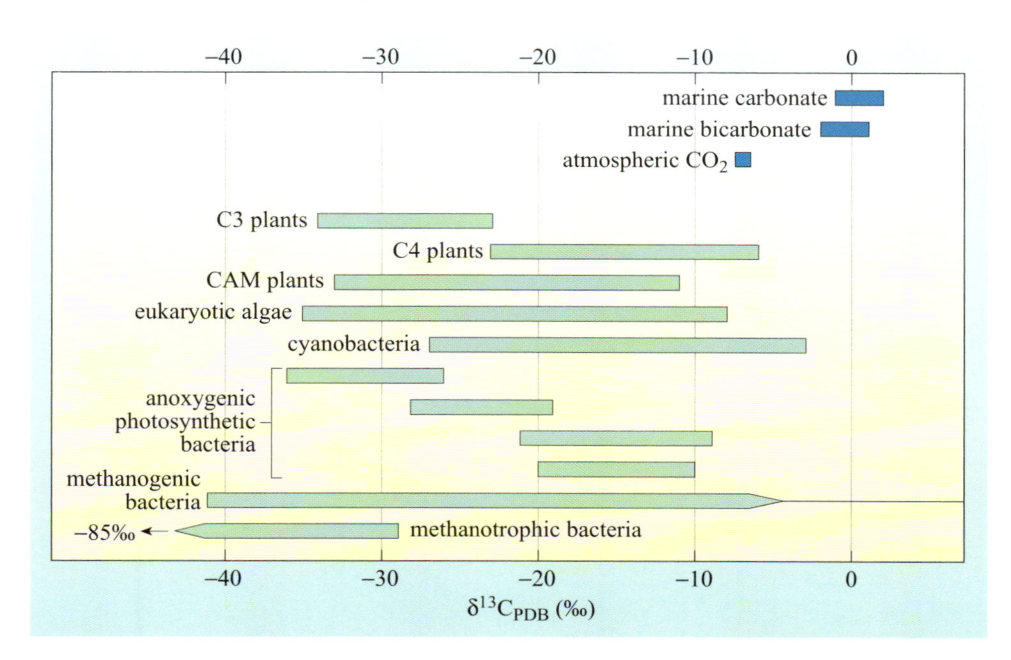

Figure 2.19 Carbon isotope ranges of major groups of higher plants and micro-organisms compared with the respective ranges of the principal inorganic carbon species in the environment. (Schidlowski *et al.*, 1983)

Figure 2.20 summarizes more than 10 000 measurements of the carbon isotopic composition of sediments of all ages. There are two groups of values shown on this figure, C_{carb}, which denotes the values obtained from carbonate rocks and C_{org}, which denotes the values measured for kerogen isolated from sediments. The difference between the two has remained roughly constant throughout the Earth's history and reflects the difference between atmospheric CO_2 and marine bicarbonate, the main sources of carbon for photosynthesis, and the organic carbon of autotrophs.

The Isua rocks at 3.8 Ga are the oldest rocks for which carbon isotope data have been obtained. Some scientists believe that the data from Isua indicate that there were biological processes operating on the Earth at that time. Figure 2.20 shows that the Isua rocks have both more negative carbonate (C_{carb}) $\delta^{13}C$ values and less negative organic (C_{org}) $\delta^{13}C$ values than those in later sediments. This could be because the rocks have been metamorphosed, i.e. subjected to high temperature and pressure, which is known to affect the carbon isotope ratios. The original $\delta^{13}C$ values, before metamorphism, may have been the same as those in younger rocks so it is possible that autotrophs were thriving some 3.8 Ga ago.

More recent analyses, though controversial due to the possibility of contamination, measured the carbon isotopic composition of small carbon inclusions in single grains of the mineral apatite from 3.85 Ga-old Isua formation rocks and gave more negative $\delta^{13}C$ values (their ranges are shown as A1 and A2 on Figure 2.20). The results gave $\delta^{13}C$ values ranging from –21‰ to –41‰, which fall well within the biological ranges shown in Figure 2.19 and are consistent with the archaebacteria, for example methanotrophs. Such organisms figured prominently in some of the Earth's oldest microbial ecosystems. For example, Figure 2.20 also shows a pronounced negative excursion in the $\delta^{13}C$ record of C_{org} at around 2.7 Ga ago with $\delta^{13}C$ values reaching as low as –60‰ in rocks from the Fortescue Formation in Australia. These excursions have been interpreted as evidence of methane-using bacteria, which often have very negative $\delta^{13}C$ values.

A methanotroph is a bacterium that can use methane as a nutrient.

Many scientists believe that the sedimentary carbon isotope record shown in Figure 2.20 can best be interpreted as representing evidence that biological fixation of carbon by autotrophic organisms had been established by 3.8 Ga ago, having

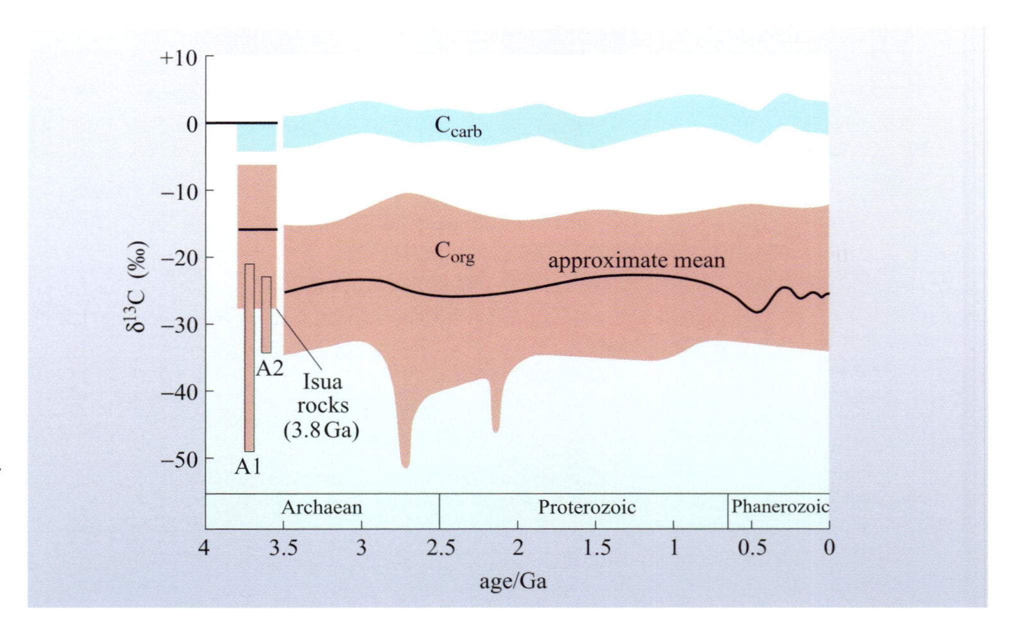

Figure 2.20 Carbon isotope compositions as $\delta^{13}C$ values for sedimentary carbonate (C_{carb}) and kerogen (C_{org}) over 3.8 Ga of the Earth's history. Superimposed on the envelope for whole rock analyses are the ranges obtained for carbon inclusions in apatite grains (A1 and A2) from two iron-rich 3.85 Ga-old Isua formation rocks. (Schidlowski *et al.*, 1983)

been fully operational by the time of the formation of the Earth's oldest sediments. However, this idea was challenged in 2002 when geochemical data suggested that some of the rocks from Greenland were not BIFs as had been previously thought, and therefore likely to preserve biologically derived carbon, but were igneous in origin and had been given the superficial appearance of a BIF rock as a result of the passage of fluids through the rock. It was argued that non-biological processes involving fluids and inorganic iron carbonate produced the carbon in these rocks. However, to fully invalidate the biological interpretation of the Earth's early $\delta^{13}C$ record will require evidence for an inorganic process operating on a global scale that can mimic, both in direction and magnitude, the isotopic fractionation associated with autotrophic carbon fixation. Evidence of the Earth's earliest biosphere remains a topic of contentious debate amongst scientists.

QUESTION 2.5

Based on what you have studied in Chapters 1 and 2 so far, outline a scenario for the emergence of life based on the following lines of evidence:

- Geological evidence that familiar geological and geochemical processes were operating on the early Earth.
- Evidence in favour of the Earth's internal heat acting as a source of energy for the first autotrophic metabolic reactions to appear as opposed to external inputs of energy into a reduced atmosphere that led to the appearance of heterotrophic organisms.
- Geological and phylogenetic evidence that suggests that hydrothermal systems were a key environment on the early Earth.

QUESTION 2.6

From the discussion in Sections 2.4.1 to 2.4.5, draw up a list of conditions on the early Earth that would have affected the potential for the emergence of life in hydrothermal systems. How do these conditions compare with those of the present day?

2.4.6 Evolving complexity

A common question that arises when considering the possibility of life elsewhere in the Universe is that of the existence of life-forms like the ones that have evolved on Earth. The evolutionary trends of life on Earth form the main basis we have for hypotheses on the nature of life elsewhere. They are marked by a series of major changes in the size, form and complexity of organisms and major expansions in diversity that have produced the enormous variety of species that populate the fossil record (Figure 2.21). This record, combined with the phylogenetic tree (Figure 1.37), form the foundation for inferences about the sequence and direction of evolution.

■ With reference to Figure 2.21, how has the size of organisms changed throughout the Earth's history?

❑ For the first 2.5–3 Ga of life on Earth, most species did not exceed a few millimetres in size and most were generally substantially smaller. In the last 600 Ma, however, the evolution of larger and more complex organisms has occurred.

Bilaterians are animals that are symmetric about a central axis.

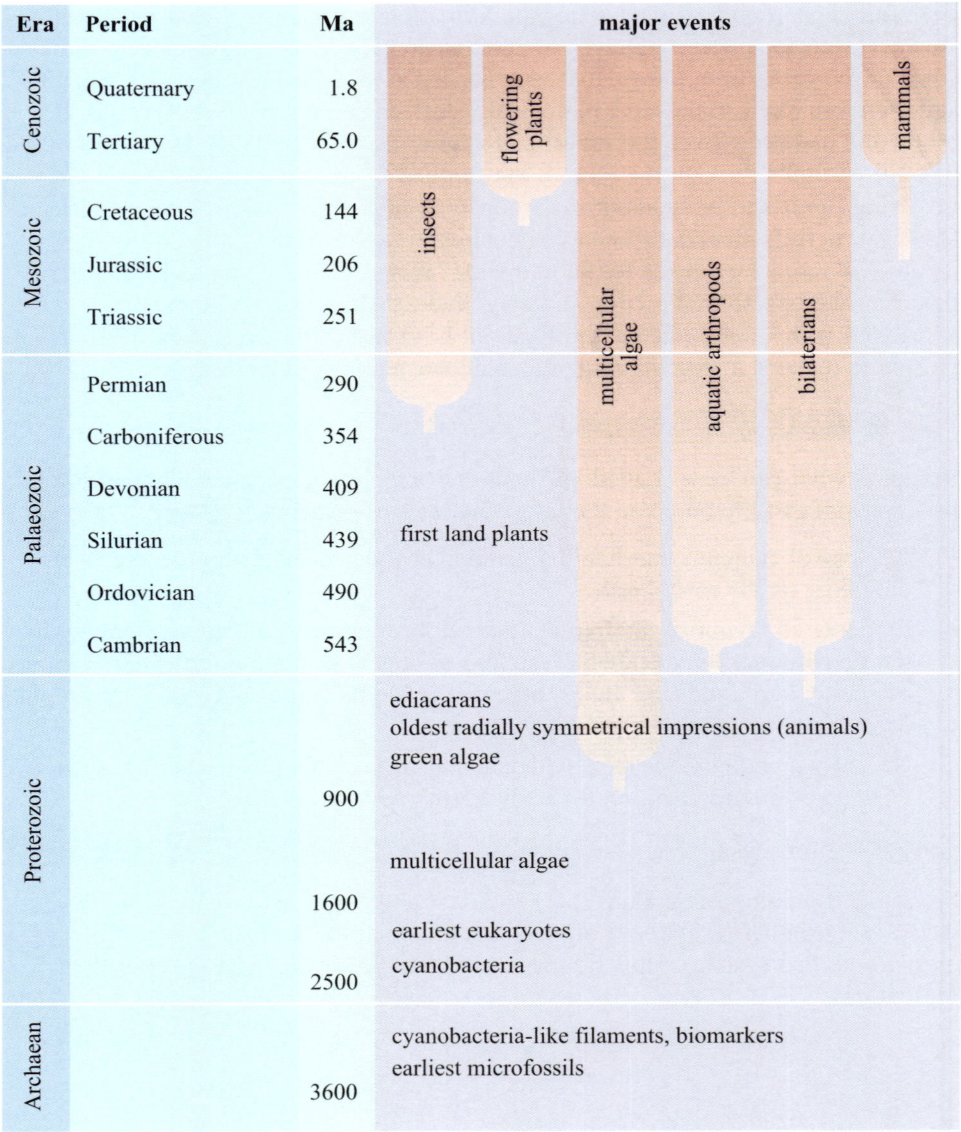

Era	Period	Ma	major events
Cenozoic	Quaternary	1.8	flowering plants; mammals
	Tertiary	65.0	
Mesozoic	Cretaceous	144	insects
	Jurassic	206	
	Triassic	251	multicellular algae; aquatic arthropods; bilaterians
Palaeozoic	Permian	290	
	Carboniferous	354	
	Devonian	409	
	Silurian	439	first land plants
	Ordovician	490	
	Cambrian	543	
Proterozoic			ediacarans oldest radially symmetrical impressions (animals) green algae
		900	multicellular algae
		1600	earliest eukaryotes cyanobacteria
		2500	
Archaean			cyanobacteria-like filaments, biomarkers earliest microfossils
		3600	

Figure 2.21 History of major evolutionary events in the geological record of Earth. (Adapted from Carroll, 2001)

The size of organisms increased significantly with the evolution of multicellular forms. In algae and bacteria, one of the simplest ways to form a multicellular organism was for the products of cell division (see Section 1.2.6) to remain together and produce long filaments. Evidence from the fossil record suggests that many early multicellular eukaryotes were indeed millimetre-sized, linear or branched filamentous forms. By around 600 Ma ago, the fossil record suggests the presence of millimetre-sized radially symmetric life-forms. From 600 Ma onwards, major changes in the size of organisms started to appear. The Ediacaran fauna are a distinctive group of fossils that arose about 670 Ma ago. They are named after Australia's Ediacara hills, where they were first discovered. They comprised tubular, frond-like, radially symmetric forms and generally reached several centimetres in size, with some as large as 1 m (Figure 2.22).

Figure 2.22 Examples of Ediacaran fauna fossils. (a) *Dickinsonia*, an elongate pancake-shaped worm; (b) *Cyclomedusa*, a jellyfish; (c) *Tribrachidium*, a bun-shaped organism with three spiral tracts on its upper surface; (d) *Inkrylovia*, an elongate bag-like form with transverse partitions. (a and b: Simon Conway Morris; c and d: Peter Crimes)

The size of organisms expanded rapidly after around 500 Ma ago, with algae and sponges reaching up to 50 cm in size. Subsequently, the sizes of animals increased by another two orders of magnitude to produce such giants as the dinosaurs and the larger mammals that have evolved in the last 65 Ma. Along with an increase in size, life on Earth has also increased in diversity since its origin. However, this has not been a steady and continuous increase. Major extinctions have caused marked reductions in the diversity of life at several periods in the Earth's history. The most recent major or mass extinction occurred 65 Ma ago and is thought to have been caused by the impact of a large comet or asteroid, a reminder of events that may have frustrated the evolution of life on the early Earth.

QUESTION 2.7

As a thinking exercise, suggest how the evolutionary trends observed for life on Earth might help us answer the question about the possible nature of life elsewhere. What assumptions do you think we need to make before undertaking such extrapolations?

2.5 Life on the edge

2.5.1 Introduction

From the preceding discussion it seems that life on the early Earth may well have arisen in an environment that we would consider extreme when compared to those that exist on much of our planet today. Evidence from both the geological record and phylogeny suggests that the first organisms on Earth may have been the heat-tolerant thermophiles or hyperthermophiles. **Extremophile** is a term used to describe the many micro-organisms that are capable of different degrees of adaptability to the extreme range of living conditions available on Earth. Such environments and organisms are likely candidates for having given rise to life on Earth. These general considerations give some support to the idea that similar ecosystems may have emerged elsewhere as well.

Must extremophiles actually 'love' (as the suffix *-phile* implies) their environment or merely tolerate it? It is certainly easier to determine in the laboratory if an organism will simply tolerate extreme conditions and it is common to find organisms in extreme environments that tolerate rather than love them. However, the number of known true extreme-loving organisms from a variety of environments is increasing, confirming that life can exist, and indeed thrive, under those conditions.

Figure 2.23 Recovery of the camera from the Surveyor 3 spacecraft by the Apollo 12 astronaut, Pete Conrad. When returned to Earth, a strain of the bacterium *Streptococcus* that was isolated from foam inside the camera was found to have survived exposure on the lunar surface. (NASA)

The ability of life to survive in extreme environments has been dramatically demonstrated by the recovery of bacteria exposed to the hostile environment of the lunar surface by the astronauts of Apollo 16 (Figure 2.23). Apollo 16 astronauts recovered a camera from the Surveyor 3 spacecraft that had landed on the Moon two and a half years earlier. When returned to Earth, foam from the inside of the camera was examined to see if any bacteria on it had survived their journey to the lunar surface. The 50–100 organisms recovered survived launch, space vacuum, 3 years of radiation exposure, deep-freeze at an average temperature of 20 K, and no nutrient, water or energy source. However, the organisms effectively did nothing while they sat on the lunar surface, they were in effect freeze-dried. These were not extremophiles, they merely survived. An important observation about extremophiles is that these organisms do not merely tolerate their lot; they do best in their punishing habitats and, in many cases, require one or more extremes in order to reproduce at all.

Studies of extremophiles are responsible for a marked change in evolutionary theory that has given rise to the phylogenetic tree you met in Section 1.9. It had been thought that living organisms could be grouped into two basic domains: bacteria, whose simple cells lack a nucleus, and eukarya, whose cells are more complex. We now know that a third group, the archaea, exists. Anatomically, the archaea lack a nucleus and closely resemble bacteria – some of their genes have similar counterparts in bacteria, a sign that the two groups function similarly in some ways. But the archaea also possess genes otherwise found only in eukarya, and a large fraction of their genes appear to be unique. These unshared genes establish the archaea's separate identity.

So what are the physical limits to life on Earth and what sort of organisms thrive under extreme conditions? Table 2.2 summarizes our present state of knowledge of the physical limits to life. We'll examine the organisms that live in these environments in the following sections.

Table 2.2 The physical limits for life on Earth, with examples of some of the organisms associated with particular environments.

Environment	Limiting conditions	Type	Example
temperature	<15 °C	psychrophiles	
	15–50 °C	mesophiles	*Homo sapiens*
	50–80 °C	thermophiles	*Thermoplasma* can reproduce at >45 °C
	80–115 °C	hyperthermophiles	*Pyrolobus fumarii* (113 °C)
radiation			*Deinococcus radiodurans*
salinity	15–37.5% NaCl	halophiles	
pH	0.7–4	acidophiles	
	8–12.5	alkalophiles	
dessication	anhydrobiotic	xerophiles	nematodes, microbes, fungi, lichens
pressure	pressure-loving – up to 130 MPa	piezophiles	
	weight-loving	barophiles	
vacuum	tolerates vacuum		microbes, insects, seeds
oxygen	cannot tolerate O_2	anaerobes	
	tolerates some O_2	microaerophiles	
	requires O_2	aerobes	*Homo sapiens*
chemical extremes	gases		*C. caldarium* (pure CO_2)
	can tolerate high concentrations of metals		

2.5.2 Temperature

Temperature presents a range of challenges to living organisms. The structural breakdown of cells caused by the formation of ice crystals in sensitive plants can be readily witnessed in those parts of the world that experience cold winters or even just the occasional frosty night. At the other extreme, high temperatures result in the structural breakdown of biological molecules such as proteins and nucleic acids, a process known as denaturation. High temperatures increase the rate at which material diffuses through cell membranes, the membrane fluidity. Temperatures of 100 °C can disrupt the structural integrity of cell membranes to the extent that they leak important cellular constituents.

Life on Earth has adapted to a surprising range of temperatures (Figure 2.24). Although the majority of organisms grow best at moderate temperatures of between 20 °C and 45 °C (the **mesophiles** in Figure 2.24), the temperature preferences of other organisms range from hyperthermophiles (able to reproduce at temperatures >80 °C) to psychrophiles where maximum growth occurs at temperatures <15 °C.

Thermophilic organisms are among the most studied extremophiles. The archaea *Thermoplasma* (Figure 1.37), for example, found in volcanic hot springs, can reproduce at temperatures in excess of 45 °C. Hyperthermophiles, such as the archaea *Sulfolobus* (Figure 1.37) have been recovered from environments where they are exposed to temperatures in excess of 100 °C. By comparison, most regular bacteria thrive at temperatures between 25 °C and 40 °C. No multicellular animals or plants are known that can tolerate temperatures above 50 °C and no microbial eukarya are able to tolerate long-term exposure to temperatures above 60 °C.

Thermophiles that are content at temperatures up to 60 °C have been known for a long time, but true extremophiles, those able to flourish in greater heat, were first discovered in the 1960s during a study of microbial life in hot springs and other waters of Yellowstone National Park in the USA. The first extremophile reported to be capable of growth at temperatures greater than 70 °C was the bacterium *Thermus* (Figure 1.37).

To date, more than 50 species of hyperthermophiles have been isolated, the most resistant of which, *Pyrolobus fumarii*, grows in the walls of black smokers on the ocean floor. It reproduces best in an environment of about 105 °C; it won't grow at all at temperatures below 90 °C. Another hyperthermophile that lives in deep-sea hydrothermal systems is the methane-producing archaean *Methanopyrus* (Figure 1.37).

QUESTION 2.8

Look at Figure 1.37. Where on the phylogenetic tree can species such as *Methanopyrus, Thermoplasma,* and *Sulfolobus* be found? How does their position 'fit' with the concept of the last common ancestor you met in Section 1.9?

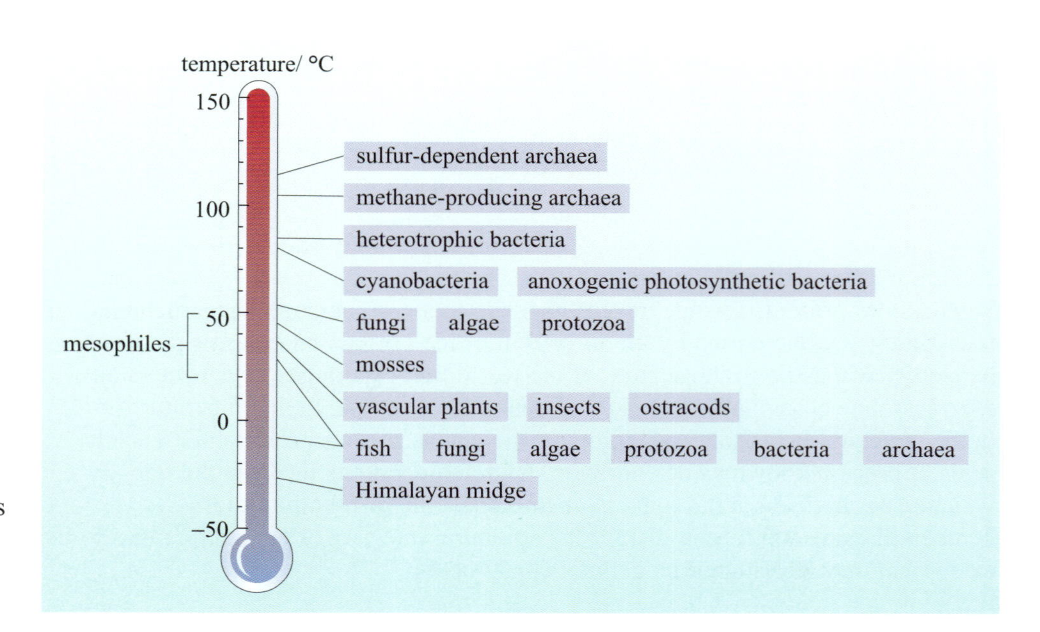

Figure 2.24 The temperature limits for major groups of organisms. (Adapted from Rothschild and Mancinelli, 2001)

What is the upper temperature limit for life? Do 'super-hyperthermophiles' capable of growth at 200 °C or 300 °C exist? At present we do not know, although it seems likely that the limit will be about 140–150 °C; this is the maximum temperature at which activity has been observed for hyperthermophile enzymes. Above this temperature, proteins and nucleic acids denature, so that a loss in the integrity of DNA and other essential molecules would probably prevent reproduction.

Denaturation involves the unfolding of the double helix of DNA.

So how have organisms adapted to these high temperatures? Since high temperatures increase membrane fluidity, one adaptation is to change the composition of the membrane to reduce that fluidity. For example bacteria will alter the ratio of different lipids in their membranes in response to the temperature at which they are grown. Thermophiles and hyperthermophiles have also evolved proteins that are better able to cope with higher temperatures. The DNA of hyperthermophiles, which would otherwise denature at temperatures above 70 °C, is more stable *in vivo* ('in life') than that of mesophiles. This reflects the fact that the G–C pair of nucleic acids is more thermally stable than the A–T or A–U pairs because of the additional hydrogen bond. Elevated G + C to A + T or A + U ratios are found in the ribosomal and transfer RNAs of thermophiles.

On Earth, cold environments are actually more common than hot ones. The Earth's oceans maintain an average temperature of 1–3 °C. However, large areas of the Earth's surface are permanently frozen or are unfrozen for only a few weeks in summer. Some of these frozen environments support life in the form of psychrophiles. Representatives of all major groups of organism are known from environments with temperatures just below 0 °C. Freezing in liquid nitrogen at a temperature of −196 °C can preserve many microbes successfully. However, the lowest recorded temperature for active microbial communities is substantially higher, at −18 °C. A typical example of a psychrophile is the bacterium *Polaromonas vacuolata*: its optimal temperature for growth is 4 °C and it finds temperatures above 12 °C too warm for reproduction. Liquid water is both a solvent for life and an important reactant or product in most metabolic processes. When water freezes the resulting ice crystals can rip cell membranes apart, and solution chemistry stops in the absence of liquid water. Freezing of the water inside cells is almost invariably lethal. The only exception to this rule reported so far, outside of cryopreservation, is the nematode *Panagrolaimus davidi*, which can withstand freezing of all body water.

As with thermophilic organisms, psychrophiles have evolved adaptations to the problems of adverse temperatures. While high temperatures increase membrane fluidity, low temperatures result in a decrease in membrane fluidity. In response, psychrophilic organisms adjust the ratios of lipids in their membranes to improve the fluidity of the cell membrane. Two principal adaptations have evolved to deal with temperatures below the freezing point of water: the protection of cells from ice formation by preventing freezing or, if ice does form, protection of the cells during thawing. One way organisms prevent ice forming is to accumulate soluble compounds that can depress the normal freezing point of water. Increased concentrations of salts and sugars can achieve this, but organisms also produce relatively inert molecules, notably glycerol, specifically for this purpose. For example, high concentrations of glycerol can enable the survival of some invertebrates to temperatures as low as −60 °C. The teleost fish, which inhabit polar regions, manufacture specific proteins which can effectively act as antifreeze agents by binding to the edges of ice crystal lattices and preventing the addition of further water molecules. This phenomenon, known as **thermal hysteresis**, depresses the freezing point of water well below its melting point, hence these proteins are known as thermal hysteresis proteins.

2.5.3 Radiation

Radiation is energy in the form of waves or particles, such as electromagnetic radiation (e.g. gamma rays, X-rays, ultraviolet radiation, visible light, or infrared radiation) and particles (neutrons, protons, electrons, or alpha particles). While very high levels of radiation do not generally occur naturally on Earth, the effects of radiation on living organisms have been well studied as a result of research on the use of radiation in medicine and on the consequences of human activity ranging from warfare to space travel. UV and ionizing radiation can cause serious damage to DNA by modifying the nucleotide bases or causing single or double-strand breaks in DNA. One organism that is known to withstand exceptional levels of ionizing radiation and that probably qualifies as a radiation extremophile is the bacterium *Deinococcus radiodurans*, first discovered in 1956. It can withstand exceptionally high doses of UV and gamma radiation. The ability to survive such extreme environments is attributed to *D. radiodurans'* ability to accurately rebuild its DNA from hundreds of radiation-damaged fragments in the absence of an intact template. This extraordinary resistance is thought to be a consequence of evolutionary adaptations to cope with extreme desiccation so that *D. radiodurans* may in fact be a xerophile (see Table 2.2).

This unique ability may result from a ring-like morphology for *D. radiodurans'* genetic material keeping free DNA fragments close together.

2.5.4 pH

pH, which ranges on a logarithmic scale from 0 to 14, measures the concentration of hydrogen ions (H^+) in solution. Biological processes tend to occur towards the middle range of the pH scale so that typical environmental pH values also fall within this range, e.g. the pH of seawater is ~8.2. Some extremophiles are known that prefer highly acidic or alkaline conditions, the **acidophiles** and **alkaliphiles**. Acidophiles thrive in the rare habitats having a pH of between 0.7 and 4, and alkaliphiles favour habitats with a pH between 8 and 12.5.

Highly acidic environments can occur naturally from geochemical activities. For example, the production of sulfur-rich gases in deep-sea hydrothermal vents and at some hot springs. However, acidophiles are not able to tolerate a significant increase in acidity inside their cells, where it would destroy important molecules such as DNA. Thus they survive by keeping the acid out. But the defensive molecules that provide this protection, as well as others that come into contact with the environment, must be able to operate in extreme acidity. Indeed, enzymes have been isolated from acidophiles that are able to work at a pH of less than 1.

Alkaliphiles live in soils laden with carbonate and in so-called soda lakes, such as those found in Egypt, the Rift Valley of Africa and the western USA. Above a pH of 8 or so, certain molecules, notably those made of RNA, break down. Consequently, alkaliphiles, like acidophiles, maintain neutrality inside their cells.

2.5.5 Salinity

Organisms can live within a range of salinities, from essentially distilled water to saturated salt solutions. **Halophiles** are organisms that require high concentrations of salt in order to live. Their optimal NaCl concentrations for growth range from twice to nearly five times the salt concentration of seawater. They are found in habitats like the Great Salt Lake (Figure 2.25), Dead Sea and salterns (evaporation basins for obtaining salt). Some high-salinity environments are also extremely alkaline because weathering of sodium carbonate and certain other salts can release ions that produce alkalinity. Not surprisingly, microbes in those environments are adapted to both high alkalinity and high salinity.

Halophiles have a particular adaptation that allows them to tolerate a high salt environment. Under normal conditions, water tends to flow across a semi-permeable membrane such as a cell wall from areas of low salt concentration to areas of higher concentration, a process known as **osmosis**. Thus a cell suspended in a very salty solution will lose water and become dehydrated unless it contains a higher concentration of salt (or some other solute) than its environment. Halophiles contend with this problem either by producing large amounts of an internal solute or by retaining a solute extracted from outside the cell. For example, the archaean *Halobacterium salinarum* concentrates potassium chloride in its interior. As with the hyperthermophiles, these adaptations will not work under more normal salinities.

Figure 2.25 The Great Salt Lake, Utah, seen from the Shuttle Atlantis. The area of the image is around 200 km by 200 km. (NASA)

2.5.6 Dessication

You saw in Sections 1.1.3 and 2.1 that its high melting and boiling points and the wide range of temperatures over which it remains liquid makes water an essential solvent for life. Water limitation therefore represents a particularly extreme environment for life. Some organisms can tolerate extreme desiccation by entering a state of apparent suspended animation known as **anhydrobiosis**, characterized by little intracellular water and no metabolic activity. It is well documented in organisms such as bacteria, yeast, fungi and plants and animals associated with environments where the water-film essential for active life is often transient and sporadic. When the film dries out these organisms appear to be dead for periods of days, weeks, or even years until moisture returns, when they 'come back to life' and resume their normal activities.

2.5.7 Pressure

Terrestrial plants and animals at the Earth's surface have evolved at normal atmospheric pressure (101 kPa = 1 atmosphere). However, hydrostatic pressure increases with depth in the oceans so marine organisms may have to deal with much higher pressures. Atmospheric pressure also decreases with altitude, so that by 10 km above sea-level, it is around one quarter of the atmospheric pressure at sea-level.

■ What effect will decreasing atmospheric pressure have on the boiling point of water?

❑ The boiling point of water decreases with decreasing pressure.

Conversely, the boiling point of water increases with increasing pressure so that water in the Earth's deepest ocean basins will remain liquid at temperatures as high as 400 °C.

Pressure presents problems to life because it forces volume changes, for example when pressure increases, the molecules in cell membranes pack more tightly, resulting in decreased membrane fluidity. Organisms that can tolerate high pressures have often adapted the compositions of their cell membranes to increase fluidity. Similarly, any biochemical reaction that results in an increase in volume, as many do, will be inhibited by an increase in pressure. Pressure-loving piezophiles (see Table 2.2) have been recovered from the Earth's deepest sea floor, the Mariana Trench, where they thrive at pressures of 70–80 MPa, but will not grow at pressures below 50 MPa.

Gravity also has an effect on the forces experienced by an organism. However, until recently, all organisms on Earth have lived at 1 *g*. The advent of space exploration means that humans have had to deal with a range of different gravity regimes, from the variable *g* experienced during launch to microgravity environments on board the International Space Station (ISS). Although most research concerned with microgravity has concerned human health, studies on board the ISS have demonstrated that gravity plays an important role in a variety of biological processes. Some effects of microgravity were expected in organisms that were adept at perceiving gravity, such as the root tips in plants. What was unknown was whether gravity played a role at the sub-cellular level where the force of gravity is almost negligible when compared with the forces governing molecular interactions. Scientists now believe that there are conditions in which the weightless environment influences the cellular machinery fundamentally resulting in specific changes to cell membranes and the reproduction of micro-organisms.

2.5.8 Oxygen

For much of its early history the Earth was an anaerobic environment. Today oxygen plays a crucial role in life on Earth and organisms inhabit environments ranging from strictly anaerobic to aerobic. Oxygen plays a key part in the mechanisms that sustain plant and animal life, photosynthesis and respiration and it is the subtle balance between the consumption of oxygen in respiration and its production in photosynthesis that is critical for the stability of the oxygen level in the Earth's atmosphere. Aerobic metabolism is far more efficient than anaerobic metabolism, but it comes at a price. Molecular oxygen can cause considerable oxidative damage to living organisms and has been implicated in a variety of human health problems, from cancer to ageing. UV radiation can produce reactive oxygen species such as hydrogen peroxide (H_2O_2) and the same species can be produced in aerobic metabolism. As a result, some organisms have evolved mechanisms to avoid or repair the effects of oxidative damage by producing antioxidants.

2.6 Extreme environments

The sheer diversity of life on Earth makes it impossible to do a complete survey of even the Earth's more extreme environments in a few pages.

> The continued discovery of new extreme environments and the organisms that inhabit them has made more plausible the search for life on other bodies in the Solar System such as Mars and Europa.

Given that many of these environments that appear to be extreme on Earth may be analogous to the normal environments for other planetary bodies, we'll examine a few of them that may have a role to play in either the origin of life or providing suitable habitats in otherwise hostile environments.

Hotsprings

Hotsprings and geysers such as those in volcanic areas of New Zealand (Figure 2.26) are characterized by hot water, steam and sometimes low pH and toxic metals such as mercury. They are, nonetheless, environments that sustain a remarkably diverse range of life. The range of colours visible in Figure 2.26 reflects different algal populations growing around the Waiotapu hot springs in New Zealand.

Figure 2.26 Hot spring in Waiotapu Park, Rotorua in North Island, New Zealand. The various colours around the edge are due to microbial mats formed by organisms that thrive in different temperature and pH environments. (Courtesy L. Thomas)

The deep sea

The deep-sea environment has high pressures and both heat and cold. In the vicinity of hydrothermal vents water temperatures may be as high as 400 °C.

- Why does water not boil at hydrothermal vents?
- Hydrostatic pressure keeps the water liquid as it raises its boiling point.

Hydrothermal vents have pH ranges from around 3 to 8 and as you saw in Section 1.7.3 were possibly critical to the evolution of early life on Earth, a conclusion supported by phylogenetic evidence that suggests that thermophiles were the last common ancestor (Section 1.9).

The fact that life can exist, and indeed thrive, at depth in the Earth's oceans without the need for photosynthesis has significant implications for the plausibility of environments for life elsewhere in the Solar System. As you'll see in Chapter 4, Jupiter's moon Europa may harbour a subsurface ocean of liquid water that lies below a layer of ice too thick to allow photosynthesis.

Hypersaline environments

Hypersaline environments include salt-flats, evaporation ponds, natural lakes (e.g. the Great Salt Lake in Utah, USA) and deep-sea hypersaline basins. Halophilic organisms are often the dominant organisms in these environments, tolerating salinities of more than 30%.

- Geological evidence, in the form of stromatolites, provides some evidence for the emergence of life possibly as early as 3.5 Ga ago. Carbon isotopes provide evidence for the operation of biological carbon fixation in rocks as old as 3.8 Ga. However, both lines of evidence are controversial.
- Life on Earth has become progressively more complex, and larger. However, large organisms evolved comparatively recently over the last 600 Ma.
- Extremophiles are organisms that have adapted to living under some of the more extreme environmental conditions on Earth such as high and low temperatures, extremes of pH and salinity, and high radiation environments. Adaptations, particularly to their cell membranes, enable these organisms to thrive in otherwise hostile environments.
- Many of the Earth's extreme environments, and the organisms that inhabit them, provide useful analogues for understanding how life may be able to exist on other planetary bodies in our Solar System and beyond.

CHAPTER 3
MARS

3.1 Introduction: Mars and life

Of all the bodies in the Solar System other than Earth, it has almost always been Mars that has dominated discussions of the possible existence of life, whether it is extinct or still present (extant). Even in ancient times, both mythology and informed thinking suggested that Mars might be inhabited. This belief prevailed through the Middle Ages even up to the modern scientific epoch. By the advent of the space age, the idea of any advanced life-forms had all but disappeared. But even in 1960, when the first space mission to Mars was attempted, the seasonal changes in colour which had been observed from the Earth were still taken by some to indicate that Mars was vegetated.

Figure 3.1 Giovanni Virginio Schiaparelli (1835–1910), Italian astronomer and director of the Milan Observatory, who discovered the relationship between comets and meteor streams in 1866. He was most famous for his meticulous observations of Mars (1877–90), including features that became known as 'Martian canals' (Figure 3.2). He continued to observe Mars faithfully until his eyesight failed. (Courtesy of Yerkes Observatory, University of Chicago)

Arguably the most notable (and possibly notorious) of observers of Mars was the 19th century Italian astronomer Giovanni Schiaparelli (Figure 3.1) who, apart from his scientific contributions, made an everlasting contribution to our cultural awareness of Mars. At a time when astronomical observations relied on attention to detail and outstanding eyesight, he would draw features which photographic plates could not record. He announced, in 1878, that he had observed extensive straight lines or streaks on the surface of Mars (Figure 3.2). Schiaparelli treated his findings with great caution and at first doubted his own observations. But successive observations convinced him of their veracity. In his description of these observations, he used the Italian word 'canali' which means 'channels' or 'grooves'. This generated enormous interest and many scientists and writers in the English-speaking world translated this word as 'canals', implying artificial waterways presumably created by intelligent beings. (Who could have imagined then that some 90 years later, spacecraft would photograph canyons and valleys

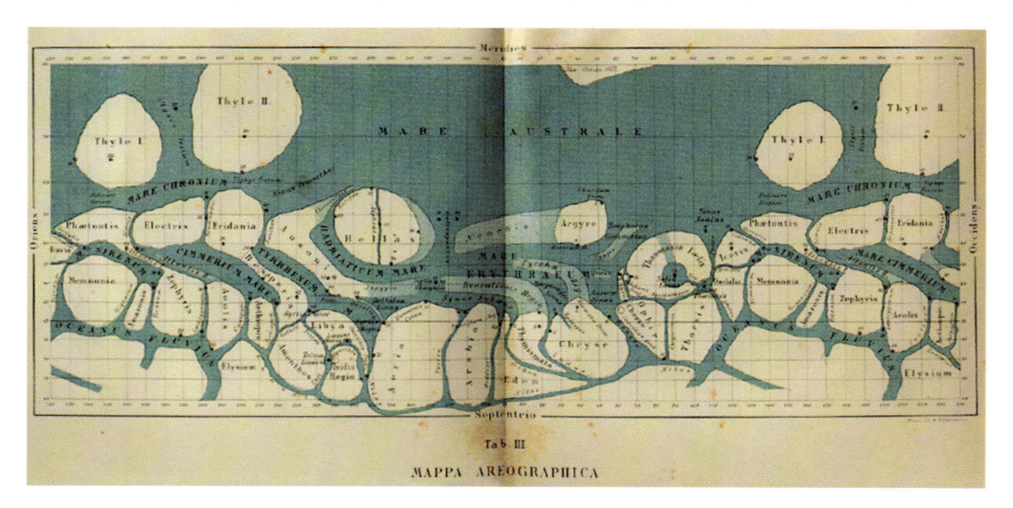

Figure 3.2 One of Schiaparelli's many sketches of Mars, this one completed in 1881, based on his telescopic observations. These led to a 'craze' amongst astronomers, both amateur and professional, to search for evidence for intelligent life and extravagant claims by some for positive evidence. Schiaparelli remained rather moderate in his assertions about 'canals', suggesting that they might be natural rather than artificial structures.

on Mars hundreds of kilometres long and produced at one time by flowing water?) For nearly a century the idea of life on Mars was embedded in popular imagination and was a fertile source for both serious literature (for example H. G. Wells' *War of the Worlds*, first published in 1898), and pulp fiction in its many manifestations (Figure 3.3). One of those most inspired by Schiaparelli's observations was Percival Lowell (1855–1916), an American polymath who devoted his energy (and fortune!) to the study of Mars (Figure 3.4). He advanced a theory in his lectures and writings that the 'canals' were a result of attempts by struggling Martian inhabitants to irrigate the planet from the melting polar ice-caps!

(a)

(b)

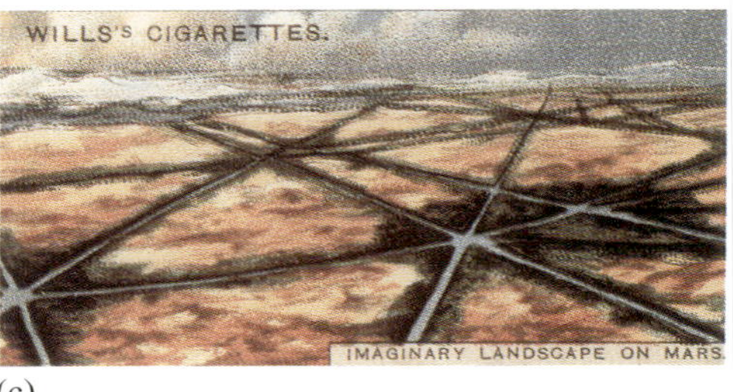

(c)

Figure 3.3 Illustrations from several examples of the treatment of life on Mars in literature. (a) The cover picture by an unknown artist of *La Guerre dans Mars,* (b) an illustration of a floating Martian city by Paul Handy in *Letters from the Planets* (1890) and (c) a cigarette card from Wills's cigarettes. (Mary Evans Picture Library)

Figure 3.4 Percival Lowell (1855–1916), a Boston-born American, who spent his life devoted to business, travel, literature and astronomy. He became widely known for his theory that the Martian 'canals' were a result of attempts by the struggling inhabitants to irrigate the planet from the melting polar ice-caps. He founded the great observatory that bears his name at Flagstaff, Arizona, initially with the exclusive intention of confirming the presence of advanced life-forms on the planet. Although his theories met with widespread opposition, he received numerous honours during his life. Nearly 14 years after Lowell's death, Clyde Tombaugh discovered the planet Pluto from the Lowell observatory, a discovery for which Lowell had paved the way with his calculations concerning the gravitational perturbations to the movement of the planet Neptune. (Copyright © Smithsonian, National Air and Space Museum)

In recent years, as a result of developments in our knowledge of life on Earth, the known abundance of the elements, the fundamentals of organic chemistry and our knowledge of the Martian environment, the belief that some form of life has at some time existed on Mars has been strengthened in the view of many (though not all) planetary scientists. However, despite enormous steps forward in both knowledge and understanding, this is still a matter of some controversy, and unequivocal evidence is still awaited. In fact, in recent times, the consensus about life on Mars has ebbed and flowed with its perceived likelihood rising and falling as the latest results and theories, both from Mars-based measurements or developments here on Earth, are digested.

But without any detailed knowledge of Mars, are there any reasons for believing that Mars might be a habitat for life?

■ Recall from Section 2.3.2 the position of the outer edge of the Sun's habitable zone. How does this relate to the position of Mars?

❑ Models for the outer edge of the present habitable zone place it between 1.37 AU (where CO_2 starts to condense) and 1.67 AU (the point at which a maximum greenhouse effect would operate). Mars is located at 1.52 AU, placing it inside the outermost estimate of the Sun's habitable zone.

There is one prerequisite for life which has dominated the issue of the existence or not of life on Mars.

■ What is this prerequisite?

❑ The need for liquid water (see Section 1.2.1).

The search for water has thus become inextricably linked to the search for life itself. Whereas it is clear that water did once exist on Mars (Figure 2.4), the question of when it disappeared, or even if it has completely disappeared, has become a dominant issue. As one of the scientists involved in this search has said:

> 'Following the water makes sense if you're prospecting for biology.
> If we could find evidence of preserved liquid water on Mars, that would be the Holy Grail.'

3.2 Background

Up to the end of the year 2002, over 30 space missions had attempted to explore Mars (Table 3.1). Starting with an unsuccessful Soviet mission (official designation Mars 1960A), which failed on launch in 1960, the space exploration of Mars was dogged by many failures, especially in the early days of the space age.

However, a series of extremely successful missions, coupled with Earth-based telescopic observations, has given us a fairly thorough picture of the basic facts about Mars. The first successful space mission was the Mariner 4 fly-by in 1964 and there followed various missions (both from the USA and the USSR) over the next 11 years, leading to the successful Viking 1 & 2 orbiters and landers. However, the results from the early missions were not encouraging for the proponents of

■ How does the surface pressure of Mars compare with that on Earth?

❑ The surface pressure on Earth is about 1000 mbar, so the Martian surface pressure is less than 1% of that on Earth.

■ Express the average surface pressure on Mars in SI units.

❑ Average surface pressure = $(6.3 \times 10^{-3} \times 10^{5})$ Pa = 6.3×10^{2} Pa.

Table 3.2 Composition[a] of the **troposphere** of Mars with sources and sinks of the components where known.

Gas	Volume ratio[b]	Major source[c]	Major sink
CO_2	9.53×10^{-1}	Evaporation, outgassing	Condensation
N_2	2.7×10^{-2}	Outgassing	Escape (as N)
^{40}Ar	1.6×10^{-2}	Outgassing	
O_2	1.3×10^{-3}	CO_2 photodissociation (3.2–3.4)	Photoreduction
CO	7×10^{-4}	CO_2 photodissociation (3.2)	Photooxidation
H_2O	3×10^{-4}	Evaporation, desorption	Condensation, adsorption
^{36}Ar	5×10^{-6}	Outgassing	
Ne	2.5×10^{-6}	Outgassing	
Kr	3×10^{-7}	Outgassing	
Xe	8×10^{-8}	Outgassing	
O_3	$(0.1 \text{ to } 20) \times 10^{-8}$	Photochemistry (3.6)	Photochemistry
NO	7×10^{-5} (at 120 km)	Photochemistry	Photochemistry

[a] Values at the surface unless indicated otherwise.

[b] The **volume ratio** is the fraction by *number* of the atoms or molecules of a species present. Chemists often refer to this as the mole fraction. It is also called the *volume mixing ratio* by some atmospheric scientists. When multiplied by the atmospheric pressure, it gives a quantity called the **partial pressure**, which may be envisaged as the fractional contribution of a component to the total pressure.

[c] Numbers in brackets refer to equation numbers in the text.

The atmosphere of Mars is composed mainly of carbon dioxide, with only a few per cent of N_2 and very minor amounts of other gases, including H_2O (Table 3.2). Being composed largely of CO_2, the atmosphere of Mars resembles that of Venus, especially at high altitude, although the **column mass** (Box 3.2) is very different (see Table A1).

In addition to these atmospheric components, reservoirs of H_2O and CO_2 are contained in the polar ice-caps (see Figure 3.5) and permafrost. So, not surprisingly, the Viking missions (see Section 3.3) observed relatively large amounts of H_2O in the atmosphere close to the north polar cap, especially during summer when the

Permafrost is a term used to describe permanently frozen soil, subsoil or other deposits.

BOX 3.2 COLUMN MASS

This parameter gives the mass of gas in a column of unit cross-sectional area (i.e. 1 m^2) extending from the surface of the planet vertically upwards to the very top of the atmosphere. If we know the atmospheric pressure, P, and the gravitational acceleration, g, at the surface, we can calculate the column mass, M_c, as follows:

Pressure = force/area; force = mass × acceleration.

Therefore, pressure = (mass × acceleration)/area.

We can identify (mass/area) with the column mass, M_c.

Therefore, we finally obtain: pressure = column mass × acceleration, or

$$M_c = P/g \tag{3.1}$$

- What units are used for column mass?
- Column mass is mass divided by area so the units are $kg\,m^{-2}$.

Figure 3.5 This image, one of the best of Mars taken from the Earth (or Earth-orbit), was acquired by the Hubble Space Telescope in 1997. It shows many features. For example, the north polar CO_2 ice-cap is clearly visible – at the time the image was taken, at the end of Martian spring, it was rapidly receding to reveal the much smaller permanent water-ice cap. It also reveals the circular, dark sea of sand dunes (Olympia Planitia) that surrounds the north pole. Another major feature is the large dark area (Syrtis Major Planitia) just below the centre. Near the southern extremity, clouds of water-ice obscure the giant impact basin, Hellas. (NASA)

cap is evaporating. Rather less enhancement of H_2O was observed at the south polar cap in its summer. The northern polar cap is believed to consist of CO_2 overlying a residual cap of H_2O ice about 600 km across which is exposed in summer. In contrast, the south polar cap appears to be composed predominantly of CO_2.

But why should the frost or ice, when its temperature is raised, turn straight into the gas or vapour state rather than liquid? To understand this, you need to make use of a **phase diagram** (Box 3.3).

Figure 3.9 Alternating layers of ice (water and carbon dioxide?) and dust (or ice and dust mixed together in differing proportions) are visible in this exotic image of part of the southern polar cap. These layers probably formed as a result of the varying axial inclination (obliquity), and they reflect climatic changes over perhaps the last billion years. This terrain has, to the best of our knowledge, no parallel elsewhere in the Solar System and its structure and appearance are not yet fully explained. The image was produced from a mosaic of images obtained by the Mars Orbiter Camera on the Mars Global Surveyor in October 1999. The region is at latitude 87° S and the image covers an area of about 10 km × 4 km. (NASA/JPL/Malin Space Science Systems/USGS Flagstaff)

Being of such low column mass, the Martian atmosphere is completely exposed to solar UV radiation right to the surface. We therefore expect that CO_2 will dissociate:

Equation 3.2 is an example of photodissociation or photolysis. In such a process, a molecule dissociates as a result of the absorption of a photon.

$$CO_2 + \text{photon} \longrightarrow CO + O \qquad (3.2)$$

However, there is a problem of timescale. In the absence of other reactions, the entire atmosphere of CO_2 would be destroyed in about 3000 years.

The key to the chemistry that maintains CO_2 in the Martian atmosphere is the intervention of H_2O. Thus, recombination of CO and O takes place mainly in the lower atmosphere through the intervention of H, OH and HO_2 (hydroperoxyl). These species occur when H_2O is **photodissociated**. As in the Earth's troposphere, OH acts to **oxidize** CO, that is it reacts to increase the proportion of oxygen in a compound (see Box 3.4). Starting with atomic hydrogen, a sequence of reactions leading to CO_2 can be deduced from laboratory studies according to Equations 3.3 to 3.5:

$$H + O_2 + M \longrightarrow HO_2 + M \qquad (3.3)$$

$$O + HO_2 \longrightarrow O_2 + OH \qquad (3.4)$$

$$CO + OH \longrightarrow CO_2 + H \qquad (3.5)$$

- What happens to the substance M that enters Equation 3.3 on the left-hand side of the equation?
- Nothing! M is an example of a catalyst so it reappears unscathed on the right-hand side.

BOX 3.4 OXIDATION AND REDUCTION

Oxygen occurs in combination with other elements in substances, such as H_2O and CO_2 that are present in the Martian atmosphere. In carbon dioxide, the ratio of oxygen to carbon (O : C) reaches its highest in compounds of these two elements. So in CO_2, the carbon is described as oxidized.

The Earth's atmosphere also contains compounds of carbon in which there is no oxygen. In one example, methane, CH_4, the ratio of hydrogen to carbon (H : C) has its highest value for compounds of these two elements: the carbon is now a **reduced** substance. When combined with the maximum amount of hydrogen, an element is said to be reduced.

Oxygen, which is responsible for the conversion of carbon-containing substances into CO_2, is an **oxidizing** substance. Hydrogen, which can convert carbon-containing substances to CH_4, is called a reducing substance. The conversion processes are referred to, respectively, as **oxidation** and **reduction**. Intermediate levels of oxidation (or reduction) are represented by substances such as carbon monoxide, CO, which also occurs in the Martian atmosphere. Oxidized and reduced are relative terms: for example, CO is oxidized relative to CH_4, but reduced relative to CO_2.

The net effect of these three reactions is the conversion of CO and O into CO_2. It is just one of a number of schemes that can be devised using known chemical reactions of OH and HO_2 that are thought to be effective in regenerating CO_2 in the Martian atmosphere. Note that this sequence involves atomic oxygen, present from the photodissociation of CO_2. Its reaction with molecular oxygen, O_2, accounts for the presence of small amounts of ozone, O_3, via Equation 3.6, one of a set of four reactions known as the Chapman scheme (named after English geophysicist Sydney Chapman, 1888–1970):

$$O + O_2 + M \longrightarrow O_3 + M \qquad (3.6)$$

Both O_3 and HO_2 are powerful oxidizing molecules. These and other oxidizing molecules have effects beyond interactions with atmospheric components, which will be considered later in the context of the Viking space mission.

Another feature of the atmosphere is dust storms of size 100 km to 1000 km which are found to be relatively common. Occasionally these can grow to planet-wide proportions, enveloping the entire planet in a shroud of dust. Dust particles raised by Martian winds are typically 1 μm across and can remain aloft for weeks before settling to the surface. The redistribution of dust by these storms plays an important role in the centimetre- and metre-scale geology of the surface.

It now seems likely therefore, that the most extreme environments on Earth in which organisms can replicate (as illustrated by the extremophiles that you met in Section 2.5) are notably less extreme than the environments that occur on the surface of Mars. A logical conclusion may be that it is very unlikely that any terrestrial organism could grow on the surface of Mars.

■ In view of what you have learned so far about the Martian environment, where might be the only sensible place to look for signs of **microbial** life?

□ There may be suitable habitats under the ground, just as there are on Earth.

The fact that both the GEX and PR experiments produced positive results even with the control sample indicates that non-biological processes were operating. Ensuing laboratory experiments on Earth involving the exposure of materials thought to be similar to Martian soil (oxides or superoxides) to UV radiation in the presence of a Martian-type atmosphere generated peroxides in or absorbed on the soil and, moreover, reproduced the results of the Viking lander experiments. Oxidized iron could act as a catalyst to produce the results seen by the PR experiment. It is likely that the surface of Mars is highly oxidized, the red colour (Figure 3.10) supporting the contention that iron is in this form. The oxidizing atmosphere of the Earth also ensures that iron is usually in similar form at the Earth's surface.

Only the LR experiment appears to have met the criteria for life detection, but it does this rather ambiguously. When the nutrient was first injected, there was a rapid increase in the amount of labelled gas emitted. Subsequent injections of nutrient caused the amount of gas to decrease initially (which is surprising if biological processes were at work) but then to increase slowly. No response was seen in the control sample sterilized at the highest temperature (160 °C). While there is still some controversy, the consensus is that the LR results can also be explained non-biologically.

Figure 3.10 A spectacular panorama taken by Mars Pathfinder a few days after landing in July 1997. The camera has a resolution of 2 mm at a distance of 2 m. The feature on the horizon at left is one of the so-called 'Twin Peaks' which are at a distance of some 1 to 2 km. (The other 'twin' is just out of shot.) The Sojourner Rover is seen analysing the rock nicknamed 'Yogi' some 3 to 4 m from the lander. The characteristic colour is suggestive of iron being in oxidized form. (NASA)

To summarize, all three Viking biology experiments gave results indicative of active chemical processes when samples of Martian soil were subjected to incubation under the conditions imposed on them. However, the experiments failed to detect any organic matter in the Mars soil, either at the surface or from samples collected a few centimetres below the surface. The indications were that strong oxidative processes were at work at the surface. Subsequent theoretical work has shown that **photochemical** processes, as well as the effects of oxidants such as hydrogen peroxide, are likely to be responsible for the destruction of all such material in the surface region.

The enigma of an active organic chemistry in the absence of life has not been fully explained and, as a result of the Viking biology experiments, the scientific community was split into those who denied (or at least strongly questioned) the existence of life on Mars, and those who did not rule out the existence of certain biotic 'oases' on Mars. However, it is clear that views of the issue of life on Mars were dominated for nearly 20 years by the results from the Viking biology experiments.

QUESTION 3.3

(a) The view of most scientists after the Viking biology experiments were performed on the surface of Mars was that they had failed to detect any positive indications of life. What arguments could be used against the notion that these experiments had ruled out all possibilities of life?

(b) These experiments employed a control sample against which results from Martian soil samples were compared. If this control sample had not been used, how might the results from the Martian samples have been interpreted?

3.4 Water, water everywhere?

Images of the surface of Mars, captured by the early space missions (Figure 3.11), showed features that resemble canyons and valleys, whereas others look distinctly like channels on Earth, indications that water once flowed as a liquid on Mars. The cratering timescale of the regions that contain these features has been used to estimate when these valleys were formed. Although different interpretations have been made of this evidence, it does appear that the valleys and gullies were not formed during a single, early event. Channels were formed during the period of late heavy bombardment, which occurred between 4 Ga and 3.8 Ga ago (see Section 1.6.1) but there is also evidence of water flow later at about 3 Ga ago, and also of very recent flow.

3.4.1 Evidence for past and recent water

Although the first direct evidence for the existence of surface water came as early as 1972 from images from the Mariner 9 spacecraft (see Figure 3.11a), the real progress in our knowledge of the existence of water on Mars has come with the results from two recent spacecraft, namely the Mars Global Surveyor and Mars Odyssey. They have produced an enormous body of evidence which reinforces the previously held idea that Mars once possessed significant bodies of water – but,

(a)

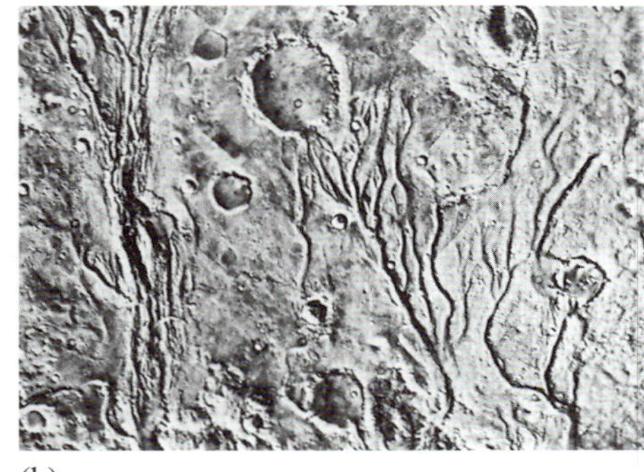
(b)

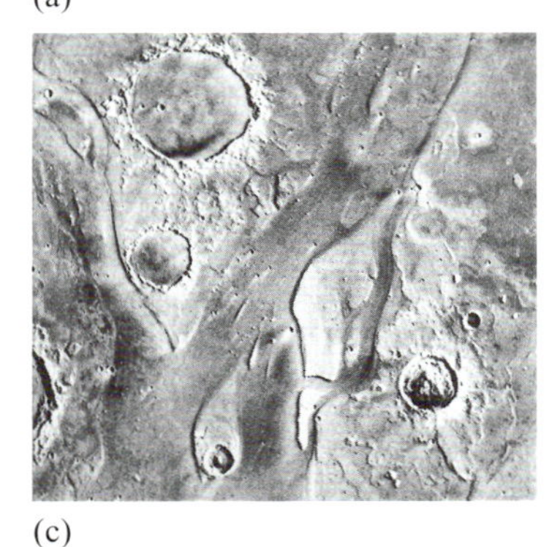
(c)

Figure 3.11 Images of the Martian surface from early space missions, showing a variety of features indicative of flowing water. (a) Part of a 700 km long channel in the heavily cratered southern highland region (20° S, 184° W) discovered by Mariner 9 in 1972. Channels like these provide firm evidence of an episode of erosion by flowing water, very early in Mars's history, before heavy bombardment had ceased entirely. (b) River valleys and impact craters. (c) A Viking Orbiter image showing the giant 'outflow channel' Ares Vallis (20° N, 33° W), a result of catastrophic flooding. The largest impact crater visible is 62 km in diameter. Clearly visible is the presence of streamlined islands with pointed prows upstream and long tapering tails downstream. (NASA)

further than this, the evidence points to this water having existed quite recently or, more speculatively, that it is even still existing. First we should look at evidence which has come from the Mars Global Surveyor (MGS) spacecraft (see Box 3.5).

BOX 3.5 THE MARS GLOBAL SURVEYOR AND MARS ODYSSEY MISSIONS

These are two of NASA's recent armada of spacecraft sent to Mars. Both are orbiting spacecraft only whose vital statistics are given in Table 3.6 below.

Table 3.6 Brief mission specifications for Mars Global Surveyor and Mars Odyssey.

	Mars Global Surveyor	Mars Odyssey
Launch	7 November 1996	7 April 2001
Arrival	12 September 1997	24 October 2001
Mass (kg)	767	758
Lifetime	The primary science mission ended on 31 January 2001, but was prolonged into an 'extended mission phase'.	The primary science mission ends in August 2004.
Primary scientific instruments	Orbiter Camera Orbiter Laser Altimeter Thermal Emission Spectrometer	Gamma-Ray Spectrometer Thermal Emission Imaging System Radiation Environment Experiment

Mars Global Surveyor became the first successful orbiter around Mars in 20 years when it entered orbit on 12 September 1997. One and a half years was spent, as planned, in trimming its orbit from an eccentric ellipse to a circle, so that its primary mapping mission started in March 1999. It studied Mars from a low-altitude, nearly polar orbit over one complete Martian year (a year on Mars is 687 Earth days or about 2 Earth years).

Mars Odyssey is targeted primarily to study the composition of the surface of Mars and to detect water and shallow buried ice. It also collects data on the radiation environment to help assess potential risks to any future human exploration and can act as a communications relay for future Mars landers. Its high-gain antenna unfurled on 6 February 2002, and its instruments began mapping Mars at the end of that month. Odyssey's Thermal Emission Imaging Sensor camera is imaging Mars simultaneously at numerous infrared wavelengths (from 8 to 20 μm) with unprecedented resolution (Box 3.6), even down to the size of a football pitch, seeking thermal and mineral 'fingerprints' hinting at 'seeps' (these are dark streaks seen in MGS Mars Orbital Camera images and possibly created by gradual seepage of liquid water), volcanic vents, or underground reservoirs.

Mars Global Surveyor

One of the great improvements offered by the instruments on the Mars Global Surveyor was the **resolution** (see Box 3.6) of its camera, the Mars Orbiter Camera (MOC). In fact, the smallest feature detectable by this instrument on Mars's surface was of length 1.4 m. Over 60 000 images have been produced and of these, over 200 of them show some very interesting features, which throw more light on the question 'Where did the water go?'

BOX 3.6 RESOLUTION ON SPACECRAFT IMAGES

Resolution is an optical term, referring to the most closely spaced objects that can be separated. Low resolution (or coarse resolution) means that closely spaced objects cannot be distinguished, whereas high-resolution images reveal fine detail. In the case of astronomical images of the sky, resolution is conventionally expressed as fractions of a degree, but resolution on planetary surfaces is more usefully expressed in terms of true distance on the ground.

In a digital image, the detail that can be seen usually depends on the size of the picture elements or *pixels* of which the image is composed. The terms resolution and pixel size are often treated as having the same meaning, although they are not strictly identical from an optical perspective. The highest resolution images of Mars obtained by the MGS MOC have pixels that are 1.4 m across, but some of the surface has been imaged with the lowest (i.e. worst) MOC resolution, namely 230 m.

QUESTION 3.4

A planetary image is produced by an orbiting spacecraft camera employing 512 pixels from top to bottom of the image and 1024 pixels from side to side. The area imaged corresponds to a scale of 4.5 km from top to bottom and 12.7 km from side to side respectively on the surface of the planet.

(a) What is the resolution (expressed as pixel size) of the imaging system in the configuration described?

(b) In this configuration, would it be possible to distinguish (or resolve) (i) 500 m-scale impact craters and (ii) 1 m-scale boulders?

(c) What happens (qualitatively) to the resolution if the same imaging system is used from a higher orbit?

They pinpointed hundreds of delicately structured gully systems (Figure 3.12). Individual gullies are just 10 metres wide (earlier missions couldn't detect such small features because of their inferior resolution) and a whole system might cover an area of only a dozen football pitches. Most are in the southern hemisphere and nearly all occur between latitudes 30° and 70°. Their sculpted terrain, cut-bank patterns, and fan-shaped accumulations of debris look remarkably similar to flash-flood gully washes in deserts on Earth (Figure 3.13). However, the headwalls of the gullies (i.e. starts of gullies) rarely have tributaries and are unlike systems fed by precipitation, so the cause appears not to be the same as flash floods on Earth.

Many (though not all) of the gully systems appear on the shaded sides of hills facing the polar ice-caps. Their geometry suggests that 'swimming-pool volumes of water could be entombed underground until suddenly it's warm enough for an ice plug to burst, letting all the water rush down the slopes,' to quote the lead scientist of the MOC team. Such a scenario is shown in Figure 3.14. In this model, underground liquid water is trapped behind an ice barrier or 'plug' which has formed on the shadowed slopes of craters and ravines. Salts dissolved in the water behind the plug could help it to remain liquid as salts have the potential to lower the freezing point of water significantly. (Remember that the phase diagram for water shown in Figure 3.6 is for *pure* water – that for salty water will be different.) Ultimately, when the dam breaks, a flood is sent down the gully, resulting in the observed patterns.

Many of the gully systems look extraordinarily recent (less than 1 Ma old) – sharply carved and crossing older, wind-scoured features. Their appearance is so fresh, in fact, that some planetary geologists think that Mars may have undergone massive, short-term climate changes, where water could come and go in hundreds of years. Indeed, scientists wonder whether liquid water might exist on Mars now, buried in some areas perhaps 500 metres underground.

MOC's findings are corroborated by data from another instrument on the spacecraft, the Mars Orbiter Laser Altimeter (MOLA). For 27 months – longer than a Martian year – MOLA gauged the daily height of the polar ice-caps, meticulously recording how much frozen material accumulated in winter and eroded (sublimated or evaporated) in summer in each hemisphere. MOLA showed that each ice-cap has a volume as great as the Greenland ice-cap on Earth (about 2.5×10^6 km^3).

Although the upper crust of the ice-caps is clearly carbon dioxide, scientists are now convinced that much of both caps' supporting mass must be frozen water because structurally, dry ice (i.e. frozen CO_2) can't support the mass of a 3 km high polar cap. MOLA and MOC measured how the polar caps shrink in each hemisphere's summer. They shrink so much, in fact, that if the observed trends were continued for just a few centuries, nearly one-third of each polar cap could evaporate into Mars's atmosphere. That would pump the atmospheric pressure up from 6 mbar to 30 or 40 mbar (remember the Earth's atmospheric pressure is about 1000 mb) which is high enough pressure for liquid water to be stable on the planet's surface under certain temperature conditions. Thus, perhaps as recently as just a century or two ago, Mars might have been clement enough for ponds of water to have dotted its surface like desert oases, and current trends suggest it might become so again. All these observations reopen the venerable question: was there – or is there – life on Mars?

(a)

(b)

Figure 3.12 Two examples of Martian gullies observed by the Mars Global Surveyor. (a) Gullies in the northern wall of the Newton crater in the northern hemisphere. The width of this image corresponds to a real distance of 6.5 km on Mars. (b) Gullies at 70° S in polar pit walls. (NASA)

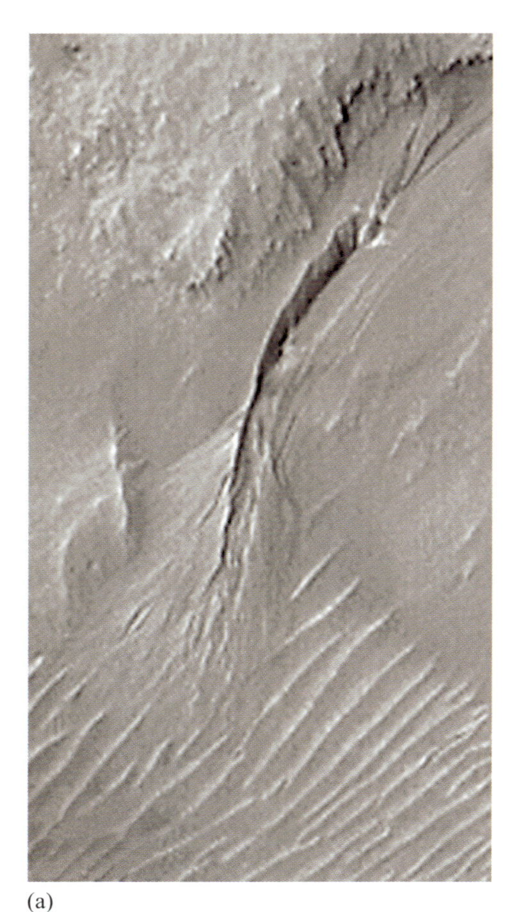

(a)

(b)

Figure 3.13 (a) The accumulated debris (or 'apron') from this gully on Mars covers sand dunes that may have formed less than a century ago. The width of this image corresponds to a real distance of 1.6 km on Mars. (b) For comparison, an apron on Earth is shown. In this example, rain water flowing under and seeping along the base of a recently-deposited volcanic ash layer (at Mount St Helens) has created the gully. (NASA)

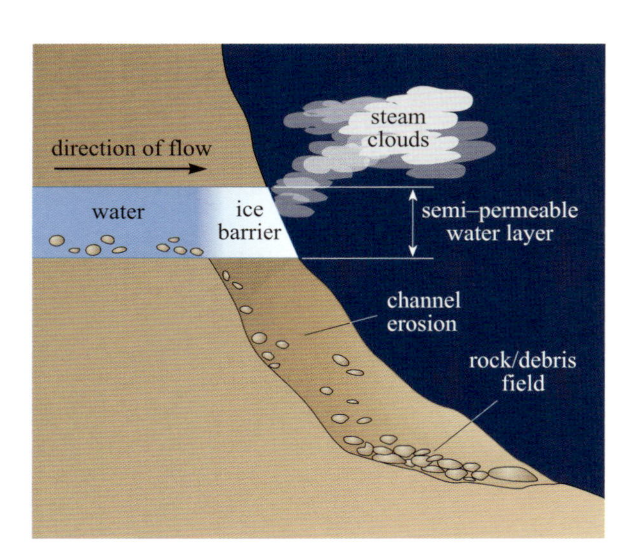

Figure 3.14 A possible model for the formation of the characteristic channels and aprons of Martian gullies.

Mars Odyssey

This mission started its main scientific tasks at the end of February 2002 and in a very short time began to produce outstanding results. We shall focus here on some very important measurements that are of fundamental importance. These are based on observations made with the Gamma-Ray Spectrometer. First we shall look at this instrument in more detail. The Mars Odyssey Gamma-Ray Spectrometer (GRS) is a suite of three different sensors that share a common electronics box and complimentary scientific objectives. The instruments are the GRS proper, the Neutron Spectrometer (NS) and the High-Energy Neutron Detector (HEND). This instrument is a follow-on instrument to the GRS that was lost when the Mars Observer Mission (Table 3.1) failed in 1993.

How GRS works

When exposed to cosmic rays (charged particles in space that come from the stars including our Sun), chemical elements in soils and rocks emit uniquely identifiable signatures in the form of gamma-rays and neutrons. The gamma-ray spectrometer analyses these signatures coming from the elements present in the Martian soil. By making these measurements, it is possible to determine which elements are present, how abundant they are and how they are distributed around the planet's surface.

The incoming cosmic rays collide with the nuclei of some of the atoms in the soil and, in some cases, they release neutrons as a result. These neutrons have high energies (and are referred to as 'fast' neutrons), and they scatter and collide with other atoms, some of which are excited to a higher energy state than usual. This extra energy can then be released in the form of gamma-rays so that the atom can return to its normal unexcited energy state. The energy E of the gamma-ray (γ) is characteristic of the atom from which it was released – in other words, it is characteristic of its parent element. Some elements such as potassium, thorium and uranium are naturally radioactive so that they don't require an external source, such as cosmic rays, to excite them.

Now the neutrons generated in the initial interactions can undergo further reactions themselves. When they collide with the nuclei of other atoms, they might lose energy, slow down, and eventually become thermalized, which means that they are moving at speeds comparable to the speed at which atoms on the surface are moving. This process is known as **moderation**; hydrogen atoms are especially important in moderating neutrons because the two have nearly identical masses. The various processes discussed here are illustrated in Figure 3.15.

'Thermal', 'epithermal' and 'fast' are terms used to describe neutrons of progressively higher energy.

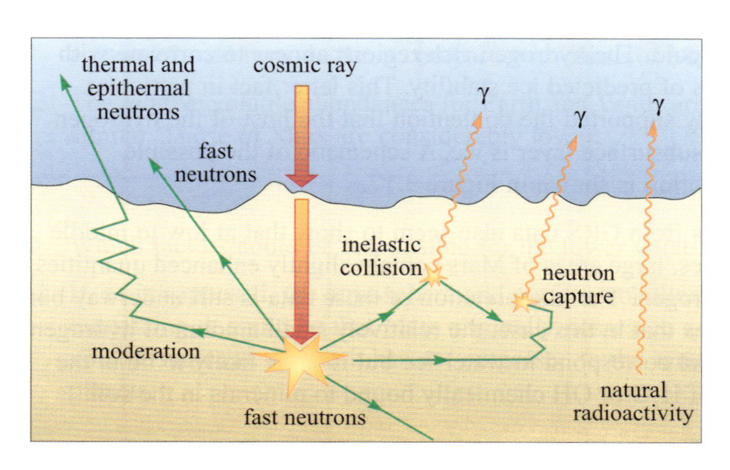

Figure 3.15 Nuclear radiation from a planetary surface produced by the interaction of incident cosmic rays with the surface. The products of the interactions described in the text are neutrons of varying energy (thermal, epithermal and fast in increasing energy) and gamma-rays (γ).

The temperature gradient in the solar nebula would militate against a low original content, since material condensing at the distance of Mars, being further from the protoSun than either Venus or Earth, and thus at a lower temperature, might be expected to be richer in volatiles. Therefore less complete outgassing seems more likely. If this is the case, the Martian atmosphere should never have been as extensive as that of Venus or Earth.

There is, however, evidence of former periods when a more substantial atmosphere existed than at present. As we have seen, the atmospheric pressure is currently so low that liquid water does not exist stably, the transition from solid to gas occurring directly. So a higher atmospheric pressure is probably needed to sustain the water which has created some of the observed water-formed features.

The most plausible explanation is that the surface temperature of Mars has been higher at some epoch in the past than it is at present. A higher temperature would result in the evaporation of some of the condensed volatiles, for example H_2O and especially CO_2. Mars would then have possessed a more substantial atmosphere. Subsequent loss of some of this atmosphere would have led to a cooling of the planet, because the surface temperature is strongly determined by the bulk of the atmosphere. Loss of water or other atmospheric components would have occurred mostly by thermal escape (but also by various other processes). This is a sufficiently important process that it is worth digressing briefly to consider it.

Ultimately, it is the strength of the gravitational field at its surface that dictates whether or not an object in the Solar System can retain an atmosphere: the stronger the field is, the stronger the gravitational forces acting on the molecules in the atmosphere. This leads to the notion of **escape velocity**, the smallest upward speed that any object (spacecraft or molecule) must have to escape from a body. The escape speed v_{esc} for a body of mass M and radius R is given by:

$$v_{esc} = \sqrt{\frac{2GM}{R}} \tag{3.8}$$

where G is the gravitational constant. Whether atmospheric molecules have sufficient speed depends on the temperature. As the temperature of a gas increases, its molecules move around more quickly, and the average speed of its molecules increases. Some fraction of the molecules will always be travelling fast enough to overcome gravitational forces, allowing them to escape to space. At low temperatures, this proportion is negligible, but at higher temperatures it becomes progressively more significant, until most molecules exceed the escape speed for the planetary body. Note that the relevant temperature is that at a level in the upper atmosphere above which the atmosphere is so thin that a molecule moving outwards has little chance of colliding with another, and so *will* escape if it has sufficient speed.

'Planetary body' is a handy term that can be used to encompass planets, satellites and asteroids.

Different gases have different molecular masses, so their average speeds are different at a given temperature.

In order for a planetary body to retain a particular gas in its atmosphere for a period of time of the same order as the age of the Solar System, the average speed of the molecules in the gas should be less than about one-sixth of the escape speed. If the average speed exceeds one-sixth of the escape speed, a significant proportion of molecules will be moving faster, and will be lost.

This condition is achieved on only a few planets and satellites. Mercury is so close to the Sun and so hot that average molecular speeds for all common gases are too great. Titan, which is a similar size to Mercury, is much less dense ($\approx 1.9 \times 10^3$ kg m^{-3}). This in turn means that its mass is less than one-half of Mercury's, and therefore its surface gravity and escape speed are lower. Titan, however, is also so far from the Sun that its temperature is only about 100 K. Thus, Titan can retain a dense atmosphere.

QUESTION 3.5

(a) The average (or, more correctly, root mean square) speed of a gas molecule is given by

$$v = \sqrt{\frac{3kT}{m}}$$

where m is the molecular mass. Based on this formula, what can you say about the likelihood of different gases being lost from a planetary atmosphere by thermal escape?

(b) In the Martian atmosphere, calculate the relative average speeds of the two most common constituents of the atmosphere.

Returning now to Mars, can the remaining atmosphere tell us anything about previous episodes of loss? Well, the answer is yes. It should carry the signature of this loss in the enrichment of heavier isotopes – lighter isotopes having been preferentially lost. Of particular interest in this context is the isotopic analysis of volatiles from Mars as measured in certain meteorites for which there is compelling evidence of Martian origin. In the following section, we shall examine this isotopic record of past climatic conditions on Mars. But first, we need to consider the evidence that these meteorites originate from Mars.

3.4.3 Meteorites from Mars

EET A79001 is a meteorite collected from Antarctica. Let's first consider the lettering and numbering system used in its designation. 'EET' refers to the collection site (which, in this instance, was Elephant Moraine in the Antarctic), 'A' designates the collection trip and '79' is the year of collection (1979). The identifying number, 001, signifies that it was the first meteorite to be classified upon return of the samples to the curatorial facility. When the collectors, who are often meteorite researchers, spot rare or otherwise unusual samples in Antarctica, a note is made to give them priority treatment during the preliminary classification procedure. EET A79001 was one such sample, and on this particular occasion the decision to promote its investigation could not have been more justified, since the meteorite is now widely believed to come from Mars.

Figure 3.18 The meteorite EET A79001. (NASA)

EET A79001 is a so-called shergottite. A photograph of EET A79001 is shown in Figure 3.18. Shergottites are linked with two other categories of meteorites known as nakhlites (pronounced '*nahk*-lights') and chassignites (pronounced '*sha*-sig-nights'). These samples, around

30 in number, are collectively referred to as the SNC meteorites (where S, N and C denote shergottites, nakhlites and chassignites). The names derive from the discovery locations, namely Shergotty (India), Nakhla (Egypt) and Chassigny (France). All of the SNC meteorites are igneous rocks formed by crystallization from magma. In appearance, the shergottites are medium-grained rocks of basaltic composition. However, one important fact distinguishes most SNC meteorites – they have relatively young formation ages, in the case of EET A79001 about 0.2 Ga ago.

- The crystallization ages of most of the SNC meteorites range from 0.2 Ga to 1.3 Ga. What are the implications of this?
- These meteorites were formed late in the history of the Solar System. Wherever they were formed, at least some part of their parent body had to have been melted 0.2 Ga ago (or remained molten until this time).

The ages of most meteorites cluster around 4.5 Ga and represent samples that formed in parent bodies of asteroidal size (i.e. a few hundred kilometres in diameter). Asteroids were heated and cooled relatively early in the history of the Solar System. Some meteorites have younger ages than 4.5 Ga, but these are the result of later impact melting which acts to re-set the radiometric dating systems. It is apparent from the textures of SNC meteorites that they are not impact-produced melts. The only reasonable environment that retains sufficient heat to produce melting 0.2 Ga ago is a parent body of planetary dimensions, so there can only be a few candidates for the source of SNC meteorites – i.e. Mercury, Venus, Earth, the Moon, Mars, or Io.

- Can you think of ways in which the SNC meteorites could be removed from the surface of the planet?
- Ejection caused by a volcanic eruption or an impact.

Volcanic processes can be rejected on various grounds. The most plausible mechanism involves removal following an impact onto a planetary-sized surface by a meteoroid or comet. A schematic representation of calculations pertaining to large impact craters is shown in Figure 3.19. In this model, the incoming projectile, in addition to pulverizing the target rocks and producing a crater, also causes a very thin layer of ejecta to be propelled away from the impact site. Some of these pieces of ejecta, which are unmelted and unshocked, can escape from the planet's surface. Other escaping ejecta components could be melted or shocked.

- If fragments ejected from another body can reach the Earth, what is the most likely source and why?
- The Moon – simply because it is so much closer than any other solid body.

Meteorites of lunar origin have now been unambiguously found on the Earth, numbering around 40 in total.

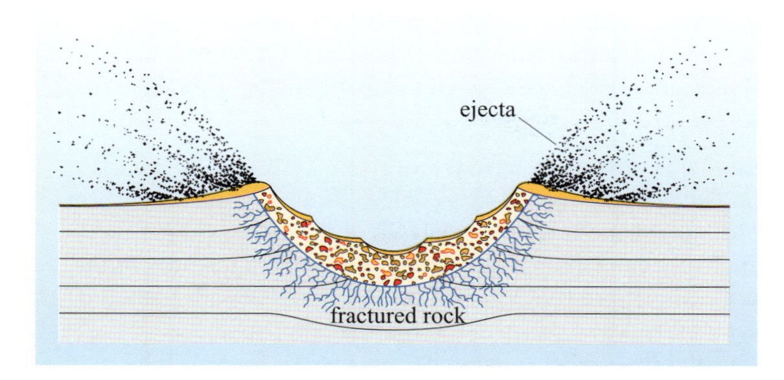

Figure 3.19 Schematic diagram of a crater-forming event on a planetary-sized body. No physical parameters are given since this is a generalized case (although imagine that it pertains to projectiles of kilometre size travelling at around 10 km s^{-1}). Note that the impact produces melting and intense fragmentation in the target rocks, with fragment size increasing away from the impact site. However, in addition, unshocked and unmelted materials can be ejected from a very thin surface layer. Close to the impact site, these may be travelling sufficiently fast to overcome the escape velocity of the planet and are, as a consequence, ejected from the planet. (Adapted from Melosh, 1989)

- Why can we be so certain that these fragments originate from the Moon?
- The Moon is the only body for which there are returned samples with which to compare. It transpired that chemical compositions, isotope ratios, minerals, and textures of the lunar meteorites are all similar to those of samples collected on the Moon during the Apollo missions. Taken together, these various characteristics are different from those of any other type of meteorite or terrestrial rock.

Having established unambiguously that certain meteorites found on Earth have originated from another body in the Solar System, we can now turn again to EET A79001. Its characteristic features bore little resemblance to any returned lunar samples so the Moon was unlikely to be the origin. In addition, lunar volcanism ceased about 3.2 Ga ago. Io's distance from Earth and its proximity to Jupiter make it dynamically very unlikely that ejecta from its surface reaches the Earth. Mercury, despite being relatively close to the Earth, is so close to the Sun that it is unlikely that material ejected from its surface could be transported to the orbit of Earth. Venus has a dense atmosphere (Table A1) and a large gravitational field (about the same as Earth) – removal of material from this planet would require speeds so high that frictional heating in the atmosphere would cause severe melting (or even complete vaporization in the case of small pieces of rock).

- Based on the above arguments, what do you think is the likely parent body of the SNC meteorites?
- Mars would seem to be a good possibility, if only by default.

In support of this it should be noted that Mars has a tenuous atmosphere (surface pressure currently 6 mb) and the gravitational field is comparatively low (less than one-half that of the Earth; see Question 3.1) – thus, solid materials could be ejected without vaporization – and it is a lot closer than Io.

However, there is some very specific evidence that quite unambiguously points to Mars as the parent body of EET A79001. We have already stated that SNC meteorites are not impact melts but are the result of igneous activity, analogous in many ways to rocks formed at or near the Earth's surface. However, SNC meteorites were subjected to an impact event, which was violent enough to accelerate them to a speed greater than the escape velocity of Mars (5 km s^{-1}).

Intuitively therefore, it may be expected that the meteorites would show some evidence of this process. Indeed, shergottites and chassignites record the effects of shock, though the nakhlites are unshocked (which poses a constraint on theoretical modelling of the impact event).

During the impact event that is thought to have removed EET A79001, localized melting occurred within the sample. These melts cooled extremely rapidly to form a glass containing trapped atmospheric gases. Analyses of these trapped gases showed them to be chemically and isotopically distinctive. The abundances of the gases contained within the shock-produced glass from EET A79001 are plotted in Figure 3.20 against the abundances of gases in the Martian atmosphere as determined by the Viking and Pathfinder missions (these are the same as the abundances shown in Table 3.2 but expressed as the number of particles per m^3 rather than as the volume ratio). The line represents points whose values are the same on both axes.

- What can you conclude from Figure 3.20 about the abundances of gases in EET A79001 and the Martian atmosphere?
- The data points show near perfect correlation between the two sets of abundances, suggesting that EET A79001 contains trapped Martian atmosphere.

The evidence is compelling and the conclusion of this evidence is therefore that EET A79001 and, in fact, all the SNC meteorites originate from Mars.

QUESTION 3.6

(a) Place in ascending order of escape velocity Mars, Venus and the Moon.

(b) What factors other than escape velocity dictate the likelihood of material from these bodies being ejected by surface impact and reaching the Earth?

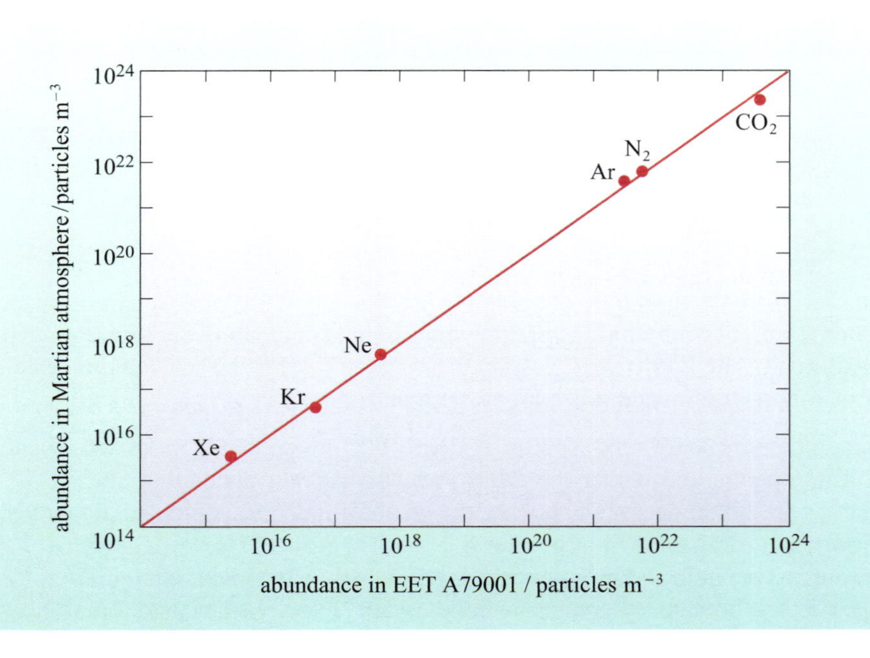

Figure 3.20 Plot of the abundances of gases in the Martian atmosphere (measured by spacecraft landers) versus the abundances in the glass from EET A79001. Both axes are logarithmic. Note the extremely good correlation, which forms part of the evidence supporting a Martian origin for SNC meteorites.

The Martian meteorite EET A79001 can give us very strong clues as to the history and evolution of water on Mars. The shock-produced glass in EET A79001 contains a variety of gases. Through analyses of the noble gases, a theoretical model has been devised to describe the early evolution of the Martian atmosphere. Before considering this further it should be noted that the isotope composition of xenon in the EET A79001 glass shows that the Martian atmosphere includes gases attributable to carbonaceous chondrites, a primitive class of meteorite (this can also be inferred from an assessment of Viking data, or by looking in detail at the chemical compositions of the SNC meteorites). It is considered that Mars received its presently observed complement of volatiles from an influx of carbonaceous chondrite-like material near the end of its formation. Outgassing from this veneer of chondritic material subsequently formed the early Martian atmosphere.

You have already seen that observations of the Martian surface show considerable evidence for the action of fluid flow (see Figure 3.11). At least two episodes of water flow and subsequent loss from the atmosphere to space have been identified. At the present time, photodissociation of water vapour results in the loss of hydrogen and oxygen. This mechanism has been operating on Mars since its formation. In earlier times, because the ultraviolet flux from the Sun was comparatively high, resulting in enhanced levels of photodissociation, the process was more efficient. Even so, the rate is too slow to account for all the water assumed to be lost. Quite clearly another process was responsible.

Over the first 100 Ma of Martian evolution, water was readily converted to hydrogen by reaction with, for instance, iron–nickel metal. The vast amounts of hydrogen produced in this way were very quickly lost from the planet by thermal escape as a rapidly moving flow of gas. During this time, gases heavier than hydrogen were swept away by the rapid flow, and so were also lost from the planet – a process known as **hydrodynamic escape**. The details of this are complicated but its imprint can now be observed in the abundances of noble gases and the isotope composition of xenon in the glass of EET A79001.

- With the knowledge that hydrogen has been lost from the Martian atmosphere, how would you anticipate its isotope composition has evolved with time, given that hydrogen has two isotopes, namely the lighter isotope of hydrogen (^{1}H, hydrogen) and the heavier isotope (^{2}H, deuterium)?

- It would be expected that ^{1}H, would be preferentially lost from the atmosphere compared with the heavier isotope ^{2}H, deuterium, resulting in an increase in the D/H ratio with time.

Measurements of water vapour in the Martian atmosphere show a D/H ratio which is about 5 times that of water in the Earth's oceans. This measurement spectacularly supports the contention of the loss of hydrogen to space, and implies that water vapour may have been carried with it by hydrodynamic escape. So our model of a Martian surface and atmosphere which at some periods was much warmer and which showed extensive surface water deposits is strongly supported.

3.4.4 Conclusions

In summary, the evidence of atmospheric evolution on Mars points to a planet in which outgassing is less complete than on Venus and Earth, and in which the atmosphere has been lost partly to space and then to the surface as the temperature consequently fell.

You have seen that the conclusions of the initial exploration of Mars by spacecraft were not particularly encouraging for those who wished to see Mars as a habitat for past or present life-forms. And these conclusions were also supported by the results of the Viking biology experiments, despite their ambiguity.

But subsequent observations by the Mars Global Surveyor and the Mars Odyssey have provided evidence of extensive bodies of frozen water below the surface as well as evidence of flowing water in the recent past – and, more controversially, the interpretation that mechanisms which produce this flowing water may be active even today. Additionally, there is separate, strong evidence from Martian meteorites for past periods of extensive water followed by escape to space. This evidence is in the form of the isotopic signature of hydrogen. So the pessimistic conclusions of most scientists following the results from Viking have not been completely vindicated. Now there is a significant body of opinion which believes that conditions in the past were conducive for simple life-forms to have existed on Mars, and even more controversially, that they may have survived in certain protected 'oases'. These arguments form the foundations for the space experiments planned and in preparation over the next decade to search for evidence of past or present life.

Before life could originate on Mars (or anywhere) certain prerequisites were needed.

■ What were the prerequisites?

❑ Water, organic materials, an energy source and a site in which to concentrate them.

Let's consider the hypothesized earliest period of extensive water flow on the surface of Mars which extended to around 3.8 Ga ago (around the same time as the end of the period of heavy asteroid bombardment). From our knowledge of the evolution of the Martian atmosphere, it seems likely that at this time the above prerequisites would have been available on Mars as on Earth. We've seen that on Earth, life would have originated between the end of the period of heavy bombardment, 4.0 Ga to 3.8 Ga ago and before 3.85 Ga to 3.5 Ga ago. This gives a 'window' of between 100 Ma and 500 Ma when life would have evolved. On Mars, this opportunity would have been somewhat shorter as the surface water was beginning to vanish rather rapidly by 3.8 Ga ago. In fact, the major uncertainty in this scenario appears to be whether liquid water was available for long enough and abundantly enough for life to arise.

3.5 The ALH 84001 story: evidence of life in a Martian meteorite?

You now know almost unequivocally that meteorites of Martian origin exist on Earth. We shall now proceed to consider the most famous of all Martian meteorites and its significance for the question of life on Mars. Until 7 August 1996, the name ALH 84001 was probably known to only a relatively small number of planetary scientists worldwide.

- ■ What is the significance of the designation ALH 84001?
- ❑ '84' signifies the year of recovery of the meteorite, namely 1984, and '001' signifies that this was the first meteorite to be classified upon return of the samples to the curatorial facility. ALH indicates the locality of recovery, which in this case was Allan Hills in Antarctica.

But on that day, everything changed. A press conference was held by a group of scientists led by David McKay, Everett Gibson and Kathy Thomas-Keptra of NASA's Johnson Space Centre to announce that certain characteristics in this meteorite, known to have come from Mars, were most likely interpreted as being the relics of ancient Martian microbial life. The effect was dramatic. Newspapers, radio and TV around the world rushed to declare 'Life found on Mars!' (or its equivalent in many languages) and with their own particular stress. It seemed that humanity's desire to find that we are not alone (or at least might not have been alone at some time in the past) had at last been answered. But had it? Well not everyone in the scientific community thinks so. In fact, it is true to say that the majority of informed opinion does not agree with the conclusions of the authors of the scientific paper that presented the full analysis of their case.

Why did a 1.9 kg potato-sized lump of rock cause such a sensation? The *recent* history of ALH 84001 started when it was discovered in 1984 in the Allan Hills region of Antarctica.

- ■ Why are so many of the world's meteorite collection found in Antarctica?
- ❑ It's a combination of the environmental conditions (e.g. uniform surface, etc) and the lack of human activity that makes Antarctica a good location for meteorite detection and collection. Of the world's collection of more than 22 000 meteorites, by far the majority have come from this source.

The Martian origin of ALH 84001 was not recognized until 1993, when it was realized that the SNC meteorites (now numbering around 30), were almost certainly from Mars. But what of the history of ALH 84001 before it was found in Antarctica in 1984? Here, we shall quote from a 1997 article by several of the team who undertook the original work:

> 'The meteorite timeline begins with the crystallization of the rock on the surface of Mars, during the first 1 percent of the planet's history. Less than a billion years later the rock was shocked and fractured by meteoritic collisions. Some time after these impacts, a water-rich fluid flowed through the fractures, and tiny globules of carbonate minerals formed in them. At the same time, molecular by-products, such as hydrocarbons, of the decay of living organisms were deposited in or near the globules by that fluid. Impacts on the surface of Mars continued to shock the rock, fracturing the globules, before a powerful collision ejected the rocks into space. After falling to Earth, the meteorite lay in the Antarctic for millennia before it was found and its momentous history revealed.'
>
> Gibson, E. K. *et al*. (December 1997) 'The Case for Relic Life on Mars', *Scientific American*, pp. 58–65.

- How old is the ALH 84001 meteorite?

- According to the quotation above, ALH 84001 formed 'during the first 1 percent of the planet's history'. Since Mars, like all the planets, formed around 4.5 Ga ago, ALH 84001 formed around 4.5×10^7 years later. This means that it has an age of nearly 4.5 Ga. This is in contrast to most SNC meteorites which have an age in the range 0.2 Ga to 1.3 Ga.

But how did they reveal this 'momentous history', namely that ALH 84001 contained evidence for the existence of living organisms? Their conclusion was based on five separate strands of evidence, each of which on its own was not compelling, but, taken together, according to the authors, were highly convincing. Almost all of their evidence comes from carbonate globules that are found on the surface of a fracture through which fluid flowed and deposited these globules. (See Figure 3.21.)

The five lines of evidence, as presented by David McKay and his colleagues can be summarized as follows:

(a) The carbonate is in the form of 'globules' comparable to crystal aggregates known to be produced by bacteria on Earth.

(b) Perhaps the most visually compelling evidence are objects that seem to be the fossilized remains of microbes themselves. Nanometre-scale carbonate structures, shown in Figure 3.22a in the globules resemble fossil spheroidal, rod-shaped and filamentary bacteria. The segmented object shown in Figure 3.22a is 380 nm long. This can be compared with terrestrial samples such as that shown in Figure 3.22b. The approximately vertical feature just to the right of centre is believed to be a minute fossil and is also 380 nm long. It was found 400 m below the Earth's surface in a formation known as Columbia River Basalt. Some curved structures in the meteorite have lengths in the range 500 to 700 nm. Others, for example ovoids (meaning egg-shaped), are as small as 30 nm in length. These are about 10 times smaller than terrestrial objects that are usually interpreted as bacteria. However, McKay *et al.* argue that typical cells often have small appendages attached, of sizes similar to these small features found within ALH 84001. The suggestion is that some of these features are fragments or parts of larger units.

(c) Inside the carbonate globules, they found fine-grained particles of magnetite (Fe_3O_4) and iron sulfide within the size range 10 to 100 nm. Using sophisticated analysis techniques involving microscopy and spectroscopy, they found that the size, purity, shapes and crystal structure of all the magnetites were typical of magnetites produced by bacteria on Earth. Such particles on Earth are known as magnetofossils. The magnetites within ALH 84001 are typically 40 to 60 nm in size. Intriguingly, some of them in ALH 84001 are arranged in chains, similar to pearls in a necklace. Terrestrial bacteria often produce magnetite in just this pattern, because as they biologically process iron and oxygen from water, they produce crystals that align themselves with the Earth's magnetic field.

(d) Another strand of evidence is that the carbonate, iron sulfide and iron oxide minerals occur together but would not be stable under any one set of physical conditions suggesting formation in the 'non-equilibrium' conditions characteristic of life.

Figure 3.21 (a) ALH 84001 shown in its entirety. (b) A cut through the rock, showing a cross-section. Just right of centre is a vertical crack through which fluid flowed and deposited globules of carbonate minerals. (c) and (d) 2 mm long fragment containing several of the globules, which are about 200 μm in size. (e) High-resolution image of one of the globules. ((a) Copyright © Proszynski I S-ka SA 1999–2001; (c) Monica Grady; (b, d, e) Douglas A. Kurtze, North Dakota State University of Physics)

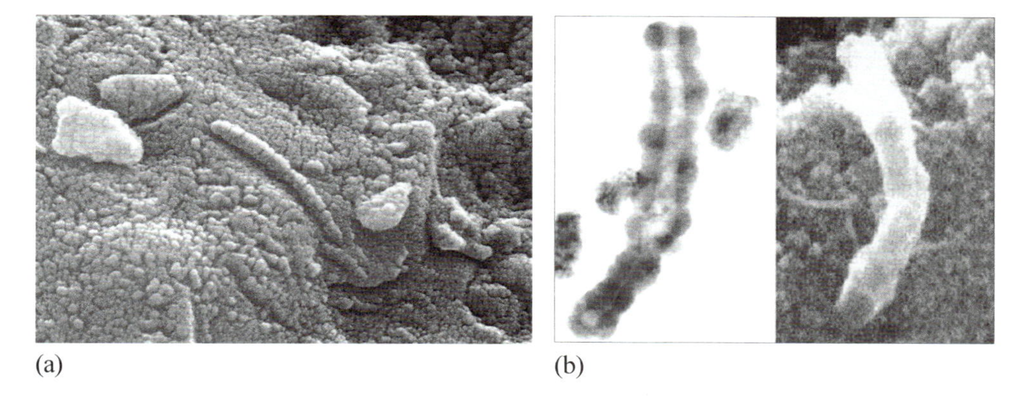

Figure 3.22 (a) The segmented object (380 nm long) was found in the Martian meteorite ALH 84001 and has been interpreted as a fossil microbe. It is claimed to resemble fossilized bacteria found on Earth. For example, the near-vertical objects in (b), which are of the same length as the object in (a), were found at a depth of 400 m below the terrestrial surface. (Everett Gibson (NASA/JSC))

(e) There is organic matter including complex hydrocarbons that could have been produced by living organisms. Now organic molecules have been found in other meteorites which are known to have come from asteroidal sources in interplanetary space, an environment hardly likely to support life, so why should this fact be significant. McKay's team argue that the type and relative abundance of the specific organic molecules are suggestive of life processes. When living organisms die and decay in the terrestrial context, they create hydrocarbons associated with coal, peat and petroleum. Many of these belong to a class of organic molecules known as polycyclic aromatic hydrocarbons (PAHs). There are literally thousands of PAHs but their presence does not necessarily demonstrate that biological processes have occurred. In ALH 84001, the PAHs are always found in carbonate-rich regions, including the globules. They contain a relatively small number of different PAH types, all of which have been identified in the decay products of microbes. Significantly, these PAHs were located inside the meteorite where terrestrial contamination is unlikely to have occurred.

All these features occur together in the carbonate veins. The team argue that these features are indigenous to the meteorite and 3.9 Ga old, and therefore that there was life on Mars at that time. There is however much scepticism in the scientific community as to the veracity of this conclusion, and much research on this subject has been instigated since 1996. Amongst the objections to these conclusions are:

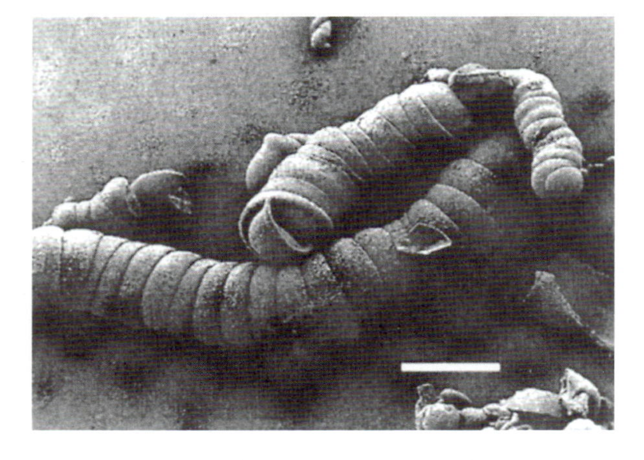

Figure 3.23 A barium carbonate crystal aggregate grown in a silica gel, in the absence of any organisms. Compare with the hypothesized microfossil in Figure 3.22a. (Courtesy of Stephen Hyde)

(a) It is always difficult to find compelling evidence of a microbial origin for such simple structures which might equally plausibly be chemical and mineralogical artefacts (see Figure 3.23).

(b) Convincing fossils of this age are extremely rare on Earth, and very hard to find even in systematic searches of well-exposed regions of well-preserved rocks. So it is very surprising that with a sample of little more than 20 rocks from Mars, one were to contain fossils.

(c) The original research compared the carbonate microstructures (evidence line (b)) to fossil bacteria but made no attempt to make comparisons with non-biogenic mineralic structures that could be mistaken for fossil bacteria.

(d) A recent study suggests that the carbonates and other materials in the veins crystallized at 200 to 500 °C from minerals melted at the moment of impact that ejected ALH 84001 into space. Experimentally produced melts of this type generate carbonate globules like those in the meteorite. If the fractures and the carbonates that fill them formed during a high-velocity impact on Mars, then the structures and compounds within them are extremely unlikely to be fossils.

(e) There are also tiny crystals of the iron sulfide mineral greigite in the carbonate globules. These are interpreted by McKay and his team as products of bacteria. On Earth, there are bacteria that gain their energy for life by converting dissolved sulfates to sulfides. The sulfide then reacts with any iron present to produce iron sulfide minerals. This is a very common process in oceanic sediments. However, there are also natural chemical processes that produce iron sulfides, so they need not be biogenic.

In fact, at the time of writing, it seems that of the five lines of evidence presented above, four can be explained without the necessity for a biogenic origin. However, one, namely (c), remains controversial. This concerns the origin of the magnetite crystals, present in abundance within the carbonate globules. In fact, within the space of a few months, scientific results on these features were published by two different groups of scientists with diametrically opposing conclusions. Both groups have studied primarily the shapes and other features of the magnetite crystals. One group claims that the Martian magnetites are physically and chemically identical to magnetites produced by a certain terrestrial bacteria strain. Conversely, the second group, using new microscopic measurements, state that 'the crystallographic and morphological evidence is inadequate to support the inference of former life on Mars'.

The original team has put forward arguments to counter the objections described above. However, the present opinion of the scientific community is that while the evidence for life is intriguing and demands further study, it is not compelling. It is probable that the argument will only be finally resolved by results from future missions to Mars. One of the most significant results from ALH 84001 was the impetus it gave to studies of life on Mars.

3.6 Planetary protection

We shall now consider an issue which is critical to all *in situ* searches for life on other planetary bodies.

- ■ Suppose you send a spacecraft to another Solar System body with an instrument designed to detect the signs of life, and that it is successful in its aim. How can you be sure that what you have detected is from the body that you have visited and not carried inadvertently by your spacecraft from Earth?
- □ Well you can't be sure unless you have been meticulous in ensuring that the chance of the spacecraft carrying micro-organisms from Earth and subsequently 'contaminating' the landing site has been eliminated.

After all, you have already come across one case where the transfer of such micro-organisms from Earth could have taken place. Figure 2.23 shows the camera from the Surveyor-3 spacecraft. On return to Earth after 2.5 years on the lunar surface, living terrestrial organisms were found in the foam inside the camera – the potential for contaminating other bodies had been clearly demonstrated.

The pursuit of preventing such contamination and all the issues relating to it are generally referred to as 'planetary protection'. There even exists a United Nations Treaty to which most space-active nations are signatories that states 'Parties to the treaty shall pursue studies of outer space.........so as to avoid their harmful contamination and also adverse changes in the environment of the Earth......'.

United Nations (1967) Treaty on Principles Governing the Activities of States in the Exploration and Use of Outer Space, Including the Moon and Other Celestial Bodies. U.N. Document No. 6347.

Remember, however, that contamination from the Earth can probably never be 100% eliminated – we can only minimize the chance of it occurring. This means that the results of all in situ experiments designed to detect extraterrestrial life must be treated with great care.

The major focus of planetary protection procedures and protocols has been on preserving other planetary environments from contamination by organisms from Earth that might grow there and thus obscure *forever* any efforts to understand the origin and evolution of life at locations other than Earth. But planetary protection is concerned not only with forward contamination, as it is called, but also back contamination, namely the return to Earth, and release into the biosphere, of potentially harmful organisms or substances. However, it is the former on which we shall concentrate here.

Clearly, different space missions to different environments and to perform different tasks do not all require the same degree of planetary protection.

■ Can you suggest what factors dictate the degree of planetary protection activity required?

❑ Factors to be considered include:

(i) the target body,

(ii) whether the spacecraft is intended to land on the target,

(iii) whether the spacecraft will return to Earth.

In the earlier days of space research, the analysis of the likelihood of contamination was implemented using a probabilistic approach. More specifically, the probability of contamination, P_c, was divided into two components, namely $P_c = P_t \times P_g$ where P_t is the probability of an organism's survival during transit from Earth surface to the surface of the target body and P_g is the probability of growth (and reproduction) of a contaminating organism on the target body.

Bioload (or bioburden) is a term used to describe the number of viable organisms present on an item.

The aim then would be, by controlling the bioload of the spacecraft and by determining the value of P_t and P_g by measurement and analysis, to ensure that the value of P_c for the spacecraft's entire bioload is less than a predefined value, often taken to be 10^{-3}.

■ What does a value for $P_c = 10^{-3}$ mean?

❑ It means that if 1000 such missions were sent to a particular planet, one of them would result in contamination.

However, this approach gradually fell into disrepute due to the difficulty of realistically estimating some of the probability terms. Instead, a simplified more robust approach has gradually been adopted.

Through a worldwide organization of scientists known as COSPAR (Committee on Space Research), the world's space-faring nations, and others, have agreed a unified approach to planetary protection. To simplify matters, all space missions that travel to a body in the Solar System are categorized into one of five categories, depending on the risk of contamination, both forward and backward. These categories, I to V in increasing order of risk, each have a different series of requirements concerning planetary protection associated with it. The categories (approved in 1984) are as follows:

- Category I missions include any mission to a target planet that is not of direct interest for understanding the process of chemical evolution. In effect, no protection of such planets (such as Mercury for example) is warranted, and so no planetary protection requirements are imposed.
- Category II missions are all types of missions to those target planets that are of significant interest for understanding the process of chemical evolution, but for which there is only a remote chance that contamination carried by a spacecraft could jeopardize future exploration. The concern is primarily over unintentional impact, since these missions are not designed to land.
- Category III missions are certain types of missions (fly-by and orbiter) to a target planet of interest for understanding the chemical evolution and/or the origins of life, or for which scientific opinion suggests a significant chance of contamination that could jeopardize a future biological experiment.
- Category IV missions are certain types of missions (mostly probe and lander) to a target planet of interest for understanding chemical evolution and/or the origins of life, or for which scientific opinion suggests a significant chance of contamination that could jeopardize future biological experiments.
- Category V missions include all Earth-return missions. The concern is for the protection of the terrestrial system as well as the scientific integrity of the returned sample.

One result of the application of the COSPAR regulations is that once the appropriate category for a spacecraft has been determined, it is possible to determine the maximum allowed bioload for that particular spacecraft. As an example, when these regulations are applied to a Category IV mission to Mars, a maximum of 300 000 organisms are allowed at launch. When the effects of the journey to Mars are taken into account, the intent is that the spacecraft is effectively 'biologically inert' when it arrives. To put the value of 300 000 into context, it should be appreciated that one sneeze disperses something like 1000 000 organisms.

■ There are three broad sources of microbiological contamination when a spacecraft is being assembled. Can you suggest what these are?

□ They are:

the environment in which the spacecraft is assembled (5%)

the materials out of which the spacecraft is constructed (15%)

the people who assemble the spacecraft (80%)

The figures in brackets represent the approximate contribution of each to the total bioload for a spacecraft built in a clean-room environment.

There is a wide range of techniques available for reducing the bioload resulting from each of these categories. For example, contact between humans and the spacecraft is kept to a minimum. When absolutely necessary, techniques are used to ensure minimal transfer of organisms by the use of barrier clothing and appropriate procedures. The environment in which the spacecraft is constructed is very rigorously controlled by the use of so-called clean-room techniques. For a Category IV mission, a typical environment would be classified as 'Class 100' which means that there are less than 100 particles (including micro-organisms) of size greater than

0.5 μm within every cubic foot of air within the controlled environment (historically, SI and Imperial units have been mixed in this definition). The main active sterilizing techniques that are used for planetary protection applications, in increasing order of complexity (and effectiveness) are given in Table 3.8.

Table 3.8 Sterilizing techniques used in planetary protection.

Technique	Notes:
Cleaning of individual components	Levels and type of cleaning will depend on the particular measurements being performed during the mission. Typical cleaning techniques include detergent cleaning, solvent cleaning and hot helium purge.
Surface sterilization	Same techniques as for individual components.
Lander sterilization	Moist or dry heat which kills micro-organisms principally by oxidation. In this technique, sufficient heat has to be delivered throughout the whole of the sample. In addition, all materials within the sample must be able to withstand high temperatures (for example, 135 °C for 8 hours or 125 °C for about 50 hours).Gas plasma sterilization in which the sample is immersed in a hydrogen peroxide plasma. It is believed that this technique is less damaging to electronic components than heating techniques. Gamma-radiation. Some electronic components and also optical glasses can be damaged by this process.

An example of the effect of moist heat sterilization is shown in Figure 3.24.

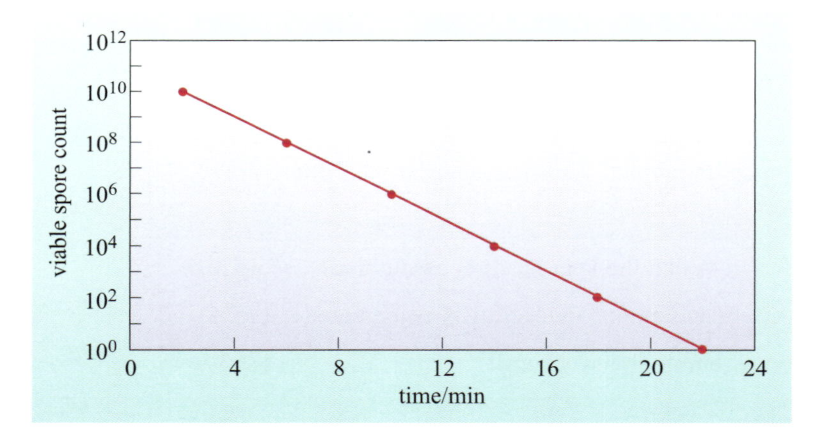

Figure 3.24 A graph showing the survival rate of spores on a sample as a function of time as a result of moist heat sterilization at 120 °C.

QUESTION 3.7

Place each of the following (hypothetical) space missions in the appropriate planetary protection category (I to V), giving your reasons in each case:

(i) cometary nucleus lander,

(ii) Mercury orbiter mission,

(iii) Jupiter orbiter with Mars fly-by en route,

(iv) Mars orbiter and lander,

(v) mission to return cometary dust to Earth.

3.7 Habitats for life

We shall end this chapter with a brief general consideration of likely Martian habitats for life.

- If we wish to search for signs of extinct or extant life on Mars, and knowing what we do about the prerequisites for life, where should we look?
- Anywhere there has been or is water.

However, there are factors, which militate against the development of life in the presence of water. The short wave ultraviolet radiation (see Box 3.7) and the oxidation due to surface peroxide means that the dry dusty soil on the surface is unlikely to be a suitable habitat for life. But terrestrial experience (see Chapter 2) suggests that microscopic life is capable of colonizing even the most extreme of occurrences of liquid water. For example, extremes of pH, salinity, temperature and pressure are not necessarily barriers. However, you have already seen that Mars has certainly experienced periods when the atmosphere was thicker and the surface warmer. Maybe therefore during these epochs, the surface was sufficiently protected for life to have developed there. So any environment indicative of water deposits might have been a suitable environment. These include lakes, thermal springs and glaciers – evidence of all these features are found on the present day surface of Mars.

BOX 3.7 MARS AND ULTRAVIOLET RADIATION

The thin atmosphere and the low concentration of ozone and oxygen means that the Martian surface is exposed to high levels of harmful solar UV radiation. Figure 3.25 shows the (measured) UV flux at the surface of the Earth and the (predicted – as it has never yet been measured directly) flux at the surface of Mars.

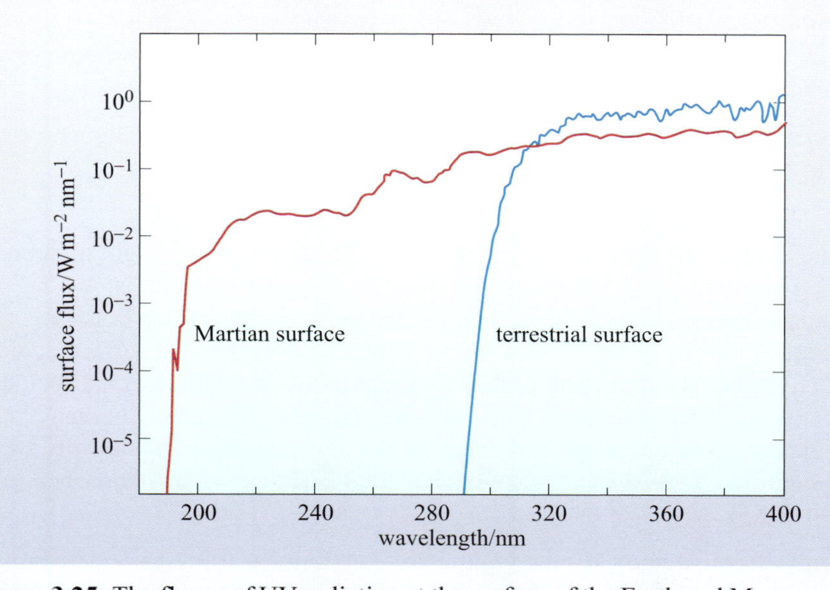

Figure 3.25 The fluxes of UV radiation at the surface of the Earth and Mars.

- What is the most obvious difference between the two spectra shown on Figure 3.25?

- The terrestrial UV spectrum 'cuts-off' at about 290 nm but on Mars it extends right down to below 200 nm.

This factor has a dramatic effect. The ultraviolet spectrum is traditionally divided into three regions, namely UV-A (315–400 nm), UV-B (280–315 nm) and UV-C (200–280 nm). Interaction of UV radiation with organic material occurs across the entire range, but varies in severity as a function of wavelength. Long wavelength UV-A is the least damaging, responsible for common effects such as tanning of the skin. UV-B (280–315 nm) is partially obscured on Earth, and is responsible for increased biological damage, causing effects such as sunburn. UV-C is the extreme case, accounting for alteration and mutation of biological organisms at the genetic level resulting in severe mutation and in some cases complete destruction. As can be seen from Figure 3.25 life on Earth is shielded from this harmful short wave UV radiation, while on Mars the atmosphere offers little protection. Environments which are protected from this radiation, such as beneath rocks or even below the surface, should therefore be favoured as habitats for life.

These environments should therefore all be amongst the targets for future searches for extinct life on Mars. Other likely targets would include evaporite deposits that are produced when a lake shrinks or disappears by evaporation – these deposits can capture other constituents including biogenic constituents. Such evaporites may have a long residence time on Mars due to the hypothesized early termination of any hydrologic cycle (involving precipitation), the absence of tectonic activity and the current dry surface conditions.

Hydrologic cycle is the term used to describe the process by which water is cyclically transferred from a planet's surface (mostly by evaporation) into the atmosphere, and back to the surface by precipitation.

For present day (extant) life, the most likely environments could be quite different.

- In view of what we know of present day conditions on Mars, where are the most probable environments for the survival of extant life-forms?

- In subsurface environments, protected from UV and the oxidizing nature of the surface.

Critically, it appears that certain extremophiles do not require sunlight to thrive – they derive their energy from chemical reactions. In fact on Earth, some populations have been found that have existed completely isolated from the surface for millions of years. This seems to open the door to subsurface ecologies on other worlds and may have profound implications for environments such as Europa (see Chapter 4). So on Mars, all subsurface environments that may harbour water should be regarded as candidates for the development of extant life, such as those identified by the Mars Global Surveyor GRS instrument (as described in Section 3.4.1).

The above discussion makes the implicit assumption that if life has existed on Mars, it was initiated spontaneously there. But we know from the very existence of the SNC meteorites that this might not necessarily be the case.

- Why are the SNC meteorites relevant to this issue?

- They show that, in principle, life from another origin (for example, the Earth) could have been transferred to Mars on ejecta from Earth by an impact. Although highly speculative, this should not be completely discounted. This could have implications for likely habitats for the development of life on Mars. Such habitats might have been therefore not the optimum ones but ones that were colonized by chance.

In addition to the above general considerations for landing site selection for any future space missions designed to search for Martian life, other more specific considerations include:

1 Concentration – It is quite important that the material indicating the presence of biota is concentrated in appreciable amounts since the first searches on the surface of Mars will probably have only limited capability for mobility and surface coverage.

2 Preservation – The preservation of the evidence is of paramount importance. Microbial material should be fossilized rapidly – this is likely to occur in an environment such as hydrothermal springs. Organic molecules and other chemical evidence may require rapid burial to avoid alteration at the surface.

3 Thin dust cover – MGS images have confirmed that the Martian surface is extensively affected by dust deposition (see also the Mars Pathfinder image in Figure 3.10 where fine-grained sediment drifts cover much of the scene, and in some places the environment has been clearly scoured and sculpted by wind). The wind-blown detritus can cover large parts of the planetary surface and the selected landing site should thus show indications of as thin as possible dust coverage.

4 Area of the target – The target for exploration should ideally be really extensive in order to be within the landing uncertainties for any lander, as it is not possible yet to land on Mars with perfect targeting precision.

These factors are not merely of academic interest. The next 20 years will see a plethora of space missions designed to try to answer once and for all the question of whether life ever has existed (or does exist) on Mars. The first of these is Mars Express/Beagle 2 and NASA's Mars Exploration Rovers. Later, we can anticipate a Mars sample/return mission in which material from Mars will be returned to Earth for the most sophisticated of analysis techniques which are available in terrestrial laboratories and ultimately a human exploration mission. By then, we can confidently hope that the question of whether life on Mars has ever existed will be finally settled.

3.8 Summary of Chapter 3

- Historical beliefs that Mars may be vegetated and inhabited were contradicted by observations from early space probes. These showed Mars to be a cold, arid habitat with a thin atmosphere unable to provide much protection against biologically harmful ultraviolet radiation and therefore superficially unlikely to be a suitable locale for life. Furthermore, it appears that water cannot exist in equilibrium under the average conditions on the surface of Mars.

- The apparently negative results from the Viking mission biology experiments appeared to kill off any speculation concerning the existence of life, despite some ambiguities. The experiments failed to detect any organic matter in the Mars soil, either at the surface or from samples collected a few centimetres below the surface. The results were interpreted as meaning that strong oxidative processes were at work at the surface.
- Evidence from subsequent space missions, such as Mars Global Surveyor and Mars Odyssey, has confirmed previous suspicions of the past existence of significant bodies of surface water and their possible existence even in recent times. They also provided strong evidence for the present existence of large deposits of subsurface water-ice.
- Evidence for earlier periods of a thicker atmosphere with a more significant presence of water also comes from analysis of the Martian meteorite EET A79001.
- Coupled with the discovery of various terrestrial extremophiles, this evidence has resulted in a re-assessment of the possibility of extant or extinct life on Mars.
- Claims for direct evidence of Martian fossils in the meteorite ALH 84001 have remained highly controversial within the scientific community with opinion as to their veracity being divided. Most scientists believe that the observed features can be explained by non-biological processes.
- Future space missions to Mars will continue to search for evidence of extinct or extant life-forms.

CHAPTER 4
ICY BODIES: EUROPA AND ELSEWHERE

4.1 Introduction

Until the 1980s, the icy satellites of the outer planets were scarcely thought of as places where life could ever have existed. Few could have imagined that one of them, Europa, would within twenty years have become the rival of Mars as a priority for astrobiological study. This chapter recounts the history of our changing perceptions of the icy satellites, examines the available evidence for their internal structures, and considers the niches offered for life to begin and to be sustained. In this context, the 'habitable zone' embraces settings devoid of both sunlight and an atmosphere. These are areas where life could survive on the energy from chemical reactions made possible by the discharge of hot chemically enriched fluids through vents on the floor of an ocean capped by a thick layer of ice.

'Ice' does not necessarily mean just frozen water. In the outer Solar System, although H_2O is usually the dominant component, ice can incorporate other frozen volatiles such as NH_3, CO_2, CO, CH_4 and N_2.

4.1.1 Satellite discoveries

All the giant planets have satellites. Jupiter's four largest satellites were discovered in 1610 by Galileo Galilei (Figure 4.1), using one of the first telescopes to be pointed at the night sky. These are now known as the **Galilean satellites**. They are much bigger than Jupiter's other satellites, the first of which was not discovered until 1892. Saturn's largest satellite, Titan, was discovered in 1655, and four more had been found by 1700.

Sir William Herschel (Figure 4.2) discovered the first two of Uranus's satellites in 1787, less than six years after he had discovered the planet itself. Neptune's largest satellite, Triton, was discovered by William Lassell (Figure 4.3) in 1846 – within three weeks of the planet being identified. Smaller and fainter satellites continued to be found. By 1950 the known tally of outer planet satellites was Jupiter, eleven; Saturn, nine; Uranus, five; and Neptune, two.

Discoveries of lesser satellites only a few kilometres across continue to be made. In the competition to be the planet with the largest number of known satellites, the lead has changed several times between Jupiter, Saturn and Uranus. However, all the satellites of the giant planets that are large enough for their own gravity to pull

Figure 4.1 Galileo Galilei, 1564–1642. Pisa-born pioneer of the experimental scientific method, whose analysis of motion paved the way for Isaac Newton's work. Galileo used one of the first telescopes to discover the four largest of Jupiter's satellites and the phases of Venus. His consequent support for the theory that the Earth moves around the Sun led to his imprisonment for heresy in 1633. (© Science Photo Library)

Figure 4.2 Sir William Herschel, 1738–1822. Born in Hanover, Herschel moved to England as a young man to work as a musician. He became an astronomer and was elected a Fellow of the Royal Society in 1781, on the strength of his lunar observations and his discovery of Uranus. Using his own 48-inch (122 cm) reflecting telescope, he discovered Titania and Oberon (satellites of Uranus) in 1787 and then Enceladus and Mimas (satellites of Saturn) in 1789. (© Science Photo Library)

Figure 4.3 William Lassell, 1799–1880. A Liverpool businessman who made his fortune in the brewing trade. He designed and built his own telescopes, including a 24-inch (61 cm) reflector, with which he discovered Triton in 1846 and two satellites of Uranus (Ariel and Umbriel) in 1851. (© National Portrait Gallery)

them into a near-spherical shape have certainly been found. For an icy body, this means the satellite must have a radius of more than about 200 km. These larger bodies are the satellites of greatest potential for astrobiology, and their basic properties are listed in Table 4.1. Two of these satellites are larger than the planet Mercury, but not so massive, because their densities are less. Four are bigger and more massive than the Moon, and a total of six are bigger and more massive than Pluto. Pluto itself (discovered in 1930) and its satellite Charon (discovered in 1978) share many of the characteristics of the large icy satellites, and so they are also listed in the table.

4.1.2 Satellite systems and their origins

The satellite systems of the giant planets have several features in common. Most satellites are in synchronous rotation, always keeping the same face towards their planet. Irregularly shaped moonlets associated with the ring system orbit closest to the planet. They travel in near-circular prograde orbits in the planet's equatorial plane. These moonlets (like the rings) are believed to be fragments of larger satellites that were destroyed by collisions or tidal forces (Figures 4.4 and 4.5). Most are bright and presumed to be icy in composition.

'Prograde' in this sense means orbiting in the same direction as the planet's spin.

Figure 4.4 Five of Saturn's innermost satellites as imaged by the Voyager spaceprobes, shown at their correct relative sizes. From left to right: Atlas, Pandora (above) and Prometheus (below), Janus (above) and Epimetheus (below). Janus is 99 km in length. The dark line across Epimetheus is the shadow of one of the narrowest of Saturn's rings. (NASA)

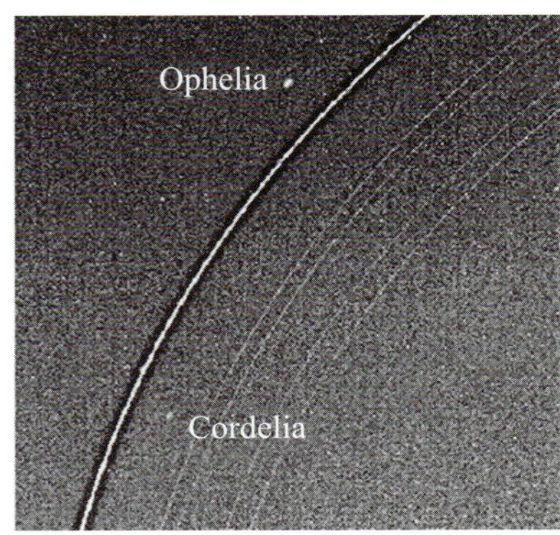

Figure 4.5 A Voyager 2 image showing Uranus's outermost and most prominent ring, which is kept in shape by the small satellites Ophelia and Cordelia (each less than about 30 km across), one of which orbits 2000 km inside the ring and the other 2000 km beyond it. The ring consists of millions of dark, dusty boulders, mostly 10 cm to 10 m in size. Four fainter rings are just visible within the orbit of Cordelia. (NASA)

Table 4.1 Basic data for the satellites of the outer planets. In the *orbital period* column, R indicates retrograde orbits. Values were up to date in March 2004, but are subject to revision. Where two or more values are given in the *radius* column, these indicate a non-spherical satellite and are the dimensions (semi-major axes) of the best-fit ellipsoids to the satellite's actual shape. The numbers of small satellites are correct as of early 2004, but are subject to change as new discoveries are made.

Planet	Satellite	Mean distance from planet/10^3 km	Orbital period/ Earth days	Radius/km	Mass/10^{20} kg	Density/ 10^3 kg m^{-3}
Jupiter	4 inner	<221.9	<0.675	<125	–	–
	Io	421.6	1.77	1821	893	3.53
	Europa	670.9	3.55	1565	480	2.99
	Ganymede	1070	7.15	2634	1482	1.94
	Callisto	1883	16.7	2403	1076	1.83
	55 outer	>7435	>130	<85	–	–
Saturn*	6 inner	<151.4	<0.695	<99	–	–
	Mimas	185.5	0.942	199	0.375	1.14
	Enceladus	238.0	1.37	249	0.649	1.00
	Tethys	294.7	1.89	530	6.28	1.00
	Dione	377.4	2.74	560	10.5	1.44
	Rhea	527.0	4.52	764	23.1	1.24
	Titan	1221.9	16.0	2575	1346	1.88
	Hyperion	1481.1	21.3	165 × 113	0.11	1.1
	Iapetus	3561.3	79.3	718	16	1.0
	Phoebe	12952	551R	115 × 105	0.007	2.3
	13 outer	>11300	>449	<16	–	–
Uranus	13 inner	<97.7	<0.762	<77	–	–
	Miranda	129.8	1.42	236	0.659	1.20
	Ariel	191.2	2.52	579	13.5	1.67
	Umbriel	266.0	4.14	585	11.7	1.40
	Titania	435.8	8.71	789	35.3	1.71
	Oberon	582.6	13.5	761	30.1	1.63
	9 outer	>4276	>267	<190	–	–
Neptune	5 inner	<73.5	<0.55	<104	–	–
	Proteus	117.6	1.12	218 × 208 × 201	0.49	1.3
	Triton	354.7	5.88R	1353	215	2.05
	Nereid	5513	360	170	0.3	1.5
	5 outer	>15686	>1874	<40	–	–
Pluto		–	–	1150	131	2.0
	Charon	19.4	6.39	586	16.1	1.9

*Saturn has three other tiny satellites: Telesto and Calypso that share the orbit of Tethys, and Helene sharing the orbit of Dione.

Orbiting further from each planet come all the satellites large enough to be spherical (or nearly so) in shape, typically in near-circular prograde orbits close to the planet's equatorial plane. These satellites probably grew within a disc of gas and dust that surrounded the planet in the later stages of its growth, mimicking in miniature the birth of the terrestrial planets from the solar nebula. Neptune's large satellite, Triton, is an exception (Figure 4.6). This has a retrograde orbit, and may be a Pluto-like Kuiper Belt object that was captured into orbit around Neptune some billions of years ago.

We have used the term 'Kuiper Belt' but you may also see it called the 'Edgeworth–Kuiper Belt'. Kenneth Edgeworth, a British astronomer, published similar ideas a few years prior to Kuiper, but this only really came to light after the term 'Kuiper Belt' had become widely established.

Beyond its large satellites each giant planet has a second collection of small irregular-shaped satellites, travelling in elongated, inclined and in many cases retrograde orbits. Most are dark bodies, rich in silicates and/or carbon compounds. These satellites are likely to be captured comets or asteroids (Figure 4.7).

Figure 4.6 A Voyager 2 image of the sunlit part of the Neptune-facing hemisphere of Triton. The south polar cap of bright nitrogen-ice is marked by streaks of sooty (carbon-rich) material erupted from geysers. The rugged surface beyond the polar cap is methane-rich ice, contaminated by nitrogen, carbon dioxide, carbon monoxide and water. (NASA)

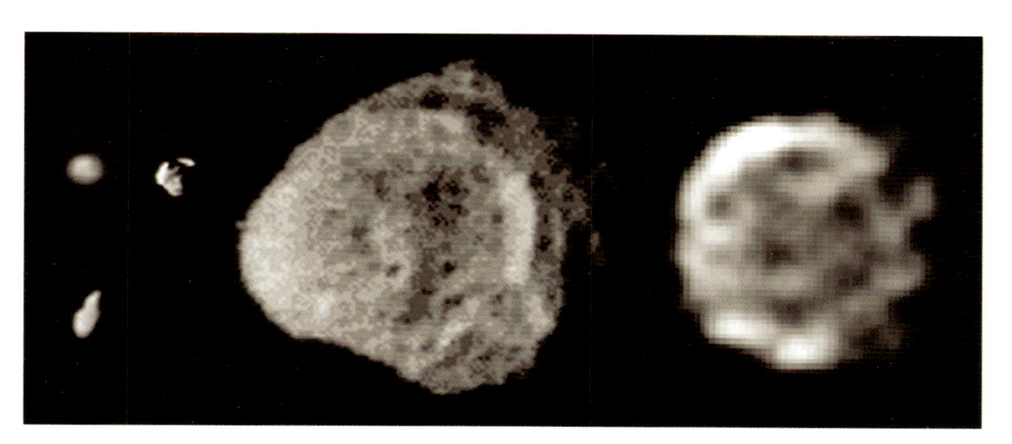

Figure 4.7 Some of the outer small satellites of Saturn shown at their correct relative sizes. From left to right: Telesto (above) and Calypso (below), Helene, Hyperion and Phoebe. Hyperion is 185 km in length. (NASA)

4.1.3 Unravelling the natures of the large satellites

Before the dawn of the space age, relatively little could be discovered about even the large satellites. Their orbits were well known, and from the subtle orbital perturbations caused by neighbouring satellites it was possible to deduce their masses. Measurements of their sizes enabled densities to be calculated to within about 20% of the currently accepted values for the Galilean satellites, and with rather less certainty for the large satellites of the other giant planets. However, it was clear that, except for Io and Europa, these bodies are not dense enough to be composed largely of rock like the terrestrial planets.

During the 1950s, spectroscopic studies by Gerard Kuiper (Figure 4.8), the discoverer of Titan's atmosphere (Section 5.2), showed that the surface of Europa is mostly clean bright water-ice, whereas that of Ganymede (which has a lower albedo) is water-ice darkened by a dusty contaminant. Spectroscopic studies have now revealed that ice dominates the surfaces of all the large satellites except Io, which is effectively a terrestrial planet in orbit about Jupiter. In the Jupiter system, the **ice** is dominantly frozen water, but with increasing distance from the Sun it becomes mixed with more volatile ices. There is indirect evidence for ammonia in the ices of Uranus's satellites, and on Neptune's large satellite Triton spectroscopic observations have detected frozen nitrogen, carbon dioxide, carbon monoxide and methane. A similar mixture to that on Triton coats Pluto's surface.

We use the term 'water-ice' where necessary to make it clear that we mean frozen water, as opposed to any other kind of ice.

Figure 4.8 Gerard Kuiper, 1905–1973. Dutch-born American planetary scientist who discovered Titan's atmosphere in 1944 and subsequently used spectroscopy to identify carbon dioxide in the atmosphere of Mars and ice on the surfaces of Europa and Ganymede. He discovered Miranda (Uranus) in 1948 and Nereid (Neptune) in 1949. In 1951 he suggested that there should be a zone of primordial debris beyond the orbit of Neptune. Although the first body in this zone was not discovered until nearly twenty years after his death, it is generally known as the Kuiper belt. (© Science Photo Library)

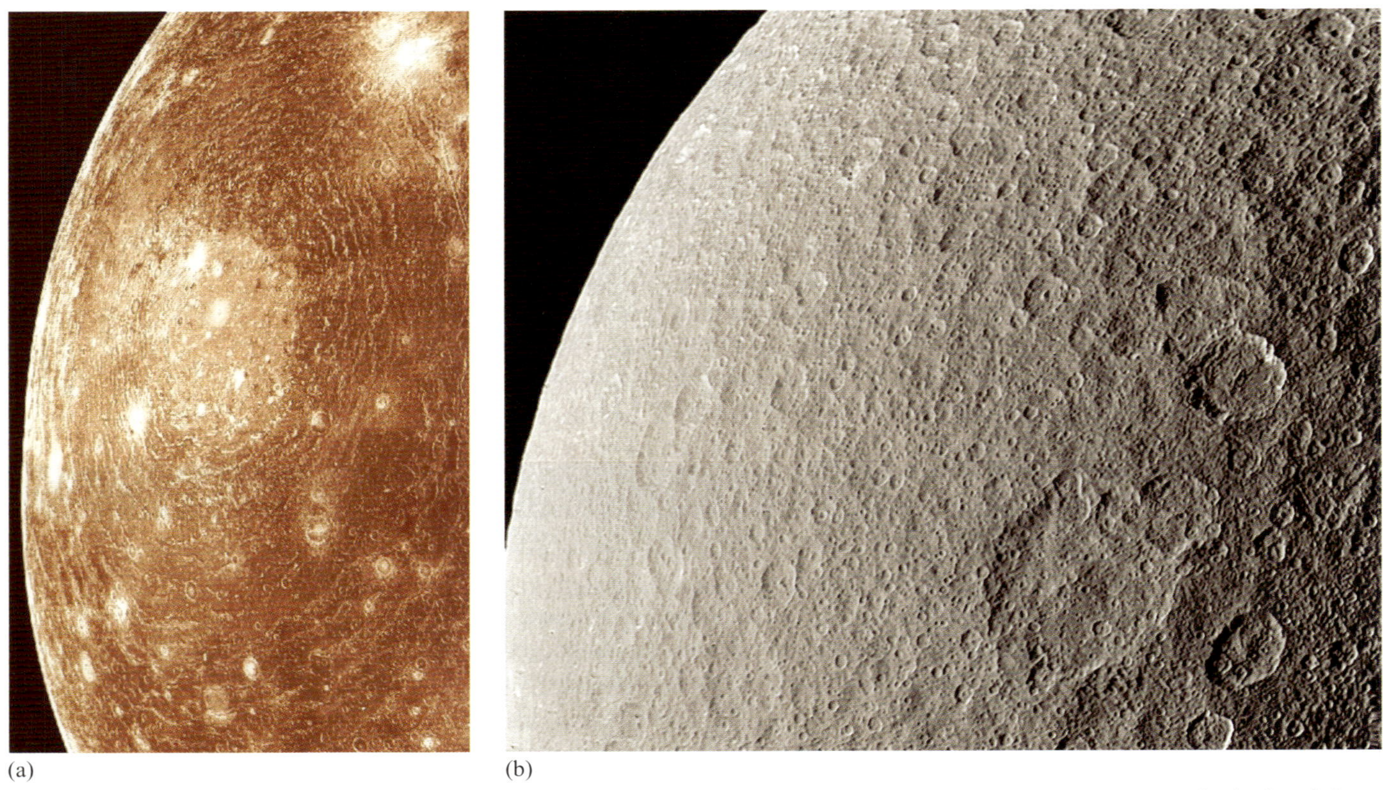

Figure 4.10 Two icy satellites whose heavily cratered appearance suggests passive worlds with little or no geological activity: (a) Callisto (image 3000 km from top to bottom) (b) part of Rhea (image 600 km from top to bottom). ((a) NASA; (b) © Calvin J. Hamilton)

With the exception of Titan, no satellite has an atmosphere thick enough to protect its surface from bombardment. The surface of an icy satellite scatters sunlight fairly evenly in all directions, which means that not even the youngest surface consists of a continuous sheet of smooth ice. Instead, any ice that was a continuous sheet originally has become broken (presumably by meteorite and micrometeorite impact) into a mass of granular fragments, with a wide range of particle sizes, in the same way that the lunar surface consists of a **regolith** of rock debris. Presumably the icy regolith is thinner (only a few particles in thickness) on the youngest icy surfaces and thickest (several metres or more) on the oldest surfaces.

'Size–frequency distribution' is a term used to describe the relative numbers of objects (in this case, craters) across a range of sizes.

Unfortunately, the resurfacing events on satellites such as those in Figure 4.11 are impossible to date. The lunar cratering timescale, which has been calibrated radiometrically (i.e., using dating methods based on the decay of radioactive isotopes), cannot be applied in the outer Solar System. This is because we can have no expectation that the Moon (1 AU from the Sun) suffered the same rate of impact bombardment as a satellite of Jupiter (5 AU from the Sun) or a satellite of Saturn at a range of nearly 10 AU. Indeed, when the size–frequency distributions of craters on the icy satellites are examined, it is found that the pattern of distribution of craters versus size range differs from the satellites of one giant planet to the next, and that each is different to that found on the Moon. This is convincing proof that *different populations* of impactors affected each region of the Solar System, so it is likely that cratering *rates* also behaved differently in each region. We can imagine a general decrease over time, but there may have been localized flurries of cratering, such as would be caused by the impact of debris originating from a nearby satellite

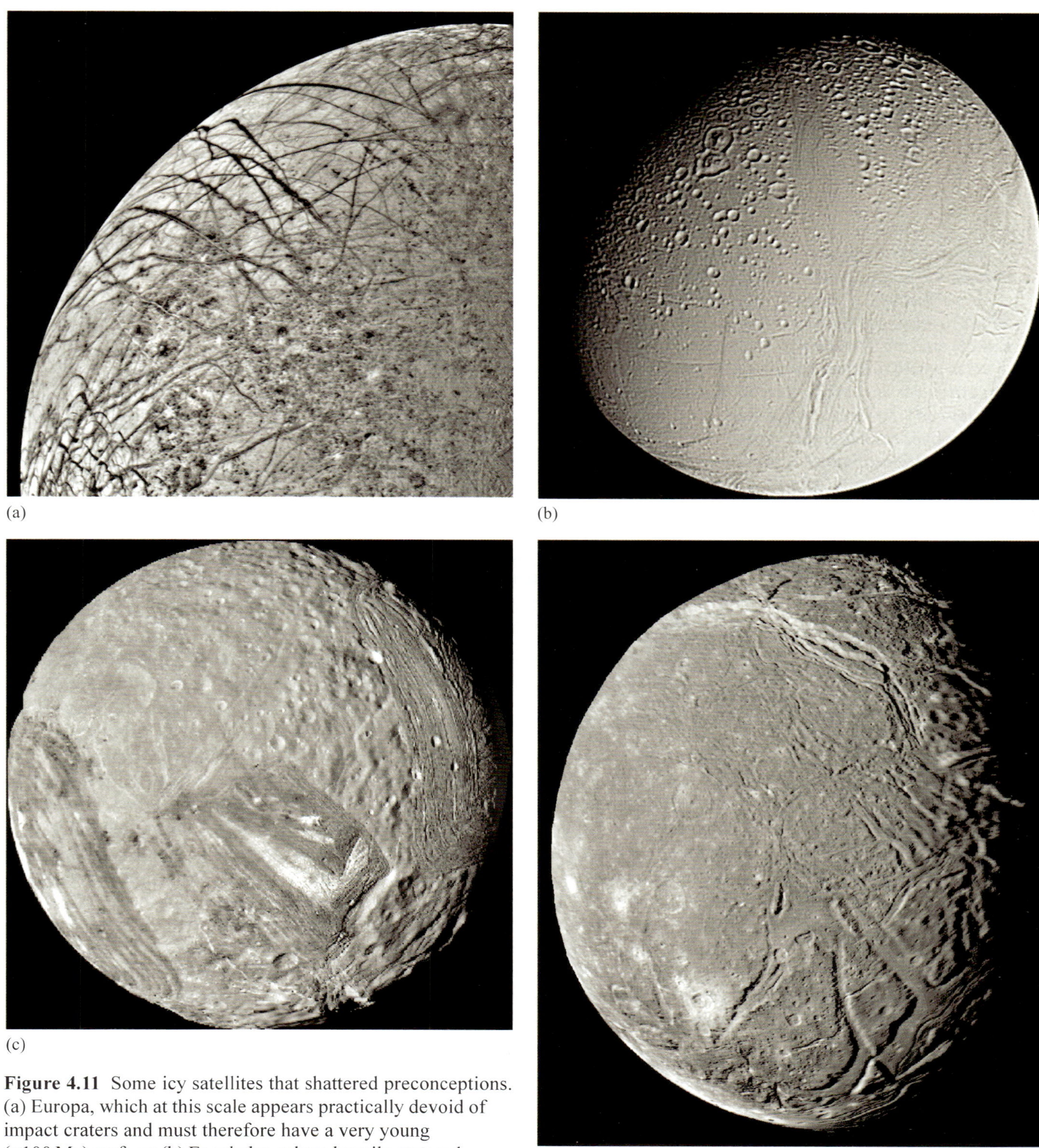

Figure 4.11 Some icy satellites that shattered preconceptions. (a) Europa, which at this scale appears practically devoid of impact craters and must therefore have a very young (<100 Ma) surface; (b) Enceladus, where heavily cratered terrain is cut by tracts of a crater-poor and therefore younger surface; (c) Miranda, whose surface is a patchwork of contrasting terrains; (d) Ariel, where the ancient heavily cratered terrain is disrupted by fault-bounded valleys, some of whose floors have become flooded by icy (cryovolcanic) 'lava' flows. (NASA)

synchronous rotation, there remain two reasons why the tidal stresses continue to vary, which allows tidal heating to continue.

1 In an elliptical orbit, the distance between planet and satellite is continually changing, and so the strength of the tidal force producing the tidal bulges varies accordingly. The bulges are slightly higher when the satellite is closer to the planet and lower when it is further away.

2 In an elliptical orbit, a satellite's speed varies with its distance from the planet (in accordance with Kepler's second law). However, the rate of the satellite's axial spin remains constant. Thus although for every orbit completed the satellite rotates exactly once, during the closest part of its orbit its rotation lags slightly behind its orbital motion, and during the furthest part of its orbit its rotation is slightly ahead of its orbital motion. Consequently, as seen from the planet, the satellite does not show exactly the same face throughout its orbit, rather it swings slightly from side to side. The tidal bulges are raised by forces acting directly on a line through the centres of the two bodies, and so their locations oscillate east and west across the satellite's surface.

So, for a satellite in an elliptical orbit, both the continual variation in the heights and the oscillation in the locations of the bulges deform the satellite's interior, and so cause heating. The reason why none of the orbits of the satellites of the giant planets has yet become exactly circular is that every satellite has neighbours. Mutual perturbations each time an inner satellite overtakes an outer (and therefore slower) satellite keep the orbits slightly elliptical, despite the tidal force from the planet.

This effect is magnified when satellites are in a situation of **orbital resonance**, i.e. where the orbital periods of satellites in adjacent orbits are simple ratios. This is particularly strong among the three inner Galilean satellites: Europa completes exactly one orbit for every two orbits by Io, and Ganymede in turn has exactly twice the orbital period of Europa. The resulting exaggerated eccentricity of the orbits, described as **forced eccentricity**, is slight (0.04 in the case of Io and 0.01 for Europa), but sufficient to power Io's volcanoes and the young (probably continuing) activity on Europa. It also explains why Ganymede shows plenty of signs of past geological activity, whereas Callisto shows few or no signs. (Although three times Callisto's orbital period is almost exactly seven times Ganymede's orbital period, this 7:3 orbital resonance does not lead to sufficient forced eccentricity of Callisto's orbit to lead to tidal heating, especially as Callisto is relatively far from Jupiter.)

Of the icy satellites, Europa (Figure 4.11a) has the youngest icy surface certainly in the Jupiter system and probably in the entire outer Solar System. Density models, supported now by more specific observations, suggest that Europa has about 100 km of icy material overlying a rocky interior. The rate of tidal heating within Europa must be less than in Io, because Europa is further from Jupiter and has a less eccentric orbit. So, after the Voyager encounters, Europa became regarded as the ice-covered equivalent of a less-active version of Io. Certainly this could explain the fracturing and resurfacing evident on Europa's surface, and speculation abounded as to whether the rate of heat transfer from the rocky part into the base of the ice would be sufficient to maintain an unfrozen ocean sandwiched between the ice and the rock. Essentially, the issue depends on which of the two alternative models in Figure 4.13 is correct. Europa and its possible ocean are the main focus of the bulk of this chapter.

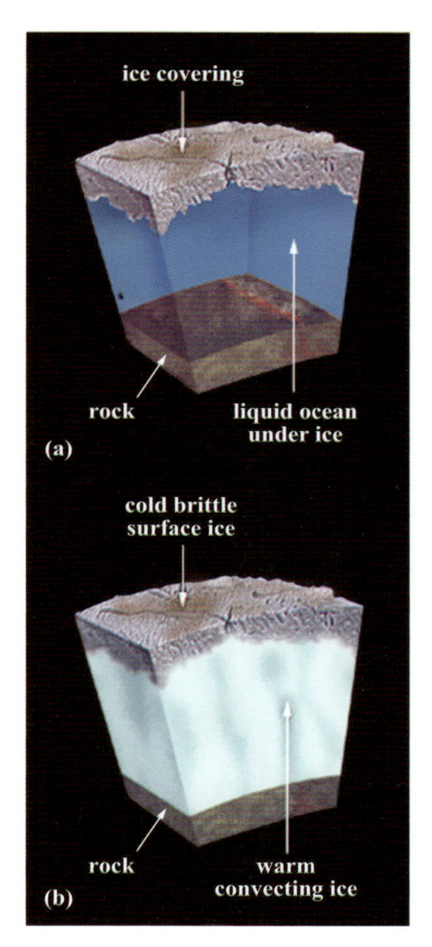

Figure 4.13 Alternative models for the nature of Europa's 'icy' layer: (a) an ocean of liquid water sandwiched between the solid ice and the rocky interior; (b) solid ice, though probably warm enough to be mobile near its base, resting directly on rock. (NASA)

The famous science fiction author Arthur C. Clarke was one of the first to realize the astrobiological implications of a tidally heated Europa, by analogy with communities around 'black smoker' hydrothermal vents on the Earth's ocean floor. In *2010: Odyssey Two* (published in 1982 as a sequel to the more famous *2001: A Space Odyssey*) he imagined an explorer's findings on the floor of the Europan ocean:

> ...the first oasis filled him with delighted surprise. It extended for almost a kilometre around a tangled mass of pipes and chimneys deposited from mineral brines gushing from the interior. Out of that natural parody of a Gothic castle, black, scalding liquids pulsed in a slow rhythm, as if driven by the beating of some mighty heart. And, like blood, they were the authentic sign of life.
>
> The boiling fluids drove back the deadly cold leaking down from above, and formed an island of warmth on the seabed. Equally important, they brought from Europa's interior all the chemicals of life. There, in an environment where none had expected it, were energy and food, in abundance...
>
> In the tropical zone close to the contorted walls of the 'castle' were delicate, spidery structures that seemed to be the analogy of plants, though almost all were capable of movement. Crawling among these were bizarre slugs and worms, some feeding on the plants, others obtaining their food directly from the mineral-laden waters around them. At greater distances from the source of heat – the submarine fire around which all the creatures warmed themselves – were sturdier, more robust organisms, not unlike crabs or spiders.
>
> Armies of biologists could have spent lifetimes studying that one small oasis.
>
> (Clarke, 1982)

It took several years for speculations such as Clarke's to become acceptable among mainstream scientists. One reason for this is that ocean-floor hydrothermal vents had not yet been recognized as one of the most likely environments where life on Earth could have originated (see Section 1.10). Another reason is that the Voyager indications of an ocean below Europa's ice were not nearly so compelling as the evidence that has become available subsequently. However, by the late 1990s NASA was presenting Europa's astrobiological potential as the main reason why the US Senate ought to provide funding for a dedicated Europa mission, which at the time of writing has a planned launch date of 2011 at the earliest.

QUESTION 4.2

Using the orbital radii given in Table 4.1, calculate the tidal force of Jupiter on Europa as a proportion of the tidal force of Jupiter on Io.

4.1.5 The Galileo mission

It was a long time before the Voyager missions were followed up by more detailed surveys of the outer planet satellites. No Uranus or Neptune missions are planned, but a mission to Saturn called Cassini–Huygens was launched in 1997 for arrival at Saturn in 2004. You will learn about this in the next chapter. However, the Jupiter system received a similar visitor first. This was Galileo, launched in 1989, which became the first spacecraft to orbit Jupiter in December 1995. It continued to function through 2002, and was scheduled to be destroyed by plunging into Jupiter's atmosphere in September 2003. This was a planetary protection measure (Chapter 3), taken to avoid the possibility of the defunct craft eventually colliding with Europa and thereby contaminating it with any unintentional bioload.

Galileo had several close encounters with each of the Galilean satellites, providing more complete and more detailed imaging than was possible during the Voyager fly-bys, using an instrument known as the solid-state imaging (SSI) camera. It also carried a near-infrared imaging spectrometer (NIMS), which was useful for determining surface compositions (and also temperatures of Io's active lava flows), an ultraviolet spectrometer, and magnetometers that revealed the satellites' responses as they move through Jupiter's magnetosphere. Perturbations to Galileo's trajectory as it passed close to the satellites placed improved constraints on their internal density distributions, indicating dense, presumably metallic, cores at the centres of Io, Europa and Ganymede. Callisto, by contrast, was proven to be only weakly differentiated, with incomplete segregation of rock and ice (Figure 4.14). See Box 4.3 for a discussion of how terms are borrowed from the terrestrial planets to describe the compositional and mechanical layers within icy satellites.

Near-infrared means the part of the infrared spectrum that is nearest to the visible. The actual spectral range covered by NIMS was 0.7–5.2 μm.

Relying largely on Galileo observations, we will now take a detailed look at Europa, to see what we can deduce about its recent history and the possibility of a life-bearing ocean below the ice.

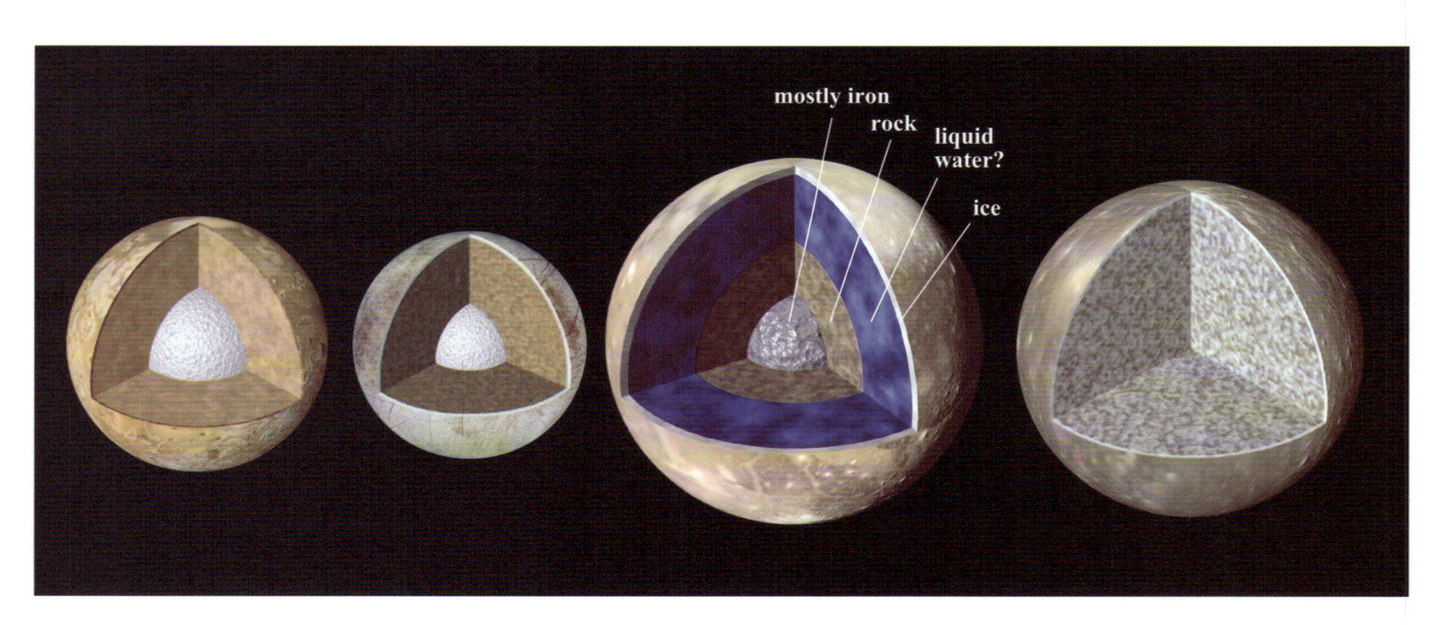

Figure 4.14 Post-Galileo models of the internal structures of the Galilean satellites (from left to right: Io, Europa, Ganymede and Callisto). (At this scale, no distinction is made between ice and liquid water in the model for Europa.) (NASA)

BOX 4.3 TERMINOLOGY FOR THE LAYERED STRUCTURE OF DIFFERENTIATED ICY BODIES

In a differentiated terrestrial planet, the term *core* is used for the dense compositionally distinct inner part, which is rich in iron. This is surrounded by a rocky (silicate) *mantle*. The extreme outer part of the rocky material is referred to as the *crust* if its composition has been altered by volcanism and other recycling processes.

In a differentiated icy body, it is logical, by analogy, to regard the rocky interior as the core (and if this is itself differentiated, with an iron-rich centre, to call that the *inner core*). The icy outer part of such a body is thus the mantle, and if the outer part of the ice differs somewhat in composition from the interior, we can call this the crust. The analogy is particularly apt because Solar System ices share many important properties with silicate rock. Among these are:

1. At the prevailing near-surface temperatures, the ice is mechanically strong and rigid, like rock near the Earth's surface.
2. When caused to melt (by heat, or in some cases a decrease in pressure) ice that is water mixed with salt or water mixed with another volatile species undergoes partial melting, just like a mixture of silicate minerals in rock. The melt and the surviving crystals have different compositions, and melting begins at a lower temperature than for pure water-ice.
3. At sufficient temperatures and pressure, ice will flow without melting, and can undergo solid-state convection, just like rock in the deeper part of the Earth's mantle.

Property 2 in this list makes it likely that the outermost part of a differentiated icy body does indeed differ at least slightly in composition from its mantle, and so we can regard this differentiated ice as a true crust. Property 3 means that we can discriminate the outer rigid ice (upper mantle plus crust) from the deeper more mobile (even though solid) ice, and distinguish these by the terms *lithosphere* and *asthenosphere*, respectively, which were originally coined for the Earth.

4.2 Europa

Europa's surface is fascinating, if often perplexing, to study. One of its special characteristics is its brightness. It has an albedo of 0.7, which is exceeded among icy satellites only by Enceladus and Triton. Overall brightness is one indicator of the youth of an icy surface: the brighter the icy surface, the younger it is. Ganymede (albedo 0.45) and Callisto (albedo 0.2) are much darker. This distinction is not usually apparent when comparing images of their surfaces (for example Figures 4.10a and 4.11a, Callisto and Europa respectively), because the brightness of each image has usually been adjusted to show features on each to best advantage.

The 'albedo' of a body is simply the fraction of the incident light that is reflected. The higher the albedo, the more light is reflected, and the brighter the body appears.

The midday temperature is about 130 K (about −140 °C) at Europa's equator and about 80 K (about −190 °C) at the poles. Europa's axis of rotation is perpendicular to the plane of its orbit, which is tilted at less than half a degree relative to Jupiter's equatorial plane. Europa experiences virtually no 'seasonal' changes in illumination during its orbit about Jupiter or during Jupiter's twelve-year orbit of the Sun, because Jupiter's axial inclination is only about 3° (so Jupiter itself virtually lacks seasons too).

Galileo detected a magnetic field about Europa, which could be generated by motion within its iron core or within a salty (and therefore electrically conducting) ocean beneath the ice. The highest-resolution images of Europa sent back by Galileo have pixels representing areas about 6 m across. Such detailed images cover only a small fraction of the total surface. 9% of Europa was imaged at better than 200 m per pixel and about half the globe was imaged at better than 1 km per pixel. This was a great advance on the coverage provided by Voyager, whose best images of Europa have pixels representing areas 1.9 km across. You will examine plenty of images of Europa shortly, but first it is worth establishing what we know about the composition of the ice.

4.2.1 Ice and salt

As noted previously, Europa's near-infrared reflectance spectrum was used as long ago as the 1950s to demonstrate that its surface is mostly water-ice. More recently, spectroscopic observations by the Hubble Space Telescope and Galileo have revealed some regions where the ice appears to be salty (see below) and have also detected traces of molecular oxygen (O_2) and smaller amounts of ozone (O_3). The oxygen and ozone almost certainly result from the breakdown of water molecules in the ice brought about by exposure to charged particles (this process is known as **radiolysis**) that are channelled onto Europa by Jupiter's magnetic field, and by solar ultraviolet radiation (a process called photodissociation or **photolysis**; Section 1.6.2). Most of the oxygen and ozone is probably held within the ice (as isolated molecules trapped within ice crystals), but some may constitute an extremely tenuous atmosphere.

■ Apart from various forms of oxygen, what else would you expect to be produced when water molecules are broken down by radiation?

❑ Given that the formula for water is H_2O, hydrogen should also be produced.

Box 4.4 shows a series of reactions that could produce oxygen and hydrogen in Europa's ice.

BOX 4.4 RADIOLYTIC AND PHOTOLYTIC BREAKDOWN OF WATER MOLECULES IN ICE

The reactions that occur to generate oxygen and hydrogen within the surface ice of an icy satellite can be summarized, in simplified form, as:

$$H_2O \rightarrow H + OH$$

$$H + H \rightarrow H_2$$

$$OH + OH \rightarrow H_2O_2$$

$$2H_2O_2 \rightarrow 2H_2O + O_2$$

Europa's ozone is likely to be the product of a chain of reactions involving radiolytic and photolytic breakdown and recombination of oxygen molecules, similar to the photolytically driven reactions that generate ozone from oxygen in the Earth's stratosphere.

Hydrogen has not yet been detected on Europa, but on Ganymede, where similar 'space weathering' of exposed ice occurs, hydrogen has been found leaking away into space.

- ■ Suggest a simple explanation to explain why there is a lot less free hydrogen than oxygen in or above Europa's surface.
- ☐ Hydrogen is a much smaller and lighter atom therefore it is easier for hydrogen to escape from within the ice. Once liberated, it is so loosely bound by Europa's weak gravity that it would be lost to space much faster than oxygen or ozone.

Hydrogen peroxide (H_2O_2), which is an intermediate product of the sequence of reactions in Box 4.4, has been identified as a trace component of the ice in reflectance spectra obtained using Galileo's near-infrared imaging spectrometer. The same instrument has also revealed distortion of the absorption bands associated with water. This indicates that, in addition to forming ice crystals, some of the water molecules are bound within hydrated salt crystals. The best match to the spectra is from a mixture of magnesium and sodium salts such as magnesium sulfate hexahydrate ($MgSO_4.6H_2O$), epsomite ($MgSO_4.7H_2O$), bloedite ($MgSO_4.Na_2SO_4.4H_2O$) and natron ($Na_2CO_3.10H_2O$). The occurrence of sulfates is supported by Galileo ultraviolet spectroscopic data that indicate the presence of compounds containing a sulfur–oxygen bond.

- ■ Although carbonates and sulfates are fairly common salts on Earth, they are not the most abundant. What kind of salt appears to be missing on Europa, compared with the Earth?
- ☐ No chlorides are in the above list – note that sodium chloride (NaCl), which is the most abundant salt dissolved in the Earth's oceans, is absent.

Actually, chlorides produce no spectral features in the available part of the spectrum, so direct observational data cannot tell us whether any chlorine salts occur on Europa's surface. What the spectral mapping by Galileo did achieve, however, was to show that the distribution of salts across Europa's surface is highly non-uniform. Large expanses are relatively salt-free, but in places where the surface has been most recently and most greatly disrupted from below, the surface salt concentration reaches 99%. You will see what these areas look like shortly.

The salts occurring on Europa's surface are unlikely to be a straightforward representation of those dissolved in any ocean beneath Europa's ice – calculations have shown that the freezing process would tend to concentrate sulfates of magnesium and sodium into the ice. This is consistent with the observed preponderance of these salts at the surface. However, the concentrations of elements dissolved in Europa's ocean are largely a matter of speculation. Two of the factors that have to be considered are the composition of Europa's rocky component, and the efficiency with which each element becomes dissolved from it into the ocean. Neither of these factors is known. Although, on average, Europa's rock is likely to be similar to carbonaceous chondrites, geochemical differentiation could mean that the rock nearest to the ice–rock interface might well be very

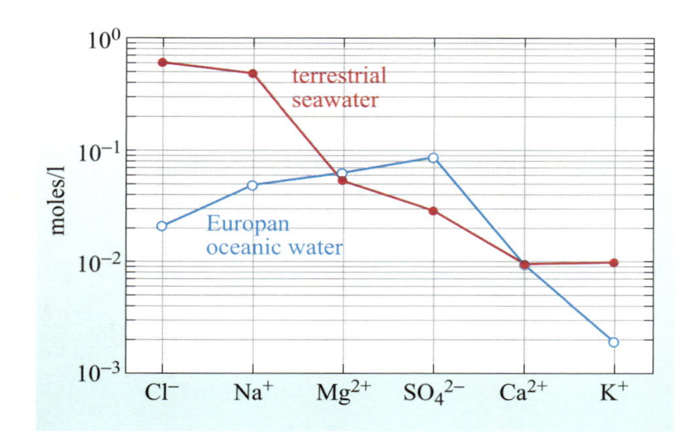

Figure 4.15 Estimated concentrations of major elements in Europan oceanic water compared with seawater on Earth.

different (as is probably the case in Io's crust, for example). The efficiency with which elements become dissolved (or sometimes reprecipitated) depends upon the temperature at which it occurs, as well as on the overall chemistry of the solution. Despite the uncertainties, attempts have been made to model the likely concentrations of dissolved elements in Europa's ocean. The results of one such model are shown in Figure 4.15.

QUESTION 4.3

According to Figure 4.15, how many more times greater is the concentration of chloride (Cl^-) in terrestrial seawater than in Europa's ocean?

4.2.2 Examining Europa's surface

It is all very well speculating about conditions in an ocean below Europa's ice, but what evidence is there that it actually exists? After all, tidal heating might not result in ice melting on a global scale, and current geophysical models of Europa's internal structure (e.g. Figure 4.14) cannot tell the difference between ice and liquid water. Fortunately, Voyager and Galileo have given us detailed images of Europa that we can use in the same way that a geologist uses aerial photographs or images from space to help decipher the processes that have shaped a particular tract of the Earth's surface.

The general view

Figure 4.16 shows a Voyager 2 image of a large region of Europa. Examine this image carefully, in order to answer Question 4.4.

BOX 4.5 LATITUDES AND LONGITUDES ON SATELLITES

As soon as the first features were discovered on the surfaces of the satellites of the outer planets, it became necessary to define co-ordinate systems to map their locations. Latitude is simple to define; it is measured in degrees north and south of the equator, which lies halfway between the satellite's poles of rotation. By convention (established by the International Astronomical Union), for a synchronously rotating satellite 0° longitude is defined to run through the centre of the planet-facing hemisphere. Longitude is normally quoted in degrees measured westwards from here, and west is always to the left when you look at a body with north towards the top.

QUESTION 4.4

(a) Study Figure 4.16 and write a short description of the kinds of features you can see on Europa's surface, noting for example relative brightness and characteristic shapes or textures. Concentrate on a simple description of appearance; we do not expect you to explain the origin or precise nature of what you can see.

(b) Try to deduce the relative ages of the features you have described.

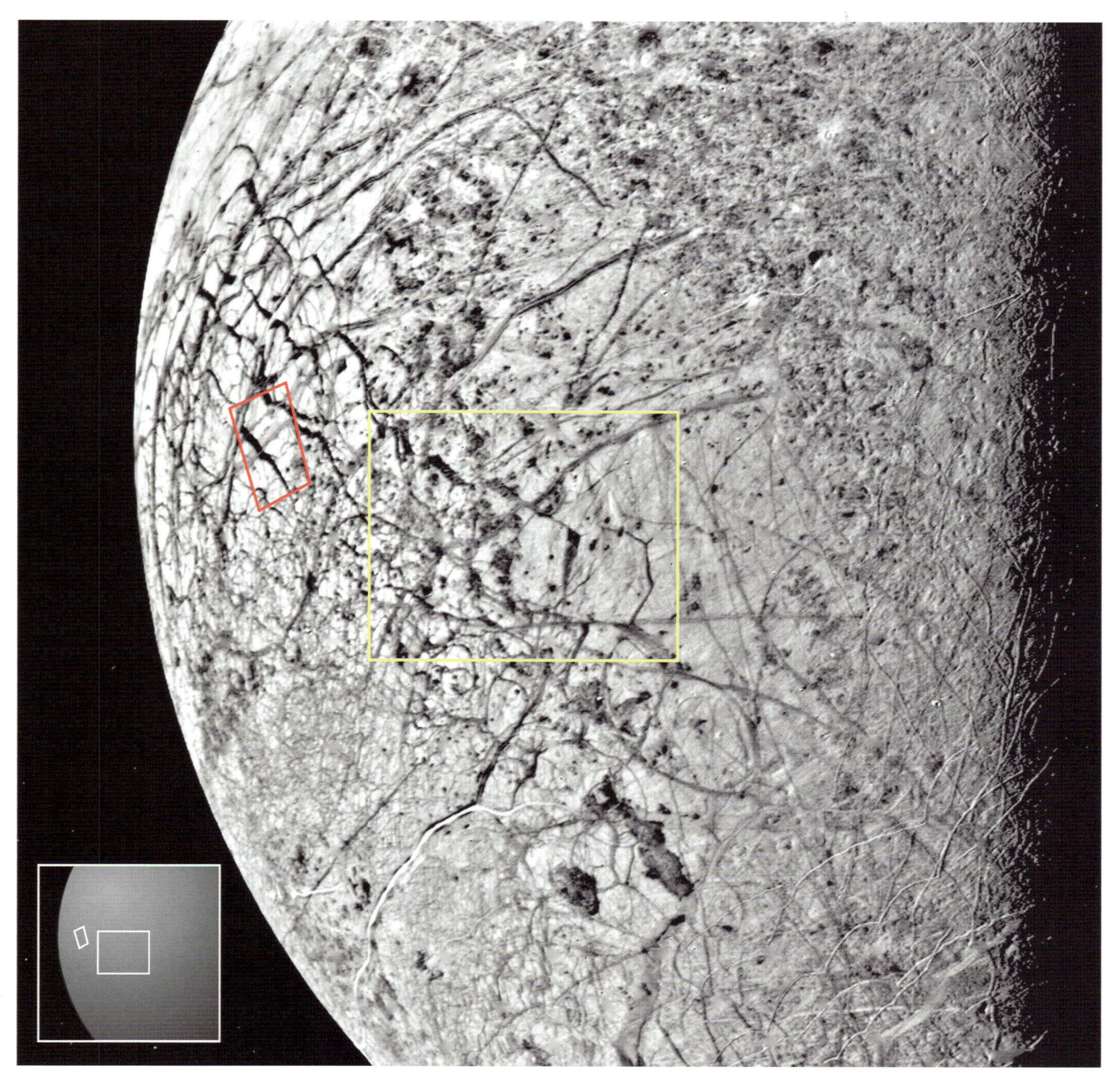

Figure 4.16 Voyager 2 image showing a region of Europa, about 3000 km across, recorded at about 2 km per pixel and centred at 10° S, 160° W (see Box 4.5 for an explanation of latitudes and longitudes on satellites). The large yellow outline shows the area covered by Figure 4.18a and the smaller red outline shows the area covered by Figure 4.19a. (NASA)

In answering Question 4.4, you should have formed an impression of an original surface that (at the scale of the image) appears relatively featureless, but was subsequently cut across by processes that produced dark bands. Later, the band-disrupted terrain was itself overprinted in places to produce mottled terrain and curved ridges. The dark bands cutting across much of Europa give it the appearance of a thoroughly cracked eggshell, but please be aware that there is no evidence in Figure 4.16 (nor on any more detailed images) that these 'cracks' are open fissures in the surface. In fact, there is very little topographic relief on Europa. The curved ridges in the lower right corner of Figure 4.16 are only about 200 m high.

The crater Pwyll

You might also have noted that there are no obvious impact craters visible in Figure 4.16. In fact there are a few. One is a bright spot, 15 km in diameter, surrounded by a dark halo of ejecta that occurs 10 mm from the top edge and 65 mm from the left-hand edge of the figure. Another is a slightly larger pale feature with a discernible central peak 20 mm from the top edge and 45 mm from the right-hand edge. The youngest large crater on Europa occurs at 26° S, 271° W, which is outside the area covered by Figure 4.16. This is shown in Figure 4.17, and is named Pwyll (pronounced 'Puh-hl' or 'Poo-eel', after a character from Welsh legend, Box 4.6). Pwyll is 26 km across, and has a dark floor and a halo of equally dark ejecta extending for about 8 km beyond its rim, which was presumably excavated from below the surface. Much brighter, finely fragmented ejecta in the form of discontinuous rays can be traced for more than 1000 km, and forms the bright region surrounding the crater in the global view in Figure 4.17a. It is the high visibility of its ejecta rays that shows Pwyll to be the youngest of Europa's large craters. Statistical arguments based on the likely frequency of comet impacts onto Europa suggest that Pwyll is very unlikely to be older than about 20 million years, and is probably about 3 million years old.

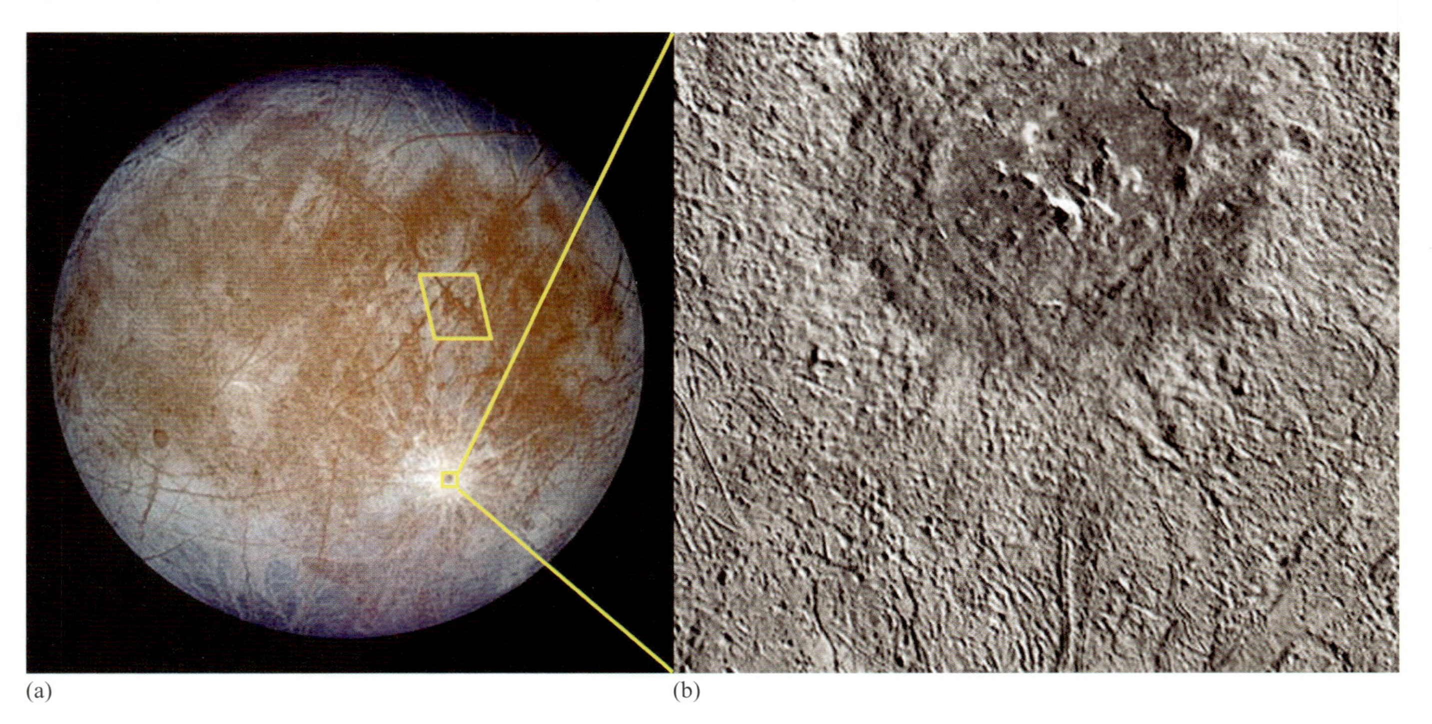

Figure 4.17 (a) Global view of Europa showing the location of the crater Pwyll, which is shown enlarged in (b), a Galileo SSI image recorded at 250 m per pixel. The outline superimposed on (a) indicates the area covered by Figure 4.21. (NASA)

BOX 4.6 NAMES ON EUROPA AND OTHER SATELLITES

In order to avoid duplication of the names of features between bodies, and to try to achieve consistency of nomenclature on each body, the International Astronomical Union has established a naming convention for each planetary body in the Solar System. Names on Europa are drawn from Celtic gods, heroes, and myths; people and places associated with the Europa myth; and place names from ancient Egypt.

The crater on Europa called Pwyll is a character from Welsh legend, who appears in the mediaeval collection of tales known as *The Mabinogion*.

By contrast, features on Io are named after gods and heroes associated with fire, sun, thunder and volcanoes, and also people and places associated with the Io myth and Dante's *Inferno*.

Incidentally, the Galilean satellites themselves and many of Jupiter's smaller satellites are named after mythological characters (of various genders and species) who, to put it delicately, became 'romantically entangled' with the god Jupiter.

QUESTION 4.5

Look at the detailed image of Pwyll in Figure 4.17b. Does Pwyll have the three-dimensional shape that you would expect of a young crater?

Expert analysis shows that most of Pwyll's rim is less than 200 m high, and that (unusually for impact craters) its floor is hardly any lower than the terrain outside. Opinion is divided as to whether the impactor responsible for Pwyll actually penetrated right through the ice, but all are agreed that the crater shows the hallmarks of an impact into relatively thin (about 20 km in thickness) and weak ice.

Thus the paucity of large craters on Europa indicates that its surface is young, and the subdued cross-sectional shape of many of those craters that do occur suggests that the ice was relatively thin when they formed.

Fracturing and motion of the ice shell

If the rigid surface layer of Europa's ice is thin (or, at least, has been thin for some of the time), and overlies either water or some kind of weak and mushy ice as indicated by large craters such as Pwyll, then we might expect to find some evidence for fracturing and motion of the rigid ice shell. This is precisely what the pattern of dark bands such as those on Figure 4.16 appears to be showing us. An area from Figure 4.16 is enlarged in Figure 4.18, with an interpretation of how plates bounded by fractures in the rigid ice shell could have moved relative to one another.

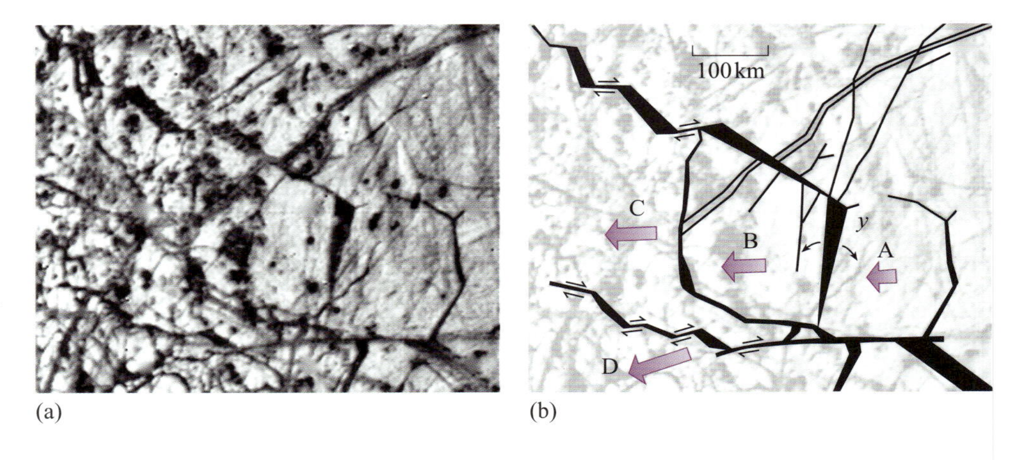

Figure 4.18 (a) An enlargement of the area that was outlined in yellow on Figure 4.16. The youngest dark bands (many of them wedge-shaped) can be seen to cross-cut and offset some of the older bands. (b) Sketch of the area covered by (a) showing how opening of the bands is consistent with shuffling and rotating neighbouring plates of ice. See text for discussion. The map has been simplified by omitting various younger blotches that appear on (a). (Copyright © 2000 David A. Rothery)

The arrows on Figure 4.18b suggest that the plates labelled A–D have all moved westwards relative to the ice at the right-hand (eastern) edge of the map. In addition, plate B has rotated about 5° anticlockwise relative to plate A (opening up the intervening wedge-shaped band that extends south from *y*); plate C has moved west relative to plate B and plate D has moved west relative to plate C.

It is tempting to make an analogy with plate tectonics on Earth, and to regard the stepped dark bands forming the north and south boundaries of plate C as lengths of spreading axis (or mid-ocean ridge) offset by transform faults. However, even if the interpretation in Figure 4.18b is correct, there are several important differences between plate tectonics on Europa and the Earth. First, Europa's jumble of overlapping dark bands (Figure 4.16) suggests that old spreading axes are abandoned and replaced by new ones after only a few tens of kilometres of spreading. However, on Earth most spreading axes last for tens to hundreds of millions of years, during which time they add hundreds or even thousands of kilometres of new lithosphere to the edges of the adjacent plates. On Earth, creation of new lithosphere at spreading axes is balanced globally by destruction of lithosphere at subduction zones.

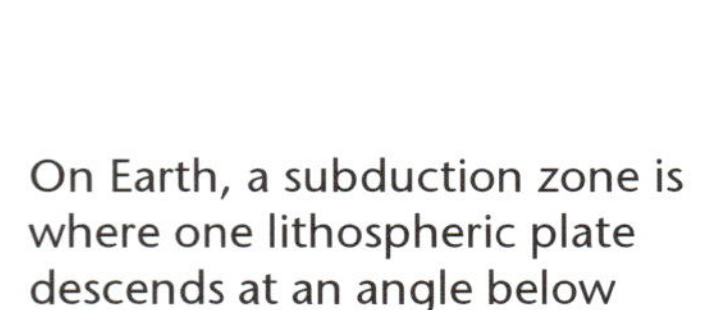
On Earth, a subduction zone is where one lithospheric plate descends at an angle below another.

There is no analogue to terrestrial subduction zones on Europa, but it is obvious that if new areas of surface ice are being added to make the dark bands then other areas must be being destroyed at an equal rate. The processes operating on Europa to achieve such a balance remained a mystery until Galileo's more detailed images became available. You will examine this evidence soon, but first it is worth exploring the extra information that Galileo images can give about the dark bands themselves. Figure 4.19a is one such image. It shows that the pale areas between the dark bands that seemed relatively featureless at the resolution of the Voyager images can be seen at higher resolution to be criss-crossed by low ridges. At this level of detail, Europa's surface has been aptly described as looking like a ball of string. Furthermore, the 'ball of string' ridges also occur within the dark bands (running parallel with their edges). When we move up to even higher resolution, as in Figure 4.19b, the 'ball of string' ridges are even more obvious (and some can be seen to have central grooves running along them), whereas the distinction between dark bands and pale terrain has become hard to see.

It is uncertain exactly how the ridges on Europa have been built. Each is probably the result of some form of cryovolcanic eruption along a crack or fissure. If this is the case, the material erupted must have been in the form of mushy ice, or perhaps

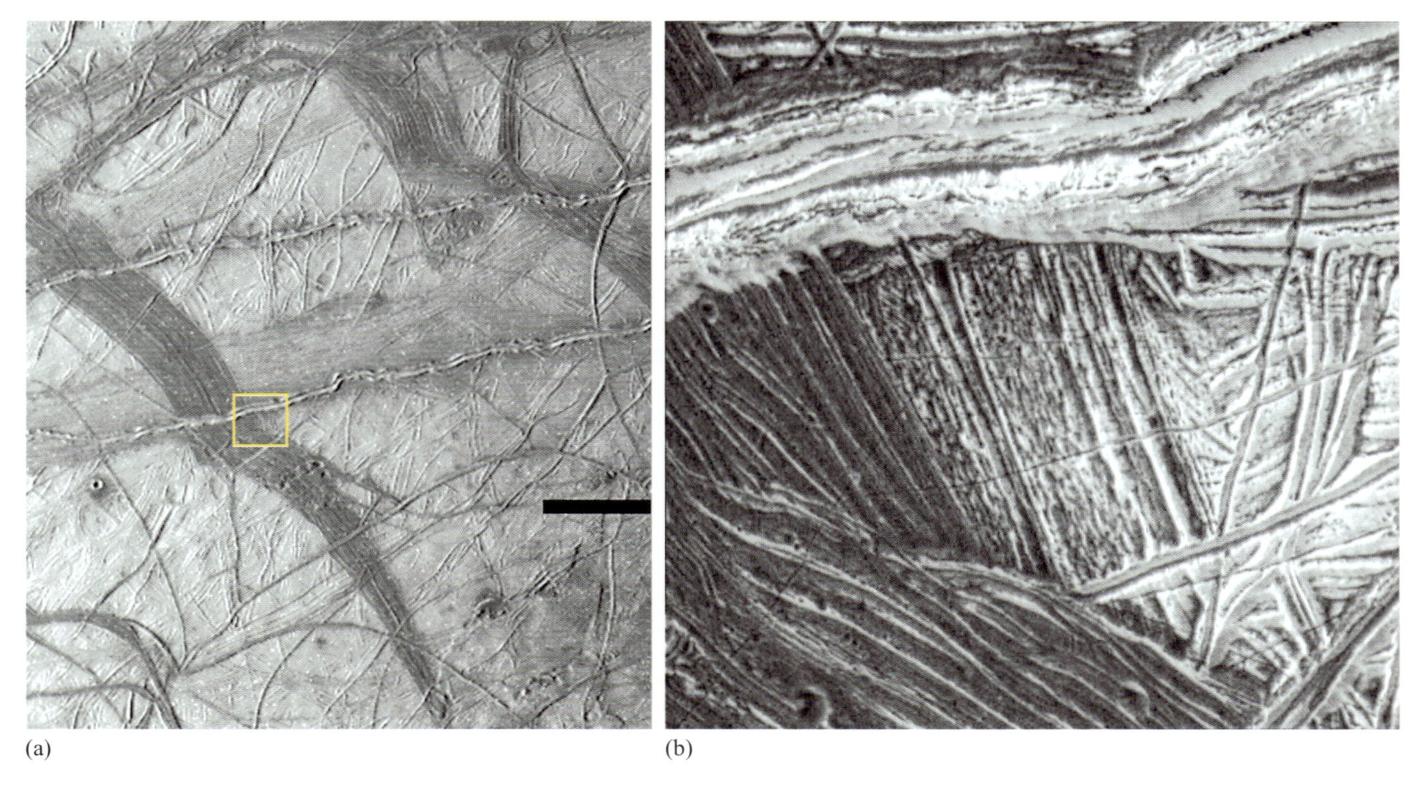

Figure 4.19 Galileo SSI images of part of Europa, near 16° S, 195° W. (a) A region, 150 km in width, recorded at 420 m per pixel. This is the area shown by the red outline on Figure 4.16 (whose shape is distorted in Figure 4.16 because of perspective). A prominent wedge-shaped dark band runs diagonally across the lower left of the image. It is cut by two narrow bright bands. The black bar at the right-hand side represents missing data. The outline indicates the area shown at higher resolution in (b). (b) Image, 20 km in width, recorded at 26 m per pixel. Apart from the bright band, which is the youngest feature shown on the image and overlies everything else, the whole surface (dark band and pale terrain alike) is seen to consist of a succession of cross-cutting 'ball of string' ridges. (NASA)

a fountain-like spray of fragmented ice, analogous to a volcanic fissure eruption on Earth (Figure 4.20) and involving the escape of gaseous volatiles during eruption. Fortunately, the details of ridge-building are not important in order to understand the general surface history and its implications for ice thickness, which appear to be as follows:

- Each 'ball of string' ridge is symptomatic of a small amount of surface extension.
- The ridges occur in sets of up to about a dozen parallel ridges, and each set can usually be seen to be cut across by a younger set. There are at least four such sets within the portion of the dark band shown in Figure 4.19b. Although not quite parallel to each other, each set runs lengthways relative to the dark band, and would in total be responsible for the kind of spreading across a dark band indicated in Figure 4.18b.
- In the older pale terrain outside the dark band the ridge sets are oriented more variably, showing a long and complex history of surface creation.
- The dark bands are the youngest parts of the 'ball of string' surface, and evidently become paler as they age. (There are many ways in which this could happen. Some involve growth or fragmentation of ice crystals over time, others depend on chemical changes caused by long-term exposure to radiation.)

Figure 4.20 A basaltic fissure eruption on Earth. The rampart is built by congealed lava on either side of the fissure. This is a possible analogue for how the 'ball of string' ridges on Europa are constructed. Human figure in left foreground indicates the scale. (USGS/Cascades Volcano Observatory)

There are two things to add to finish the story of surface creation in the area covered by Figure 4.19. First, the bright bands cut across the 'ball of string' texture and so are clearly younger than it. These bands may be a slightly different kind of cryovolcanic feature – their feathery edges, seen at the highest resolution (Figure 4.19b), could represent debris shed downslope from a central high. Second, there are some very narrow grooves (barely visible in Figure 4.19b) that also cut both dark and pale 'ball of string' texture, one of which widens towards the east where it becomes an otherwise unremarkable contributor to the texture. Many features such as these are probably cracks where extension occurred without an accompanying eruption. Others are evidently the surface expressions of faults with sideways (instead of extensional) movement (as you will see shortly).

The dark bands and the intervening tracts of pale terrain were constructed by a long and complicated series of events, each of which was associated with spreading on a local scale.

More surface disruption

Now let's examine some detailed images of the region of Europa's northern hemisphere that was indicated on Figure 4.17. A medium resolution image is shown in Figure 4.21, and higher resolution images from within this area are shown in Figures 4.22–4.24.

QUESTION 4.6

Study Figure 4.21. How would you classify the majority of the surface in this region, including the part of it shown in more detail in Figure 4.22?

QUESTION 4.7

Look at the two features labelled A and B in Figure 4.22. A is a groove with a slightly raised rim on either side, running diagonally down to the right from the top of the image. B is a ridge, with a groove down its centre, running almost directly down from the top. Try to account for the relationship between these two features where they cross, and deduce which of these two features is the younger.

These images provide clear evidence that tectonics on Europa involve relative sideways movements as well as simple spreading apart of the surface.

If you are familiar with plate tectonics on Earth you will probably not be surprised by this. Now turn your attention to the other parts of Figure 4.21, notably those covered by Figures 4.23 and 4.24.

Figure 4.21 Galileo SSI image of part of Europa, 200 km in width, centred at 10° N, 270° W and recorded at 180 m per pixel. This image was made by combining near-infrared, green and violet images in red, green and blue, respectively, and then exaggerating the resulting colours. In this rendering, blue is the general colour of the icy surface, and ice-poor (probably salty) areas show up red. The white patches are thin sprinklings of ejecta from the crater Pwyll, which is 1000 km to the south. The yellow outline locates Figure 4.22. Other yellow marks indicate the lower corners of Figure 4.23 and all four corners of Figure 4.24. (NASA)

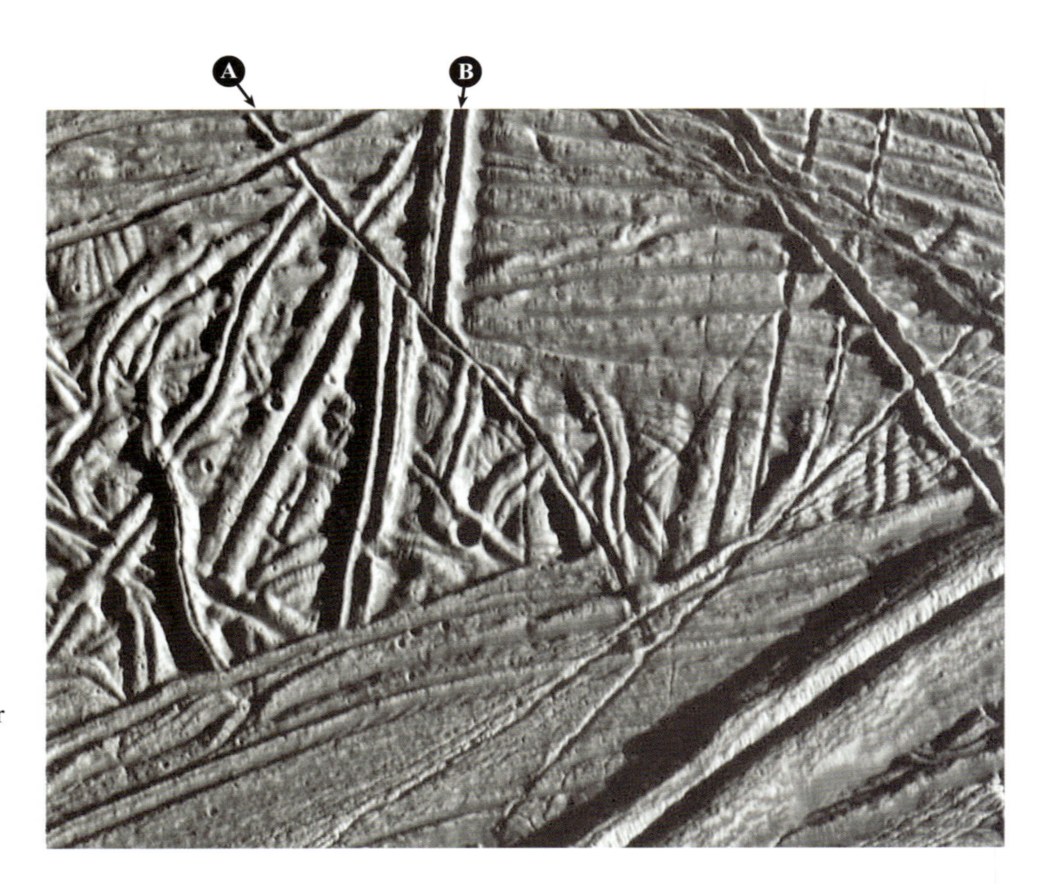

Figure 4.22 Galileo SSI image, 15 km in width, recorded at 20 m per pixel. Solar illumination is from the right. See Figure 4.21 for location. Letters A and B are referred to in Question 4.7. (NASA)

QUESTION 4.8

What has happened to the 'ball of string' texture in (a) Figure 4.23, and (b) Figure 4.24?

The usual explanation for the dome features in regions such as Figure 4.23 is that they are places where warm buoyant material (which could be warm ice, slush or water) has risen from below. Where injected as an intrusion at shallow depth, the result is a subtle dome (e.g. the example in D4) over which the surface may have been sufficiently stretched to rupture. In more extreme cases the 'ball of string' surface appears to have melted, exposing the risen material, which is surrounded by cliffs that drop down to the new surface (e.g. the example in D/E–5/6). Elsewhere the risen material seems to have spread out across the top of the old 'ball of string' surface (e.g. the example in D/E–1/2). Regions such as the one covering Figure 4.24 are essentially just more extensive versions of the D/E–5/6 situation, and demonstrate the effects of heating events on a regional scale. There are several examples of this type of terrain on Europa, which is described formally as **chaos**. Conventionally, within a chaos region, the slabs of 'ball of string' surface are referred to as 'rafts' and the low-lying hummocky material in between is called 'matrix'.

The matrix is most simply interpreted as the (now refrozen) surface of an ocean that was exposed when the overlying ice was removed, presumably by melting caused by an injection of heat from below. Near the edges of chaos, rafts have broken away from the continuous ice sheet and drifted inwards by relatively small

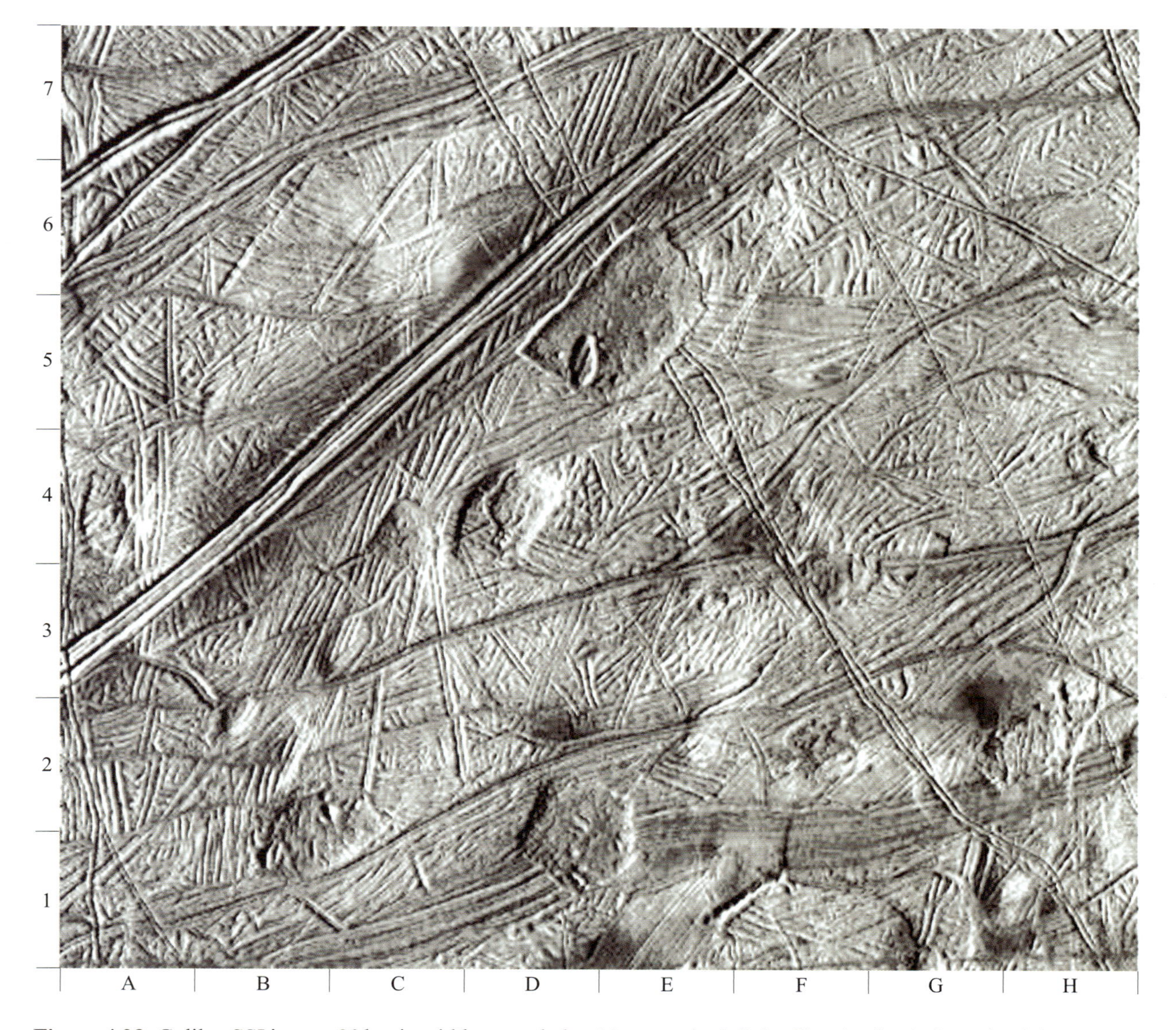

Figure 4.23 Galileo SSI image, 80 km in width, recorded at 54 m per pixel. Solar illumination is from the right. See Figure 4.21 for location. The letters and numbers around the edge define 10 km × 10 km squares for reference in the answer to Question 4.8. (NASA)

distances, and in many cases their original configurations can be deduced. Near the centres of chaos, rafts are less abundant and it is not usually possible to see how they once fitted together. The rafts are analogous to ice floes formed in the Earth's oceans when floating pack-ice breaks up in the spring. The even height of the cliff at the edge of most rafts shows that these rafts are lying horizontally. However, there is one raft, 5 km × 2 km in size, just to the northeast of the centre of Figure 4.24 with (to judge by the cliff's shadow) an exceptionally high cliff on its northwest side but no sign of a cliff on its southeast side. This raft is shown enlarged in Figure 4.25. It looks as though the raft has been tilted down towards the southeast. Some of the knobbly hills sticking up out of the matrix may be the corners of smaller or more steeply tilted rafts, the most obvious being a triangular hill with an exceptionally long shadow immediately to the south of the tilted raft in Figure 4.25.

Figure 4.24 Galileo SSI image, 45 km in width, recorded at 54 m per pixel. Colours are constructed in the same way as in Figure 4.21. Solar illumination is from the right. See Figure 4.21 for location. The outline shows the area shown at very high resolution in Figure 4.26. (NASA)

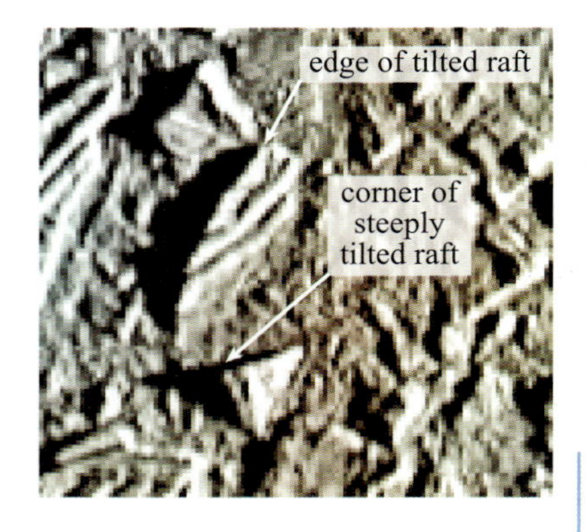

Figure 4.25 Enlargement of an area 8 km in width just northeast of the centre of Figure 4.24, showing a gently tilted raft and the corner of a more steeply tilted raft. (NASA)

Chaos makes up about a third of the 9% of Europa's surface that has been imaged at adequate resolution (less than about 200 m per pixel) to make identification certain. On low-resolution images, both chaos and dome-disrupted regions such as Figure 4.23 appear as mottled terrain like the area in the northeast of Figure 4.16. The region covered by Figure 4.24 is named Conamara Chaos, and as you can see on Figure 4.21 it extends for about 100 km north to south and about 80 km east to west. The largest chaos region on Europa is more than a thousand kilometres across, and the small end of the size spectrum is exemplified by the resurfaced area at D/E–5/6 in Figure 4.23.

- If large chaos regions are places where 'ball of string' surface has been destroyed, what could be their significance for the global tectonics of Europa?
- Destruction of surface in chaos regions could balance the spreading implied by the creation of the 'ball-of-string' texture. We noted earlier that such spreading could not occur unless it was matched globally by the surface being destroyed at an equal rate. On Earth, this is achieved where plates are subducted deep into the mantle.

Conamara Chaos is named after a region in the west of Galway, Ireland (usually spelt Connemara in English). The name derives from *Conmaicne mara*, meaning the seaside land of the descendants of Conmac. In Irish legend, Conmac was a son of Fergus Mòr, king of Ulster and Maedhbh, queen of Connacht.

It is hard to imagine how we could prove that formation of chaos in one part of Europa is accompanied simultaneously by addition of new ridges and grooves to 'ball of string' textured regions elsewhere, unless we actually could see it happening. However, recognition of chaos does at least show how a balance could be achieved between the creation and destruction of surface.

So how old is Conamara Chaos?

- Look carefully at Figure 4.24. Is the white ejecta from Pwyll visible on top of the matrix as well as on top of the rafts and, if so, what does this tell us about the relative ages of the chaos and Pwyll?

- Both rafts and matrix in the western part of Figure 4.24 are white, in contrast to redder surfaces in the east. This is because raft and matrix surfaces alike have been overlain by a sprinkling of ray ejecta from Pwyll. This is also apparent on Figure 4.21. The fact that the matrix has ejecta on top of it means that the chaos existed before Pwyll was formed. As noted earlier in this section, Pwyll is probably about 3 million years old, so this is the likely lower age limit for Conamara Chaos.

In addition to the white ray ejecta, there are quite a few craters less than 1 km in diameter in this region. These are more common within the rays, and so are almost certainly secondary craters produced by impact of the largest blocks of ejecta expelled from Pwyll. On Figure 4.24, these craters appear more common on the rafts than in the matrix. This difference could be apparent rather than real, because the jumbled surface texture of the matrix would make the craters difficult to see. However, even on the highest resolution images such as Figure 4.26 craters appear scarcer on the matrix. Some of the small craters on the rafts must pre-date the break-up into chaos, but if most of the small craters we see are Pwyll secondaries then some of those that formed on the matrix since its creation would seem to have been erased. One way this could have happened is if the matrix remained mobile and continued to deform for some millions of years *after* its surface froze, whereas the surfaces of the rafts were rigid for the whole time.

So, the evidence so far points to Conamara Chaos having formed (probably at least 3 million years ago) by melting of a patch of ice some tens of kilometres across, accompanied by break-up of the adjacent floating ice into rafts, some of which drifted inwards across the temporarily unroofed ocean. A skin of ice or slush would have rapidly covered any exposed water but, depending on the amount of heating from below, this skin could have remained sufficiently pliant to allow the rafts to plough through it for up to several million years. Before the matrix became fully rigid, ejecta from the Pwyll impact was distributed across the area, with the accompanying formation of secondary impact craters. Continuing deformation or local resurfacing of the matrix was sufficient to erase a significant proportion of the Pwyll secondary craters on the matrix.

Figure 4.26 Galileo SSI image, 7 km in width, recorded at 9 m per pixel. Solar illumination is from the lower right. See Figure 4.24 for location. (NASA)

But that is not quite the end of the story.

QUESTION 4.9

Examine Figure 4.24 again, and locate a groove that runs diagonally across the image from just below the northwest (top left) corner. Look closely at the units this groove cuts. Deduce the implications for the age of this groove and the nature of the material that it cuts.

You have now seen the last major piece of the puzzle. After its matrix has become sufficiently rigid, chaos on Europa begins to experience brittle fracturing, and new grooves form that look similar to some of those on ordinary 'ball of string' terrain. Perhaps, given sufficient time, rafts and matrix alike in a chaos region will become thoroughly overprinted by additional generations of ridges and grooves, and the rafts so split up by successive spreading increments across each crack that they lose their integrity. The entire area will take on the appearance of a ball of string – in fact, it will actually be 'ball of string' terrain. For all we know, areas such as those in Figures 4.19 and 4.22 could be former chaos of which no recognizable traces remain.

Chaos areas, and in particular the drifted rafts, are compelling evidence that, at least at the time of chaos formation, the 'ball of string' textured surface ice was floating on a liquid. This would not have to be an ocean of global extent, because the underlying liquid would not need to cover an area much wider than the overlying chaos.

In Section 4.2.4, we will argue that whether the ocean is global, local, permanent or ephemeral is of no great importance for the existence of life. However, first let's see if we can work out how thick the ice is.

4.2.3 How thick is Europa's ice?

You learned in Section 4.1.4 that geophysical data show the 'icy' outer part of Europa to be about 100 km thick, but that the information is inadequate to distinguish between the extreme possibilities of solid ice all the way down to the bedrock and a floating sheet of ice supported above a liquid ocean (Figure 4.13). The subdued topography of craters such as Pwyll and our interpretation of chaos regions both strongly suggest that the latter is more likely. We can determine how thick the ice was at the time of chaos formation, provided we are willing to take the present surface of the matrix to be at roughly the same height (relative to raft surfaces) as the surface of the ocean when it was exposed. The lack of any obvious disturbance of the matrix adjacent to the blocks even in the very high-resolution image in Figure 4.26 indicates that this is a reasonable assumption. If this is correct, then the height of a raft surface above the matrix carries important information.

- ■ Look at the rafts in Figure 4.24. Do you get the impression that each raft has its surface at a different height above the matrix?
- ❑ With the exception of the tilted rafts, they all appear to be at about the same height.

This is just a crude visual impression. However, there are various ways to determine relative heights on spacecraft images. The best way is to use the stereoscopic information contained in two images of the same area taken from different perspectives. Unfortunately, Galileo did not obtain high-resolution stereoscopic images of Europa. Instead, we can measure the widths of the shadows cast by the rafts onto the matrix, and combine this information with knowledge of the angle of the Sun above the local horizon to estimate the height of the cliff. This shows that most of the cliffs at the edges of rafts in Figure 4.24 are about 100 m high.

- Why would the surface of a raft (or the top of any object floating in a fluid) be higher than the surface of the matrix (or the fluid in which the object is floating)?
- The only simple explanation is that the rafts are less dense than the fluid in which they were floating.

This is certainly true of ice floating in the Earth's oceans, and gave rise to the metaphor 'only the tip of the iceberg', which refers to the small fraction of something that is apparent when most of it is hidden. On Europa, the height difference can tell us the total thickness of the rafts, if we know the densities of the raft and the ocean. The principle behind this is known to geologists and geophysicists as 'isostasy' (see Box 4.7).

Isostasy is really just another name for buoyancy.

BOX 4.7 THE THICKNESS OF A FLOATING RAFT

Figure 4.27 shows a tabular raft floating at equilibrium (i.e. at its position of neutral buoyancy) in a liquid. In this situation, the pressure at the base of the raft must be the same as the pressure in the liquid immediately adjacent to the base of the raft. The formula for pressure, P, at depth d beneath a substance of density ρ is given by:

$$P = \rho g d \tag{4.2}$$

where g is the acceleration due to gravity. In the situation illustrated in Figure 4.27, identical pressures occur at the base of the raft, which occurs below a total raft thickness of $(h + w)$ and at a depth w in the liquid. The difference in any atmospheric pressure between the raft surface and the liquid surface is negligible, so we can ignore this and write:

$$P = \rho_1 g(h + w) = \rho_2 g w$$

As we are interested in determining the raft thickness, $(h + w)$, we can divide by g, to get:

$$\rho_1(h + w) = \rho_2 w \tag{4.3}$$

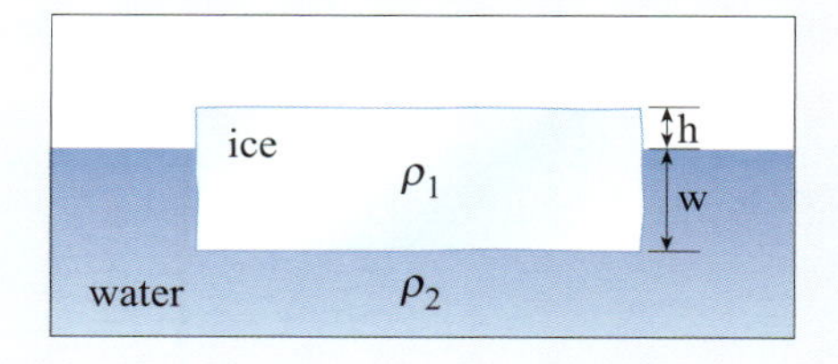

Figure 4.27 Cross-section of a raft (density ρ_1) floating in a liquid (density ρ_2). The height of the raft's surface above the liquid is h and the depth below the liquid surface to the base of the raft is w.

We do not actually know the density of the raft (impure ice) or of the liquid (likely to be a salt solution, rather than pure water). However, we can assume a reasonable range of values, given that we can be fairly sure that the raft is mostly H_2O ice and that the liquid is some kind of salty water. The density of water rich in dissolved sulfates of magnesium and sodium (for example of a composition close to that in Figure 4.15) would be about $1180\,\text{kg}\,\text{m}^{-3}$. Ice freezing from such a solution could have a density as high as $1126\,\text{kg}\,\text{m}^{-3}$ if rich in these salts or as low as $927\,\text{kg}\,\text{m}^{-3}$ if salt-free.

QUESTION 4.10

(a) Rearrange Equation 4.3 to find an expression for w.

(b) Use this rearranged equation to determine the maximum and minimum depths to the base of the rafts, and hence the raft thicknesses in Conamara Chaos, given that h is 100 m, ρ_1 is not less than $927\,\text{kg}\,\text{m}^{-3}$ and not more than $1126\,\text{kg}\,\text{m}^{-3}$, and ρ_2 is $1180\,\text{kg}\,\text{m}^{-3}$.

If you were to assume pure ice floating in pure water, this method would give a raft thickness intermediate between the extremes you calculated in Question 4.10. Thus, the heights of the cliffs at the edges of rafts show with a fair degree of confidence that when the ice broke up to create the rafts its thickness was not less than a few hundred metres and not more than a few kilometres.

This is not necessarily the long-term ice thickness on Europa. Clearly, it is possible that the local heating event responsible for chaos generation might have melted quite a lot from the base of the continuous ice sheet before this finally broke up. On the other hand, the method we have used to calculate the thickness of the rafts relies on the ice of the re-frozen matrix being both thinner and weaker than the raft ice, at least until cooling-related thickening and ridge and groove development has turned the matrix into 'ball of string' terrain. So we can imagine regions of ice on Europa both thinner and thicker than the values you calculated in Question 4.10.

4.2.4 Heat and life

The weight of evidence in the case of Europa points strongly towards ice overlying salty water, at least within the past few millions years although not necessarily today. There are signs that localized heating episodes have melted and fractured the ice. The intensity of tidal heating has probably waxed and waned in step with fluctuations in the amount of forced eccentricity of Europa's orbit, but we can anticipate that conditions on Europa would have varied through a broadly similar range during much of the Solar System's lifetime. What, then, are the prospects for life on Europa?

Let's consider the surface ice first. You learned about cold-tolerant (psychrophilic) organisms in Chapter 2. On Earth, active microbial communities have been found within Antarctic sea ice at temperatures as low as $-18\,°\text{C}$. Here, algae and other organisms survive by photosynthesis in summer that is possibly supplemented when there is less light available by metabolising dissolved organic matter, but these are probably survivor species that need liquid water for part of their life cycles.

- Can you suggest four reasons why Europan surface ice is unlikely to be so hospitable for life as Antarctic sea ice?
- Firstly, Europa's surface temperature of −140 °C even at the equator is far lower than in Antarctic sea ice, and we know of no way for water-based metabolism to proceed in such cold conditions. Secondly, liquid water would occur here far less frequently than within the Antarctic ice shelf. Thirdly, Jupiter is 5.2 AU from the Sun, so (according to the inverse-square law) the sunlight available for photosynthesis on Europa is some 27 times weaker than on Earth. Fourthly, unless there is a thriving ecosystem elsewhere on Europa, there would be no dissolved organic matter food source to supplement the energy available from photosynthesis.

Thanks to the escape of tidal heat, the temperature within Europa's ice is likely to increase with depth. However, even on Earth the light intensity is too low for photosynthesis to continue more than about 20 m deep within the ice. This is only a tiny fraction of Europa's ice thickness. There could be no ice warmer than −20 °C at a shallow enough depth for photosynthesis, except within very young matrix ice of chaos regions, or in the walls of fissures for brief periods during fracturing or ridge building eruptions. It is faintly conceivable that primitive photosynthetic organisms may lie entombed and dormant within Europa's near-surface ice for periods of millions of years, and become active only during relatively brief episodes of local heating (full-blown chaos generation, or above warm dome-forming intrusions as in Figure 4.23, or within an active fissure). This would be a pretty marginal existence. It is perhaps to the energy and nutrients that could be provided by hydrothermal vents that we must appeal if we wish to find the basis of a robust and persistent ecology of the kind imagined by Arthur C. Clarke (Section 4.1.4).

Whether hydrothermal vents exist on Europa, and, if so, their abundance and their power, depend upon how deep within Europa the tidal heating occurs. This has not been determined, because it depends on unknown factors such as the strength and other properties of Europa's ice and rocky interior. At one extreme, virtually all the tidal energy could be dissipated within the icy shell (in which case chaos formation would be a result of direct heating of the ice). This would mean that the ocean was kept warm largely because of heat from above. Any hydrothermal vents on the ocean floor would be scarce and weak, and powered only by the feeble leakage of radiogenic heat from Europa's rocky interior. On the other hand, if tidal heating were concentrated in Europa's rocky part, flow of heat from the rock into the overlying ocean would be much stronger. As on Earth, ocean water would soak into the underlying hot rock, where it would become heated and react chemically, eventually escaping back into the ocean via hydrothermal vents. A static rocky substrate would not be very favourable for sustaining life because the ocean would deplete the available chemicals over million-year timescales. However, if tidal heating were sufficient to cause partial melting within Europa's rock, hydrothermal circulation would be especially strong over sites where igneous rock was being intruded at shallow depth, and strongest of all at any places where volcanic eruptions occurred onto the ocean floor. Moreover, the repeated arrival of new igneous rock at or a little way below the ocean floor would mean that the chemistry was continually renewed, so that some of the circulating water would always find something with which to react.

- Thinking back to what you learned in Chapters 1 and 2 about the relationship between life and hydrothermal vents, can you suggest why the presence of hydrothermal vents on Europa could be particularly important for the origin of life on Europa?
- Phylogenetic evidence, in particular the ribosomal RNA tree (Figure 1.37), suggests that thermophylic autotrophic microbes dependent on chemosynthesis are the last common ancestor for life on Earth. Therefore life on Earth may well have begun at hot vents. If it did, then it could perhaps have begun with equal ease at hot vents on Europa.

An ocean of global extent would not have been necessary for life to begin. Relatively small pools of water sandwiched between ice and hot rock would have been enough. However an ocean, or at least an extensive body of water, would certainly make it easier for life to survive. Life that was trapped in a single pocket of water would have no escape when the hydrothermal vent that had been feeding it cooled down and ceased to flow. It would have to survive in a frozen state until the unlikely eventuality of a new vent starting up nearby. However, an ocean, or at least an extensive seaway, would mean that organisms (including free-floating larval stages of any multicellular life) could drift from vent to vent, allowing species to survive – even though individual colonies would meet their demise with the extinction of their vent.

The primary producers at hot-vent ecosystems on Earth derive their energy from a redox (oxidation–reduction) reaction. Typically, they exploit a reaction whose equilibrium position depends on temperature. For example if a high temperature (such as where hot fluids react with rock during hydrothermal circulation) drives the reaction in one direction but a low temperature (where vent water mixes back with ocean water) tends to drive the reaction the other way, then an organism can extract energy by getting involved in this 'reverse' reaction. This is only effective when the low-temperature ('reverse') reaction is kinetically inhibited, which provides the opportunity for a biological catalyst to become involved.

A chemical reaction is 'kinetically inhibited' when there is a significant energy barrier to be overcome to enable the reaction to proceed.

An example of this in ocean-floor hydrothermal systems on Earth is the biological production of methane ('methanogenesis'). During hydrothermal alteration of newly created oceanic crust iron reacts with water. The iron is oxidized and the water reduced to hydrogen. Carbon is discharged in vent fluids as carbon dioxide, arising partly from oxidation of crustal and mantle carbon and partly from breakdown of carbonate rocks that have been drawn into the mantle at subduction zones. Thus, hot vent fluids are rich in carbon dioxide and hydrogen. In solution, these gases are related by the equilibrium reaction:

When a chemical reaction is written this way, (aq) signifies something in aqueous solution, (l) signifies a liquid, (s) signifies a solid and (g) is a gas

$$CO_2(aq) + 4H_2(aq) \rightleftharpoons CH_4(aq) + 2H_2O(l) \qquad (4.4)$$

At high temperatures the equilibrium lies well to the left, so that in a hot solution carbon dioxide and hydrogen are stable. At lower temperatures, including those in seawater, the equilibrium position lies well to the right, but in a lifeless ocean an energy barrier would inhibit the reaction from moving in this direction. However, with biological mediation most of the carbon dioxide and hydrogen can react to form methane and water as the temperature falls. This is the reaction that methanogenic bacteria exploit as their source of energy.

$$2CO_2(aq) + 6H_2(aq) \rightarrow (CH_2O)_n + CH_4(aq) + 3H_2O(l) \quad (4.5)$$

$(CH_2O)_n$ indicates carbohydrate in biological cell material, and the subscript n indicates that the real formula is more complicated than simply CH_2O.

In principle, this reaction could be used by Europan equivalents of methanogenic bacteria at hot vents. There are reasons, however, why this particular reaction may not be a viable source of biological energy on Europa. One reason is that without (so far as we know) subduction of oxidized species, Europa's hydrothermal fluids are likely to be considerably more reducing than the Earth's. This would lead to vent fluids being naturally rich in methane rather than carbon dioxide, which would therefore deprive methanogens of their energy source. Another reason is that high pressure drives the reaction in Equation 4.4 towards the right. Now do Question 4.11, which compares the pressure on the Earth's ocean floor with that at the floor of Europa's ocean.

QUESTION 4.11

The pressure on an ocean floor is given by the expression $P = \rho g d$, which you have already met in a slightly different context in Box 4.7, as Equation 4.2. For our current purpose, ρ is the average density of the overlying ocean, g is the acceleration due to gravity on the planetary body concerned, and d is the depth of the ocean. On Earth, we can take ρ to be $1030\,\mathrm{kg\,m^{-3}}$, g to be $9.8\,\mathrm{m\,s^{-2}}$, and d to be 3.0 km (the approximate depth of a mid-ocean ridge). On Europa, treating the ice thickness as negligible relative to the ocean thickness, we can take ρ to be $1180\,\mathrm{kg\,m^{-3}}$, g to be $1.3\,\mathrm{m\,s^{-2}}$, and d to be 100 km. Use these values to calculate the pressure at the exit of a hydrothermal vent on:

(a) the Earth's ocean floor

(b) Europa's ocean floor.

Thus, the pressure on Europa's ocean floor is about five times that at a mid-ocean ridge hydrothermal vent on Earth. This may not seem a big difference, and would be unlikely to have any adverse effect on, say, biological cell structure. However, it would affect the equilibrium in Equation 4.4, so that carbon tended to be outgassed as methane rather than carbon dioxide. The situation would be even less favourable for methanogenic life if Europa's subduction-deprived mantle is more reduced than the Earth's, because this would make the methane to carbon dioxide ratio very high in the first place. It would also mean that Europa is unlikely to provide favourable habitats for analogues of terrestrial SLiME (subsurface lithautotrophic microbial ecosystems) of the kind you read about in Section 2.6.

Perhaps, then, biological methanogenesis is not viable on Europa. In an extreme case, Europa's hydrothermal fluids could be so reducing that the only plausible oxidants that could provide an energy source for life would be oxidized metals, such as ferric iron (Fe^{3+}). A suitable reaction is represented by:

$$2Fe(OH)_3(aq) + H_2(aq) \rightleftharpoons 2FeO(s) + 4H_2O(l) \quad (4.6)$$

in which the iron in vent fluids is reduced by reaction with hydrogen. Alternative reducing agents could be hydrogen sulfide or even methane. In all these cases, biological organisms could feed off the energy released during reduction of Fe^{3+} to Fe^{2+}.

On the other hand, it is conceivable that Europa's ocean may actually be moderately oxidizing in character.

- Can you recall from earlier in this chapter a mechanism whereby molecular oxygen is known to be generated on Europa?
- In Section 4.2.1 you learned how exposure to charged particles and solar ultraviolet radiation in the near-surface ice leads to radiolytic and photolytic breakdown of water molecules to produce oxygen and hydrogen.

The hydrogen escapes relatively easily to space, but much of the oxygen is held within the ice crystals. These processes are only effective in the upper few micrometres (μm) of the ice, but 'gardening' by micrometeorites and slightly larger impacts can be expected to mix the products to a depth of about 1 m in the regolith. We do not know how efficiently, if at all, such oxygen is eventually mixed into the ocean, but obviously this could occur from time to time when melting, especially during chaos formation, reaches the surface.

There is actually a radiolytic mechanism whereby oxygen could be generated from either ice or liquid water at *any* depth below the surface. This is because one of the common elements thought to be dissolved in the Europan ocean has a radioactive isotope.

- Look back at Figure 4.15, and see if you can recognize which of these elements has a radioactive isotope.
- The element with a radioactive isotope is potassium.

The radioactive isotope is ^{40}K, which on Earth, and presumably Europa too, makes up about 0.012% of the total potassium today, and would have been about ten times more common shortly after Europa was formed. β-particles and γ-radiation are emitted by ^{40}K as it decays and both can radiolytically break water into hydrogen and oxygen, by means of the series of reactions indicated in Box 4.4.

This process could yield about 10^{10} moles of oxygen per year in Europa's ocean. Provided there is sufficient carbon available and a suitable reaction pathway, this would be enough to support about 10^7–10^9 kg yr^{-1} of biomass production. However, the limited availability of carbon in the right form and right place almost certainly means that the actual rate (if any) of biomass production in Europa's ocean is probably less than this. A likely value, allowing for a modest amount of hydrothermal energy, is about 10^5–10^6 kg yr^{-1}.

QUESTION 4.12

Rates of biomass production on the Earth today are about 5×10^{13} kg yr^{-1} by photosynthesis on land, a similar amount by marine photosynthesis (mainly by microscopic plankton), and about 10^{10} kg yr^{-1} by chemosynthesis at ocean-floor hydrothermal vents.

How do these rates of biomass production compare (by orders of magnitude) with the value proposed for Europa?

It is important to remember that the estimates for Europa are very uncertain, and could be underestimates by two or three orders of magnitude – or ridiculous overestimates if Europa supports no life at all. However, Europa could offer sites that are just as favourable for life to have originated as those on the early Earth, and equally hospitable for so-called extremophiles to flourish today, albeit in smaller quantities than on Earth. Europa's small mass (so small that it cannot hold onto an atmosphere) and its distance from the Sun beyond the 'habitable zone' have prevented it from developing a photosynthesis-dominated biosphere, but the viability of any chemosynthetically supported biosphere is independent of this. Thus, according to some assumptions at least, the prospects for life on Europa appear encouraging.

Whether any life has remained at the level of simple single-celled autotrophs or diversified into multicellular forms, and whether any heterotrophic organisms have evolved to prey on these (as imagined by Arthur C. Clarke) remains to be seen.

We will conclude our discussion of Europa with a brief look at plans to gather further data on this intriguing world.

4.2.5 How can we find out more about Europa?

The next mission to be launched to the outer Solar System is likely to be NASA's Jupiter Icy Moons Orbiter (JIMO). This is tentatively scheduled for launch in 2011 at the earliest. On arrival at Jupiter it would go into orbit first round Callisto, then Ganymede and finally Europa.

The main objectives of JIMO at Europa are to:

1 Determine the presence or absence of a subsurface ocean.

2 Characterize the three-dimensional distribution of any subsurface liquid water and its overlying ice layers.

3 Understand the formation of surface features, including sites of recent or current activity, and identify candidate sites for future lander missions.

QUESTION 4.13

JIMO is an ambitious mission with ambitious objectives. What techniques do you think a spaceprobe in orbit could use to meet these objectives?

The answer we gave to Question 4.13 covers all we expected you to come up with, but there are other techniques that are also likely to be useful. At the time of writing, the actual instrument package for JIMO has not been finalized. However, it is expected to include an imaging system with spectroscopic capability, a laser altimeter, and an ice-penetrating radar. The laser altimeter will map Europa's topography, and in particular it will determine the height of Europa's tidal bulges. The bulges should be only about 1 m high if the ice is solid throughout, but about 30 m high if there is 10 km of ice floating on water, so altimetry is a neat way of addressing the presence or absence of a subsurface ocean. The radar will be directed directly downwards with the intention of recording echoes from the ice–water interface. Unless the ice is particularly salty, which would tend to attenuate

Figure 4.28 A Europa orbiter in action (not JIMO, but a simpler mission that has now been cancelled). The blue beam illuminating the surface is a schematic indication of the ice-penetrating radar beam, which is intended to map the ice thickness with a depth resolution of about 100 m. (NASA)

the signal, the radar should detect the ice–water interface wherever it lies at less than about 10 km depth, as is likely to be the case in young chaos areas (Section 4.2.3). This, plus any further visual clues to ice thinness or recent activity from the imaging system, will be the main means of selecting landing sites for future lander missions.

An artist's impression of a Europa-orbiting mission in action is shown in Figure 4.28.

The next mission to Europa is likely to arrive several years after JIMO. The current ambition is to equip such a mission with a miniature robotic submarine (a 'hydrobot') capable of exploring the ocean to seek for signs of life. In order to deploy this, a way has to be found to make an access hole in the ice, which presumably must be done either by mechanical drilling or by using heat to melt a borehole. Even after landing on the thinnest ice, the technological challenges of making such a hole would be severe.

There is also another problem, which is the planetary protection issue (Chapter 3) of how to prevent contamination of Europa's biosphere with organisms inadvertently carried from Earth. It would be foolish to send a sophisticated suite of instruments to Europa unless we could be as certain as possible that any signs of biological activity were not attributable to microbes carried to Europa by the same mission or any previous spacecraft. Contamination of Europa's biosphere (or the accidental establishment of a biosphere where none had previously existed) would undermine any conclusions about the independent origin and evolution of life that could otherwise be drawn following the discovery and study of Europan life. Most investigators would recognize an ethical duty to safeguard the integrity of future studies of Europa's biosphere and to protect against potential harm to any Europan organisms. This duty is codified in legal form by the 1967 United Nations' *Treaty on Principles Governing the Activities of States in the Exploration and Use of Outer Space, Including the Moon and Other Celestial Bodies*.

Very few terrestrial microbes would survive a journey to Europa, and of these only a tiny proportion would be likely to be able to feed and reproduce on Europa or in its ocean. However, just one viable organism delivered to the right (or wrong!) place that was then able to feed and multiply would do incalculable harm. With this in mind, a report on preventing biological contamination of Europa published in 2000 by the US National Academy of Sciences recommends that the bioload of any Europa-bound mission should be minimized by using levels of cleanliness during assembly and subsequent sterilization that are at least as stringent as those currently agreed for Mars missions.

Illuminating lessons about preparing to penetrate through a thick ice cover into a body of water may be learned from the case of Lake Vostok. This is a large lake that has been trapped beneath the Antarctic ice for possibly several million years, and is suspected of housing a sealed-in ecosystem. Exploration of Lake Vostok and the proper implementation of anti-contamination protocols are widely held to be realistic rehearsals for exploration of Europa's ocean, as discussed in Box 4.8.

BOX 4.8 LAKE VOSTOK – AN ICE CONUNDRUM

In 1974, Russian scientists began drilling deep into the ice at their Vostok research base, situated at the geomagnetic south pole in Antarctica. Samples of ice and the gases and other trace materials trapped within it provide a valuable and continuous record of climate changes and large volcanic eruptions during the past 400 000 years. Incidentally, viable micro-organisms were found entombed within the ancient ice too. It was not until 1994, by which time the borehole had reached a depth of about 3 km, that seismic and other studies revealed that the ice overlies the largest subglacial lake in the world, covering about 2×10^5 km^2, which is the same area as Lake Ontario. This is known as Lake Vostok (Figure 4.29).

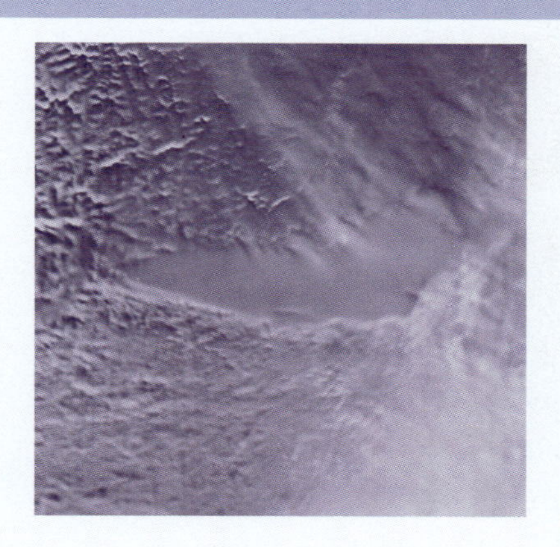

Figure 4.29 Satellite radar image showing the ice surface of part of Antarctica. Lake Vostok is the elongated flat area near the centre. Image is about 600 m across. (NASA)

In places the water depth reaches about 1 km. The oldest ice overlying the water is less than a million years old, but the ice sheet as a whole is slowly flowing across the lake, so the lake itself may have been sealed off from the surface for as long as 14 million years. The lake is suspected of supporting its own ecosystem, subsisting either by a meagre rain of organic matter at places where the overlying ice melts or by chemical energy at suspected hot springs.

These realizations united scientists from many nations in plans to bore through the base of the ice in order to sample the lake water and deploy a probe into the lake. One method suggested to keep the hole sealed and to prevent contamination was that the hole should be drilled to within a few metres of the roof of the lake. A cylindrical probe would then be lowered to the base of the hole that could sterilize itself while waiting for the hole above to freeze over, and then melt its way down into the lake. It would pay out a tether behind itself as it travelled, which would act as a communications link to the surface (Figure 4.30).

However, two serious objections emerged that put at least a temporary halt to these schemes. First, the self-sterilization techniques for the probe were untested. Secondly, when the Russians had begun drilling back in the 1970s, they were anxious to stop the hole freezing shut behind the drill bit so they pumped a mixture of aviation fuel and antifreeze (Freon) into the hole. There is now 60 tonnes of this toxic chemical mix in the hole, and no one can be sure that none of this will leak into the lake if the hole is continued. Pressure from a coalition of environmental groups caused drilling to stop in 2001 (Figure 4.31), and plans for any kind of penetration into the lake were put on hold for maybe a decade.

Figure 4.30 Artist's impression of a probe released into Lake Vostok from the base of the borehole. (© Rob Wood/Wood Ronsaville Harlin, Inc.)

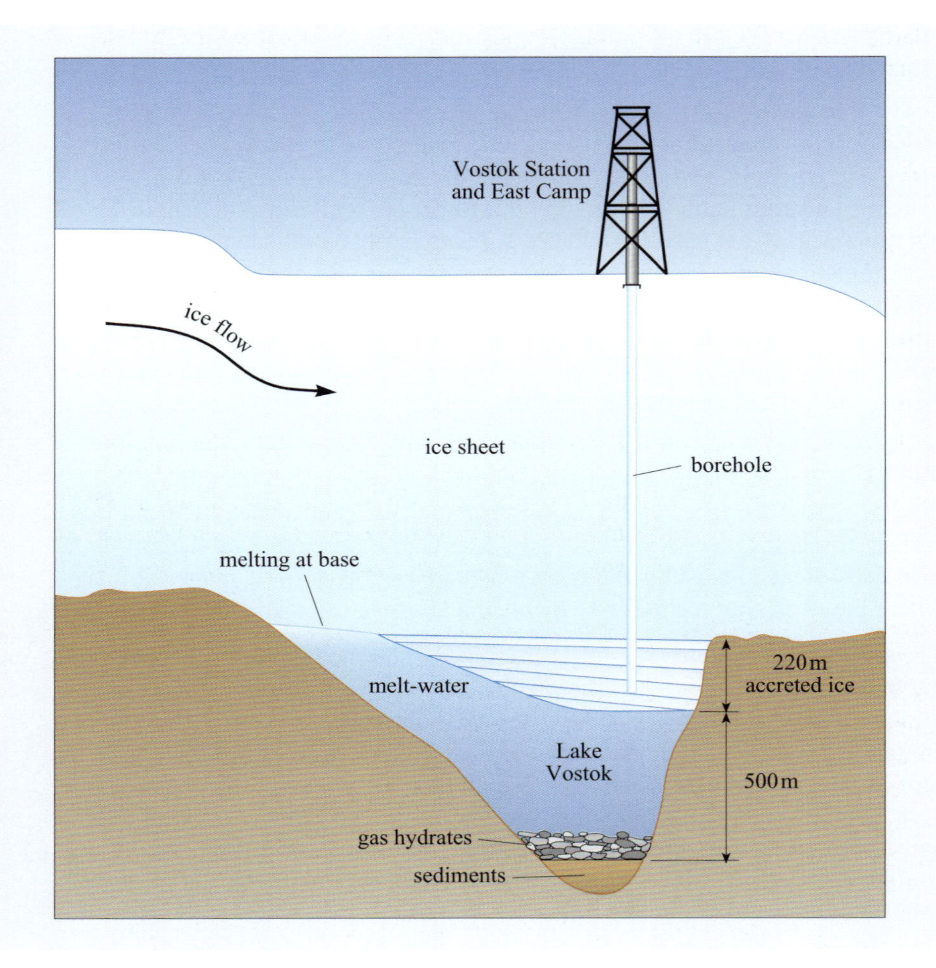

Figure 4.31 Schematic cross-section through Lake Vostok and the overlying ice (not to scale). The borehole stops less than 100 m before the roof of the lake. It has penetrated the full thickness of the ancient ice cap, and terminates within ice that has frozen more recently onto the roof.

4.3 Other icy bodies as abodes of life?

You have seen that Europa offers arguably the most promising habitat for present-day life in the Solar System, other than on the Earth itself. This is because ice or water overlying warm rock can lead to hydrothermal circulation, offering hot springs where life could originate in the first place and subsist on chemical energy thereafter. There would not actually have to be a global ocean below the ice, but this would help by allowing organisms to spread from one vent to another.

- Can you suggest places in the Solar System where conditions might now be, or might formerly have been, sufficiently similar to Europa for life to have got underway there too?

- Any of the tidally heated icy satellites would seem promising, especially those that were at any time heated sufficiently for a global ocean to form below the ice. Of those illustrated earlier in this chapter, Callisto and Rhea (Figure 4.10) do not seem promising, whereas Enceladus and Ariel (Figure 4.11b and d) would seem the likeliest.

Enceladus is a particularly intriguing proposition. Voyager obtained useful images of less than half its surface. Although parts of Enceladus are fairly heavily cratered, other areas (e.g. the lower right of Figure 4.11b) show no craters at all even on the highest resolution (2 km per pixel) images. The Cassini mission (described in the next chapter) is scheduled for several close fly-bys of Enceladus during its 2004–2008 tour of the Saturn system. These new images may reveal when and how the youngest regions became resurfaced, help us to understand Encaladus's tidal heating history, and provide a basis for more informed speculation about its astrobiological potential.

The surface of Triton (Figure 4.6) shows great variety, with plenty of evidence of cryovolcanic resurfacing. We know from spectroscopic evidence that the surface ice is a mixture of nitrogen, methane, carbon dioxide, carbon monoxide and water, and it is likely that there is some ammonia too. This is probably a true differentiated crust in the geochemical sense, overlying a mantle that is richer in water-ice. Triton's bulk density suggests that a rocky core begins at a depth of about 350 km. There is a fair degree of superimposed impact cratering on all terrain types, so widespread cryovolcanism appears to have ceased – probably at least hundreds of millions of years ago. Apart from seasonal changes in the sizes of the polar caps of frozen nitrogen, and what appear to be solar-powered geysers rupturing the south polar cap, no current or recent activity has been identified. This is consistent with the lack of a known tidal heat source. However, in the aftermath of its capture by Neptune there would have been a period of probably about a billion years while tides acted to force Triton's orbit to become circular. This is probably when most of the cryovolcanism took place. During this period there could have been a Europa-like ocean below the ice with plenty of time for life to become established. If so, life could be clinging on thanks to feeble radiogenic heat – or perhaps future explorers will find nothing but the fossilized remains of an extinct biosphere.

The tilt of a planet's axis is conventionally measured relative to the perpendicular to its orbital plane. Pluto's tilt of >90° signifies that its rotation is retrograde.

Apart from Enceladus, the icy satellite that may be experiencing the greatest rate of tidal heating today is one that you probably did not consider at all. This is Charon, the single known satellite of Pluto (Table 4.1). We know much less about Charon than the other large icy satellites, because no spaceprobe has been there, so we have to rely on telescopic data, such as spectroscopic information and the albedo maps in Figure 4.32. Charon orbits in Pluto's equatorial plane, and their rotations are mutually tidally locked so that they permanently keep the same faces towards each other. There would seem little scope here for on-going tidal heating. However, Pluto's axis (and hence Charon's orbit) is tilted over at an angle of 119.6°. This leads to competing tidal pulls on Charon by Pluto and the Sun in such a configuration that, according to some models, there could be substantial tidal heating in Charon's interior today.

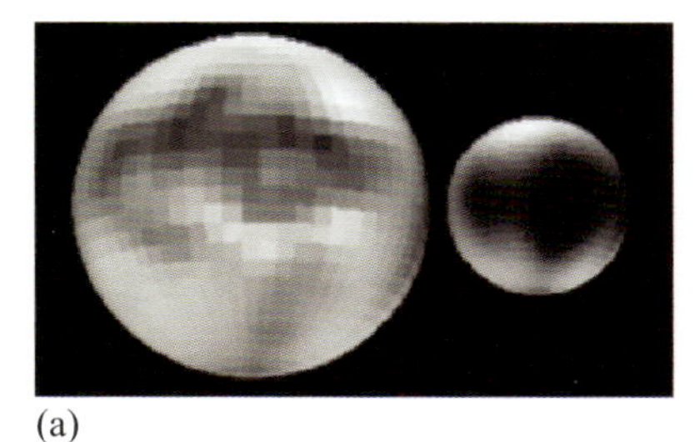

(a)

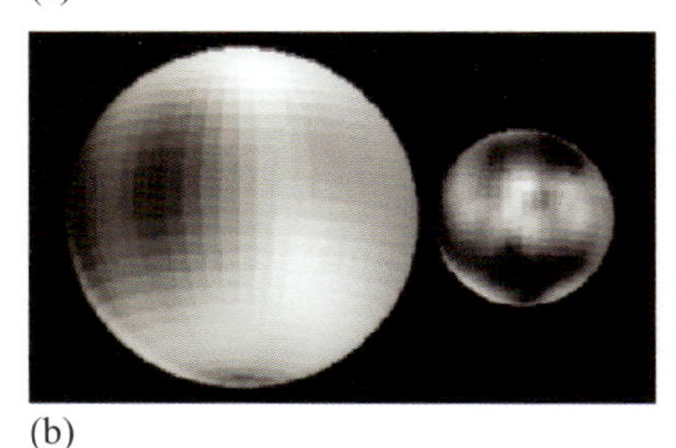

(b)

Figure 4.32 Maps of surface albedo patterns on Pluto (left) and Charon (right) calculated from variations in brightness as they rotate, and a series of mutual occultations that occurred during 1985–1990. (a) Charon-facing side of Pluto and anti-Pluto side of Charon; (b) anti-Charon side of Pluto and Pluto-facing side of Charon. (Image courtesy of Marc W. Buie/Lowell Observatory)

Pluto itself is probably not being heated tidally, but spectroscopy has revealed its surface to consist of at least as rich a cocktail of ices as on Triton. It is likely to be fully differentiated, especially if Charon owes its origin to a giant impact event similar to that which formed the Moon. An ocean, with life-bearing potential, could have persisted below the solid ice for a considerable period until most of the accretional heat from such a collision had leaked away. This heat would have been stoked up by tidal forces until Pluto's rotation became synchronous with Charon's orbital period. Speculation is likely to continue relatively unbounded until a spaceprobe visits Pluto and Charon. The first will probably be a much delayed

NASA mission currently called New Horizons. The earliest date this could now be launched is 2006, and its Pluto fly-by cannot happen until 2015 at the earliest.

In an icy satellite with no tidal heating, we would have to look to other sources of geothermal heat to power hydrothermal vents, such as radiogenic heating or heat left over from the accretion process. Any icy satellite with a differentiated structure must have experienced at least some water–rock geochemical interaction, but this may not necessarily have been strong enough or sufficiently prolonged to favour life.

Finally, what about Jupiter's outer two Galilean satellites?

- Does the evidence in Figures 4.10a and 4.14 make Callisto look like a favourable site for a Europa-style ocean?
- Figure 4.10a shows a uniformly heavily cratered and, therefore, ancient surface, and Figure 4.14 shows an interior that is only weakly differentiated. Both these factors suggest that an ocean is unlikely.

It is not thought likely that Callisto experienced significant tidal heating after its rotation became synchronous. Ganymede, however, may have been affected by tidal heating episodes (though not so strongly as Europa) when its degree of forced eccentricity fluctuated as a result of mutual orbital interaction with Europa and Callisto. It bears signs of this in the way that its ancient heavily cratered surface is transected by belts of younger terrain (Figure 4.33). However, even the youngest belts of cross-cutting terrain on Ganymede have numerous impact craters superimposed on them, and are likely to exceed a billion years in age.

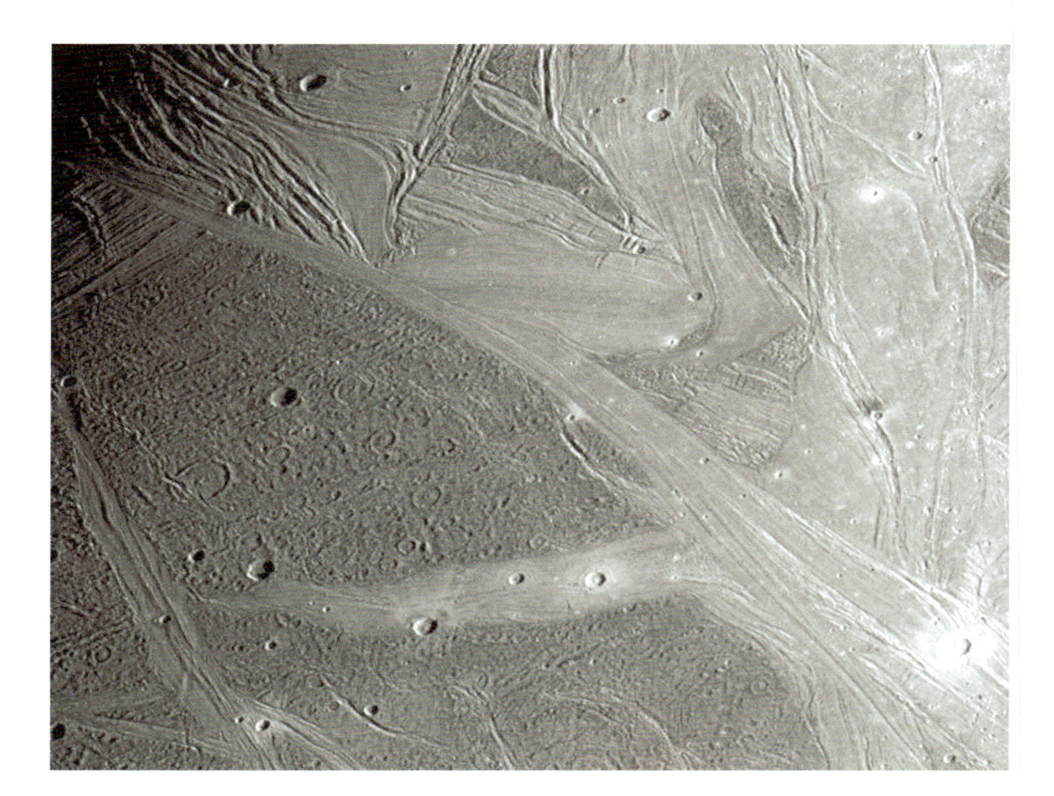

Figure 4.33 A Galileo SSI view, 600 km in width, of part of Ganymede. Several generations of ridged and grooved pale terrain cut across an older and more heavily cratered terrain. (NASA)

The imaging data would seem to be giving us a clear story. However, the Galileo Orbiter measured magnetic fields apparently induced within both these satellites by their orbital passage through Jupiter's magnetosphere, which complicate the issue. In the case of Ganymede, the induced field could originate in an iron-rich core, but what we know of Callisto's internal density distribution shows fairly robustly that it can have no such core. That being so, the only reasonable explanation remaining seems to be that Callisto has an electrically conducting ocean at least 10 km thick and no more than 100 km deep. A similar ocean could also account for Ganymede's magnetic field.

The proposition that there are relatively shallow oceans beneath the surfaces of Ganymede and Callisto seems totally at odds with the ancient appearances of their surfaces, and allows us to end this chapter with a caution.

Where there is an ocean there could be life, but we understand far too little about any of the icy satellites. Although there appear to be many reasons why some of them could harbour life, and that most could have done so at times in the past, it may be a long time before we know for sure.

Now test some of the knowledge and skills you have developed in this chapter by answering the following questions.

QUESTION 4.14

Examine again the groove in Figure 4.24 that you looked at in Question 4.9 (the same groove that you can see near the right-hand edge of Figure 4.26). On Figure 4.24, locate where the line of this groove passes between adjacent rafts about 5 km northwestward of the edge of the outline indicating Figure 4.26 (conveniently, 5 km is approximately the length of a short side of this outline).

(a) What evidence is there for the relative ages of this groove and the matrix between the rafts at this location?

(b) What are the implications for the time period over which the matrix was mobile in Conamara Chaos as a whole?

(c) How can we deduce that the ejecta from the Pwyll impact overlies the matrix at this location, and what does that tell you?

QUESTION 4.15

In Question 4.10b you calculated a raft thickness in Conamara Chaos based on the assumption that the heights of the cliffs at the raft edges were a result of rafts floating in quite a dense brine. However, perhaps at the time when the topography became 'frozen in' the rafts were actually floating in some kind of slush, which would be less dense than the brine. What is the implied raft thickness for a raft density of 1126 kg m^{-3} and a slush density 1140 kg m^{-3}?

QUESTION 4.16

Imagine it is the year 2100, and that the fifth of a series of probes into Europa's ocean has at last detected life in the form of micro-organisms that appear to be based on the same sort of DNA as on Earth. List the alternative implications for the establishment of life on Europa that could be drawn from this discovery, and how you would hope (eventually) to deduce the truth.

4.4 Summary of Chapter 4

- Many of the large icy bodies in the outer Solar System are internally differentiated. Thanks largely to tidal heating, some, especially Europa, are likely to have an ocean sandwiched between the icy exterior and the rocky core. Others may have had such an ocean in the past.
- Wherever water rests on warm rock, water must percolate into it and become heated. This will cause hydrothermal convection to begin. Hot, chemical-rich water will emerge at vents, where the resulting local chemical disequilibrium provides an opportunity for living organisms to extract energy by acting as mediators (biological catalysts) for redox reactions.
- If it is true that life on Earth originated at hydrothermal vents, then it is equally likely that life could have become established around similar vents at the 'ice'-rock interface on icy bodies.

CHAPTER 5
TITAN

5.1 Introduction

Saturn has at least 30 satellites, so you may wonder why we've devoted an entire chapter to one of them: Titan. Titan is unique in that it is the only satellite known to possess a dense atmosphere and, as you will see in Section 5.4 its surface may present us with a very exotic environment. Although it's not generally believed that Titan has ever harboured life, it does have a role to play in our understanding of the development of life. This is because the photochemical processes presently occurring in its atmosphere are believed to result in the formation of a wide range of organic molecules. In this chapter, you will look at our present knowledge of Titan, the theoretical models that explain some of these observations and finally the prospects for improving our knowledge of Titan in the near future.

5.2 Observations

Titan was discovered by the Dutch physicist and astronomer, Christiaan Huygens (1629–1695), pronounced 'hoi-gens' in English, (Figure 5.1) some 45 years after Galileo discovered the four large moons of Jupiter (the Galilean satellites). However, Titan was not named until the mid-1800s when British astronomer John Herschel (son of William Herschel who had himself discovered two satellites of Saturn) suggested that Saturn's satellites should be named after Saturn's brothers, collectively called the Titans, and Saturn's sisters, the Titanesses. These were the mythological giants who were believed to rule in the heavens before Jupiter conquered them. Because the satellite discovered by Huygens was so much larger than the rest, astronomers chose to name it Titan rather than naming it after one of the individual Titans.

Figure 5.1 Christiaan Huygens (1629–1695), sometimes spelled Christian Huyghens, Dutch mathematician, physicist and astronomer, who discovered Titan in 1655, using a telescope (built by himself and his brother Constantijn) of far better quality than those used by Galileo. Within a year of this discovery, he successfully explained the nature of Saturn's rings. Also amongst many other achievements in his career, he made fundamental contributions to the understanding of the nature of light and invented the pendulum clock. The ESA Probe that is due to land on Titan's surface as part of the Cassini mission has been named in his honour. (© Royal Astronomical Society Library)

'Limb' is a term used by astronomers to describe the apparent edge of the Sun, Moon or any other celestial body with a detectable disc.

Not much more was learnt about Titan until the early years of the 20th century when the Spanish astronomer Comas Sola published some observations which mentioned, almost in passing, that he had detected the phenomenon of **limb darkening** when observing Titan, a phenomenon in which a planet or star appears darker at its limb than at its centre.

The exact cause is not important to us here, what is vital though is that this phenomenon requires the existence of an *atmosphere*. Limb darkening, therefore, provided the first indication that Titan possessed an atmosphere. The next major step forward came in the early 1940s when Gerard Kuiper (Figure 4.8), using a spectrometer on the new 82 inch (2.08 m) McDonald Observatory telescope in Texas, identified the characteristic signature of gaseous methane in the near-infrared light coming from Titan. This essentially confirmed the existence of an atmosphere around Titan. No other satellite of any of the planets has been found to have anything other than a minute trace of an atmosphere.

With improving telescopes and instruments, observations of Titan became more sophisticated over the following decades, but it was the space age that provided the 'quantum leap' in our understanding of Titan. The first spacecraft fly-by was by Pioneer 11 in September 1979 but with its relatively unsophisticated collection of instruments and a closest approach of some 363 000 km, little was learnt about Titan. It was in November 1980 when the Voyager 1 spacecraft flew by Titan at a distance of 4394 km that our knowledge of Titan increased enormously.

■ What are some of the benefits of making observations of a planet or satellite from a fly-by spacecraft compared to ground-based Earth observations?

❑ The observations aren't constrained by limits imposed by the Earth's atmosphere (which means you can observe right across the infrared and ultraviolet parts of the electromagnetic spectrum where much of the information about the composition and structure of an atmosphere lies), the spatial resolution (i.e. smallest detail discernible) is much improved (compared with observing from the Earth) and the intensity of the emitted radiation (or any other phenomenon) is generally much higher because of the closeness of the observing point.

When the Voyager 1 instrument suite was turned towards Titan, several hundred images were taken with its on-board imaging system. A typical example is shown in Figure 5.2a. You might find this image slightly bland and disappointing – well so did the waiting scientists over 20 years ago! Indeed all the Voyager 1 images showed an almost featureless orange haze that covers all of Titan. However, images from Voyager 2, from a much greater distance, revealed more features. An example is shown in Figure 5.2b where you can see a faint dark band around the north pole and a slight contrast between the northern and southern hemispheres – the northern hemisphere was observed to be about 20% darker at blue (i.e. shorter) wavelengths.

Despite extensive image processing carried out on the images, there were no signs of any gaps in the haze to give even a fleeting view of the surface. Neither did they show any feature that could be described as a cloud, which could be tracked to give an indication of wind velocity in the atmosphere.

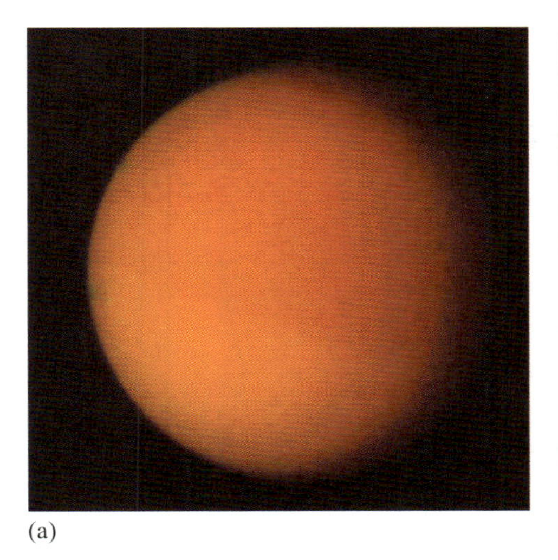
(a)

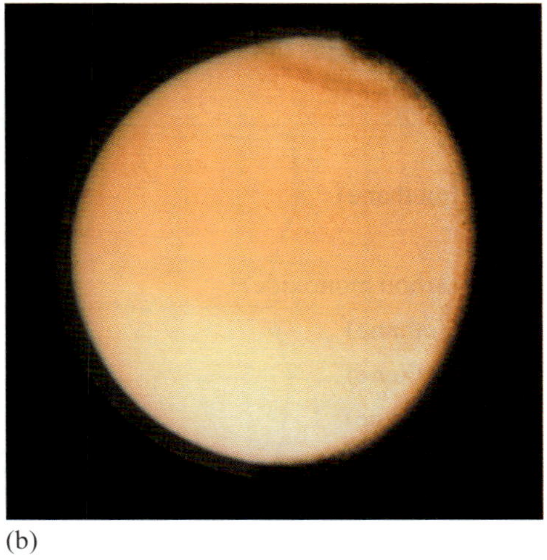
(b)

Figure 5.2 (a) Voyager 1 and (b) Voyager 2 images of Titan acquired during the fly-bys in 1980 and 1981. The only discernible features are a dark polar 'hood' and a slight contrast between the hemispheres in (b). (NASA)

However, the Voyager spacecraft carried an array of scientific instruments, including a set of spectrometers covering the infrared and ultraviolet parts of the electromagnetic spectrum. When the radiation from Titan was analysed with these instruments a set of complex spectra with a wealth of features was obtained, including spectra indicative of particular gases in Titan's atmosphere.

Figure 5.3 shows a typical spectrum from Titan obtained by the Voyager infrared spectrometer (IRIS – Infrared Radiometer Interferometer and Spectrometer). Each major feature is marked with an identification of the gas in the atmosphere that is responsible for emitting that particular wavelength. The height or intensity of each feature can be used to calculate the relative concentration of that particular gas in Titan's atmosphere. The calculated relative concentrations of all the detected gases from the Voyager spectrometers and later Earth-based telescopic studies are shown in Table 5.1.

- What are the main similarities and differences between the compositions of Titan's atmosphere and Earth's atmosphere?

- The main similarity is that each atmosphere has nitrogen as its main constituent. A striking difference is that Titan's atmosphere is rich in hydrocarbons.

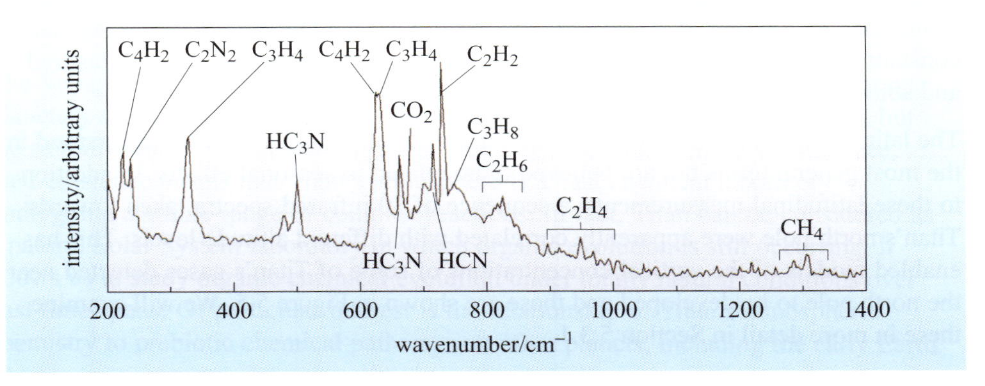

Figure 5.3 Voyager infrared spectrometer (IRIS) data from Titan's atmosphere.

Before considering aspects of the atmosphere in detail, we should consider some of the other data that Voyager was able to measure. The Voyager Radio Science System (RSS) used an onboard radio transmitter to send a signal back to the Earth. By arranging for the line-of-sight back to the Earth to pass through Titan's atmosphere (Figure 5.6), it was possible to measure various properties of the atmosphere such as temperature and pressure profiles (Figure 5.7). Measurements were possible down to the surface of Titan since the haze layer that prevented visible images of the surface from being obtained, did not adversely affect radio waves. The surface temperature is 94 K, which drops to 71 K at the tropopause level of about 45 km.

- Should we be surprised by how cold Titan's surface temperature is?
- No – because of Titan's distance from the Sun, namely about 9.6 AU, the solar input to Titan is almost 100 (actually $9.6 \times 9.6 \approx 92$) times weaker than at the Earth. Therefore, unless there is another source of heat (i.e. internal), you would expect the temperature to be very low.

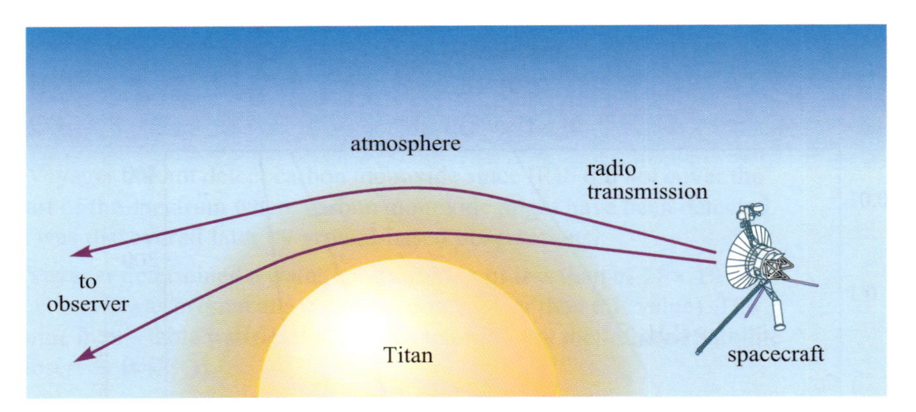

Figure 5.6 Schematic diagram showing the use of the Voyager RSS to determine atmospheric properties. When the path from the spacecraft to the Earth passes through an atmosphere, the signal is deflected, as shown. In addition, the strength of the signal is reduced and its polarization changed.

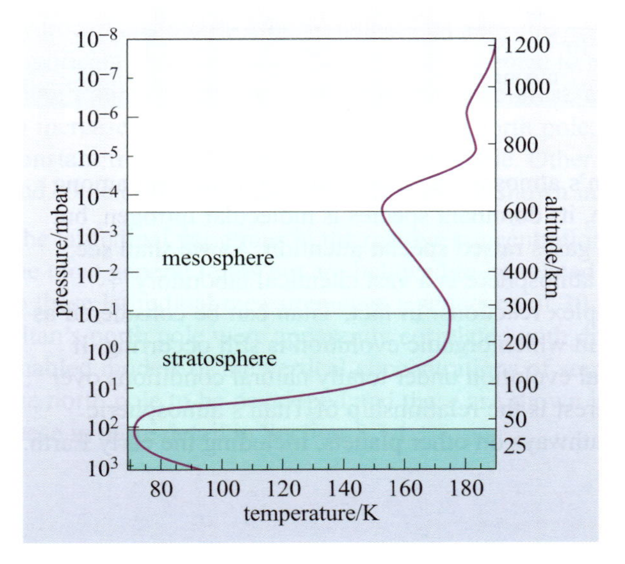

Figure 5.7 Temperature profile for Titan, derived from Voyager RSS data. Also derived from RSS data was the pressure profile that can be constructed by comparing the left-hand (pressure) and right-hand (altitude) axes. Also shown is the designation of various regions of the atmosphere, named by analogy with Earth, although 2 regions have been left blank because you will identify them in Question 5.1.

QUESTION 5.1

(a) Add the labels 'troposphere' and 'thermosphere' to the temperature profile for appropriate sections of the curve in Figure 5.7.

(b) By measuring from Figure 5.7, estimate the lapse rate (rate of decrease of temperature with altitude) for the lowest region of the atmosphere. How does this compare with the value measured by Voyager 1 and quoted in Section 5.4?

If you examine the pressure information in Figure 5.7, you will see that the surface pressure on Titan is between 1 bar and 1.5 bar, i.e. 50% greater than the surface pressure on Earth. Coupled with the fact that Titan's surface gravity is only about 15% of that on Earth, this means that we are dealing with a very substantial and dense atmosphere.

- Can you recall a parameter that is useful for comparing the 'quantity of atmosphere' on a planetary body?
- The column mass, M_c (see Box 3.2) defined as:

$$M_c = P/g \tag{5.1}$$

where P is the atmospheric pressure and g is the gravitational acceleration at the surface.

- Using data from Table 5.2 and Equation 5.1, calculate the column mass of Titan's atmosphere. (*Hint*: Think carefully about units.)
- From Table 5.2, P = 1496 mbar ≈ 1.5 bar and g = 1.35 m s^{-2}. But we need to use the SI unit of pressure, namely the pascal (Pa), with 1 bar = 10^5 Pa.

So $M_c = 1.5 \times 10^5\ \text{Pa}/1.35\ \text{m s}^{-2} = 1.1 \times 10^5\ \text{kg m}^{-2}$.

This figure should be compared with the atmospheric column mass value for Earth, namely 1.0×10^4 kg m^{-2} thus confirming that Titan's atmosphere is more substantial than Earth's.

With the data obtained from the Voyager 1 encounter in November 1980, the Voyager 2 encounter some 9 months later (with a fly-by distance some 170 times greater than for Voyager 1, i.e. 663 385 km as opposed to 4394 km) and previous data, we are now in a position to examine some of Titan's basic parameters. These are shown in Table 5.2 and several points are worth noting. First, we see that Titan is larger than the terrestrial planet Mercury. It is the second largest planetary satellite, only just smaller than Ganymede. Titan, in common with most other planetary satellites, has its axial rotation period 'locked' to the orbital period. This is known as *synchronous rotation* and results from tidal forces generated when a major satellite orbits a planet (Box 4.2).

For many years, Titan was believed to be the largest satellite before it was realized that the dimensions of the visible image included the haze layer above the solid surface.

- Can you think of another satellite that shows this effect?
- The Moon. This is why the same face is always turned towards the Earth.

Table 5.2 Titan's vital statistics.

Equatorial radius	(2575 ± 0.5) km
Mean density	1.88×10^3 kg m^{-3}
Distance from centre of Saturn	1.22×10^6 km
Mass	1.346×10^{23} kg
Surface gravity	1.35 m s^{-2}
Orbital period	15.95 days
Axial rotation period	15.95 days
Orbital eccentricity	0.0292
Surface temperature	(94.0 ± 0.7) K
Surface pressure	(1496 ± 20) mbar
Main atmospheric constituents	N_2, CH_4, H_2, CO
Mass of atmosphere	4×10^{17} kg

QUESTION 5.2

Using the data for the mass and radius of Titan in Table 5.2, confirm the value for the mean density of Titan. How does this value compare with other solid bodies in the Solar System and what does this suggest to you about the composition of Titan?

QUESTION 5.3

The surface gravity on a solid body can be calculated from the expression $g = GM/R^2$, where G is the gravitational constant, and M and R are the body's mass and radius respectively. Using the data for mass and radius in Table 5.2, confirm the value for surface gravity quoted in the same table.

5.3 Titan's atmosphere

The atmospheres of the giant planets are in constant turmoil. Recorded wind speeds can be very high, and there are vast storms. In chemical terms, however, it is safe to assume that below the cloud layers the various atoms and molecules are in chemical equilibrium. Indeed, the very turbulence of the atmosphere helps the gases to reach equilibrium by ensuring that they are well mixed.

■ What extra participant in atmospheric chemistry must we consider above the clouds that is less significant lower down?

❑ Sunlight.

Although in the dark outer reaches of the Solar System much less sunlight is received per m^2 than on Earth, the absorption of light still plays an important role in the atmospheric chemistry of Titan. This chemistry is rich and complex and, rather than trying to cover all the important reactions, we will consider two problems that arise from the observed concentrations of different molecules in Titan's atmosphere:

1 Carbon-containing molecules occur in Titan's atmosphere that would not be expected were the atmosphere in chemical equilibrium (Box 5.1).

2 The element nitrogen occurs on Titan as the diatomic nitrogen molecule, N_2, whereas it occurs as the compound ammonia, NH_3, in the atmospheres of the giant planets.

5.3.1 Hydrocarbons

As you saw in Table 5.1, the atmosphere of Titan contains a wide range of organic molecules, in particular hydrocarbons.

Modelling of Titan's atmosphere indicates that if it were in **chemical equilibrium**, the predominant hydrocarbon would be methane, CH_4, with only negligible amounts of the other hydrocarbons present (Box 5.1). So why do we observe hydrocarbons such as ethene, ethane, etc? These molecules are the result of photochemical reactions in the outer atmosphere. In those regions where solar radiation (particularly ultraviolet) penetrates, atmospheric composition is determined not by chemical equilibrium but by the interaction of molecules with radiation.

BOX 5.1 CHEMICAL EQUILIBRIUM

If we take a box containing a mixture of chemicals, say nitrogen, hydrogen and ammonia and leave it for a very long time, ensuring that the temperature and pressure remain constant and that no chemicals or radiation leave or enter the box, then the chemicals will reach equilibrium. At equilibrium, chemical reactions will be taking place but the total rate of production of each compound equals the total rate of its destruction so that the amounts of the various compounds present will stay the same. For any chemical reaction occurring, the relative amounts of the compounds involved that are present at equilibrium are given by the **equilibrium constant**, *K*, for that reaction. The value of *K* varies with the temperature, but does not depend on the total amount of chemicals present.

Let us take as an example of chemical equilibrium the reaction between carbon, C, and hydrogen, H_2, to form methane, CH_4:

$$C + 2H_2 = CH_4 \quad (5.2)$$

We can put into our box any amount of carbon, hydrogen and/or methane. It does not matter whether we start with a mixture of carbon and hydrogen, or with methane, or with a mixture of hydrogen and methane, so long as both elements are present in sufficient amounts. After a very long time, we will have carbon, hydrogen and methane present in equilibrium amounts. If we measured the concentrations of the three compounds at equilibrium we could obtain the equilibrium constant, *K*. For the reaction in Equation 5.2, the equilibrium constant is given by

$$K = \frac{[CH_4]}{[C] \times [H_2]^2} \quad (5.3)$$

where the square brackets [ ] denote concentrations. The value of *K* is such that if the concentration of hydrogen is very much higher than that of carbon then most of the carbon will be converted to methane. However, in the atmosphere of Titan, several molecules containing just carbon and hydrogen are observed. We have to include equilibrium constants for the reaction between carbon, C, and hydrogen, H_2, to form methane, CH_4, ethane, C_2H_6, ethene, C_2H_4, and ethyne, C_2H_2. Given the equilibrium constants we can calculate the relative amounts of hydrogen, methane, ethane, ethene and ethyne. But in addition we have to consider the equilibria between carbon, hydrogen and other elements such as nitrogen and oxygen to form compounds such as carbon monoxide, ammonia and water. All these equilibria are linked so that a large number of equations have to be solved to obtain the abundances. Luckily this is just the sort of problem that computers are good at.

From our knowledge of how chemicals react, we can choose for any planetary atmosphere a set of reactions involving the most likely molecules formed from the most abundant elements.

For example, assuming chemical equilibrium, models for the chemical composition of Titan's atmosphere predict fractional abundances of ethane, C_2H_6, and ethyne, C_2H_2, that are negligible. However, if the effect of ultraviolet radiation is included, then the models predict fractional abundances of 10^{-5} and between 10^{-8} and 10^{-6}, respectively. These figures agree reasonably well with the observed fractional abundances in Table 5.1.

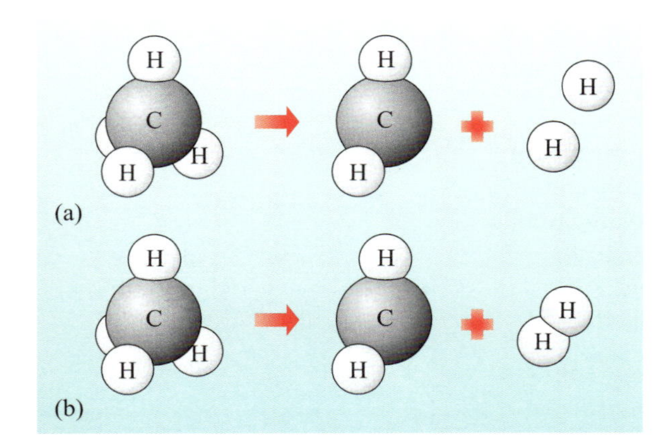

Figure 5.8 Schematic representation of the photodissociation of methane: (a) corresponds to Equation 5.4 and (b) corresponds to Equation 5.5.

How, then, can the observed variety of hydrocarbons be produced through the action of solar radiation on methane? The methane molecule has a central carbon atom joined to four hydrogen atoms, which can be thought of as lying at the corners of a tetrahedron (see Figure 5.8a). It can absorb ultraviolet radiation and break up or dissociate into smaller fragments in several ways, a process known as photodissociation. For example, one bond could acquire so much vibrational energy that the bond breaks and a hydrogen atom separates off. However, the C—H bonds do not necessarily vibrate in isolation, and two, three or four bonds can vibrate together. The most common photodissociations of methane in the atmospheres of Titan (and the giant planets) are those in which two hydrogens are lost. These are illustrated in Figure 5.8, and in Equations 5.4 and 5.5.

$$CH_4 + \text{photon} \longrightarrow CH_2 + H + H \tag{5.4}$$

$$CH_4 + \text{photon} \longrightarrow CH_2 + H_2 \tag{5.5}$$

A photon is a particle representing a basic unit (with a distinct energy) of visible light or other electromagnetic radiation.

In the atmospheres of the giant planets the most abundant molecule is H_2. This reacts with the carbon-containing product of Equations 5.4 and 5.5, CH_2. The two main products of this reaction are methane and CH_3 (methyl). Since the methyl molecule is not bonded to four hydrogen atoms it has a spare electron in its outer shell (note the molecule still has an equal number of protons and electrons so it carries no charge). Were it to make a chemical bond this electron would be paired-off with an electron from another atom. However, since it is unpaired, this makes the methyl molecule highly reactive.

Molecules with no charge but which have an unpaired electron that can take part in forming chemical bonds are known as **radicals** and they are common products of photochemical reactions in which a bond is broken.

The formation of methane, of course, just takes us back to the starting material, however, the two methyl radicals will readily combine to form ethane, C_2H_6, as in Equation 5.6:

$$CH_3 + CH_3 + M \longrightarrow C_2H_6 + M \tag{5.6}$$

where M indicates a third molecule that does not take part chemically in the reaction but removes some of the energy of the reacting molecules.

A particularly interesting set of photochemical reactions occurs when methane loses three hydrogens to form the highly reactive radical CH (Figure 5.9). This happens about 8% of the time when methane absorbs ultraviolet radiation; it's an important reaction as it leads ultimately to the production of hydrocarbons containing long chains of carbon atoms.

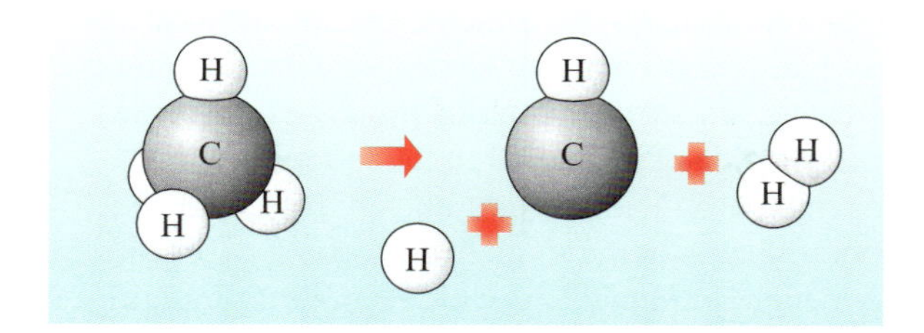

Figure 5.9 Schematic representation of the photodissociation of methane to form CH.

The CH radical produced by the loss of three hydrogen atoms reacts with methane to form a molecule in which two carbon atoms are bonded together, ethene, C_2H_4:

$$CH + CH_4 \longrightarrow H_2C{=}CH_2 + H \qquad (5.7)$$

The ethene produced readily absorbs ultraviolet radiation and loses hydrogen to form ethyne (HC≡CH) almost as soon as it is produced and so there is very little of it around. The observed abundance of ethene on Titan (and Jupiter) is lower than those of ethane ($H_3C{-}CH_3$) and ethyne.

C−C, C=C, and C≡C denote carbon atoms bound together by single, double and triple bonds respectively.

On Jupiter, ethyne is quite an abundant molecule in the atmosphere since when it is broken up by radiation, the products rapidly react with hydrogen to re-form ethyne:

$$HC{\equiv}CH + \text{photon} \longrightarrow HC{\equiv}C + H \qquad (5.8)$$

$$HC{\equiv}C + H_2 \longrightarrow HC{\equiv}CH + H \qquad (5.9)$$

- Are these reactions (Equations 5.8 and 5.9) likely to happen on Titan?
- No. The major constituent of Titan's atmosphere is not hydrogen but the less reactive gas nitrogen.

Thus on Titan HC≡C does not re-form ethyne, instead it goes on to produce larger molecules. One of the simplest ways it can do this is illustrated by Equations 5.10 and 5.11.

$$HC{\equiv}C + HC{\equiv}CH \longrightarrow HC{\equiv}C{-}C{\equiv}CH + H \qquad (5.10)$$

$$HC{\equiv}C + HC{\equiv}C{-}C{\equiv}CH \longrightarrow HC{\equiv}C{-}C{\equiv}C{-}C{\equiv}CH + H \qquad (5.11)$$

Successive reactions of this sort can produce very long molecules with hundreds of linked carbon atoms.

- Write an equation for the formation of the next molecule in the series after HC≡C−C≡C−C≡CH (i.e. one with eight carbon atoms).
- $HC{\equiv}C + HC{\equiv}C{-}C{\equiv}C{-}C{\equiv}CH \longrightarrow HC{\equiv}C{-}C{\equiv}C{-}C{\equiv}C{-}C{\equiv}CH + H$ (5.12)

It is not known how far the production of carbon chains may have proceeded on Titan, but it is possible that some of the longer-chain molecules may be responsible for the haze observed in the atmosphere (see Section 5.3.3). The haze, therefore, is sometimes referred to as a photochemical haze or smog. However, the formation of very long chains may be inhibited by the presence of hydrogen atoms in the atmosphere, which would tend to promote the reverse reactions in Equations 5.10 and 5.11.

To obtain a detailed picture of the chemistry of Titan's atmosphere, we would need to consider all possible reactions. By building up our knowledge of which reactions are important, it is possible to construct a model of the major processes that take place. To do this, individual chemical reactions are studied in the laboratory to determine their rates and to deduce which are the important ones in the conditions that prevail in the particular atmosphere that is being modelled. In some cases, there are just a small number of reactions which dominate and which can describe fairly well the state of any particular atmosphere. In other cases, one has to consider a very large number of inter-related reactions in order to get a fair representation of the state of an atmosphere.

The photochemistry of methane is modified by the presence of large quantities of nitrogen on Titan (see Table 5.1). As nitrogen is relatively unreactive, it does not participate directly in Titan's atmospheric chemistry. However, N_2 molecules will dissociate under the action of solar photons, galactic cosmic rays and electrons from Saturn's magnetosphere to form atomic nitrogen, N. Sunlight is believed to be the dominant cause of the photodissociation of N_2 above an altitude of around 700 km in Titan's atmosphere. At lower altitudes dissociation by galactic cosmic rays is thought to be more significant with electrons from Saturn's magnetosphere also playing a role in the altitude range of 500 km to 750 km. The nitrogen atoms, however they are formed, do have an effect on the methane photochemistry already described as well as playing a role in the production of **nitriles**, a class of organic compound containing the group CN.

- ■ Why should photodissociation be a more significant process above 700 km than below it?
- ❑ The strength of sunlight, which drives photochemical reactions, increases with greater height as solar radiation is absorbed in an atmosphere.

This is illustrated in Figure 5.10 which shows a prediction of how the intensity of solar radiation varies with height above Titan's surface. This suggests that the intensity near the surface is perhaps one-seventh of that at high altitude. This will clearly have a large effect on the rate of photochemical reactions – the rates are likely to be far higher in the stratosphere than at lower levels.

Following the acquisition of detailed compositional information on Titan's atmosphere by the Voyager fly-bys in the early 1980s, a series of increasingly sophisticated chemical models, based on the processes discussed above have been developed. To date, the most complete chemical model has considered a total of 122 reactions and 37 dissociation processes. These models have to take account of variations with height in an atmosphere. Factors that vary with height include the temperature and density, chemical abundances and the strength of sunlight. The most recent model for Titan's

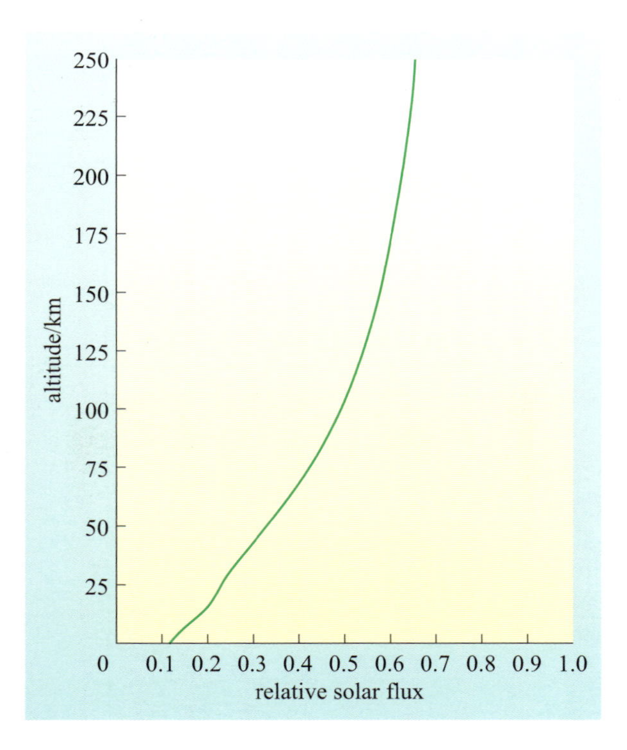

Figure 5.10 The predicted variation of relative solar flux with altitude above Titan's surface. Relative solar flux is used to indicate that full allowance has been made at all altitudes for the amount of solar radiation that is absorbed, scattered and re-radiated.

atmosphere has had some success in describing the observed properties, however difficulties still remain that will require further observations and analyses.

The atmospheric models also provided some interesting insights into what has become known as the 'carbon monoxide problem'. Carbon monoxide, was not detected by either of the Voyagers. However, scientists had expected to find it since it is a common species and because it can be produced from the photolysis of CO_2 in the upper atmosphere. Carbon dioxide was detected by Voyager 1 in infrared spectra. Subsequently, CO was detected in 1983 in the near-infrared from ground-based observations. Carbon monoxide and carbon dioxide were then the only oxygen-bearing gases known in Titan's atmosphere. However, the observed abundances of CO and CO_2 imply that photodissociation of CO_2 is not the only process by which CO is produced in Titan's atmosphere. Carbon monoxide can also be produced by the reaction of the products resulting from the dissociation of water and methane molecules.

Photolysis or photodissociation refers to chemical reactions produced by exposure to light.

The water could be derived from icy particles from comets or from Saturn's ring system. The potential role of water in Titan's atmospheric chemistry was confirmed in 1998 when the European Space Agency's (ESA) Earth-orbiting Infrared Satellite Observatory (ISO) detected water. The superior spectral resolution of ISO's infrared spectrometer gave spectra with considerably more detail (Figure 5.14) than had been observed by Voyager. This enabled some of the fine spectral features characteristic of water to be detected. Even with the new data from ISO there still does not seem to be enough water in Titan's atmosphere to explain the observed amounts of CO and CO_2. This discrepancy will probably have to wait for its solution until the arrival of the Cassini–Huygens mission to the Saturnian system (Box 5.2).

BOX 5.2 THE CASSINI–HUYGENS MISSION

The Cassini–Huygens project is collaboration between The European Space Agency (ESA), NASA and the Italian Space Agency (ASI). The aim of the project is to place a spacecraft in orbit around Saturn and to deliver a probe to the surface of Titan. The former is the Cassini Orbiter provided by NASA and the latter is the Huygens Probe, the contribution from ESA. ASI is responsible for the spacecraft's 4 metre high-gain radio antenna and part of the communications system. The Orbiter is named after the Italian astronomer Jean-Dominique Cassini (Figure 5.11).

Cassini–Huygens was launched on 15 October 1997 by a Titan IVB/Centaur launch vehicle, and became the heaviest spacecraft (about 5630 kg at launch) to be launched towards the outer Solar System.

Figure 5.11 Jean-Dominique Cassini (born Giovanni Domenico) (1625–1712), the first of four generations of Italian astronomers who served as Director of the Paris Observatory. He discovered four Saturnian satellites (Iapetus, Rhea, Dione and Tethys) and in 1675 the distinct gap in Saturn's ring system, which now bears his name. The spacecraft (and mission) which will orbit the Saturnian system has been named in his honour. (Painting by Duragel, courtesy of the Observatoire de Paris)

Using the currently available propulsion technology, it isn't possible to launch a spacecraft of this mass directly to Saturn. Instead, a series of gravity assists had to be used. This technique, first employed by the Mariner 10 spacecraft at Venus in 1973 (which allowed it to reach Mercury), involves flying a spacecraft close to a planetary body in order for the spacecraft to gain energy. This requires the spacecraft to fly past the planet at just the right distance and angle. If successful, the technique reduces the launch energy requirements for a particular mission and enables otherwise impossible missions to be undertaken; it can also provide the means to carry a much larger payload than would otherwise be possible. In the case of Cassini–Huygens, four gravity assist manoeuvres have been used, at Venus (twice), Earth and Jupiter (Figure 5.12). These gravity assists have increased the spacecraft's velocity relative to the Sun by about 20 km s^{-1} and the spacecraft is due to reach Saturn in July 2004, releasing the Huygens Probe into Titan's atmosphere in December 2004.

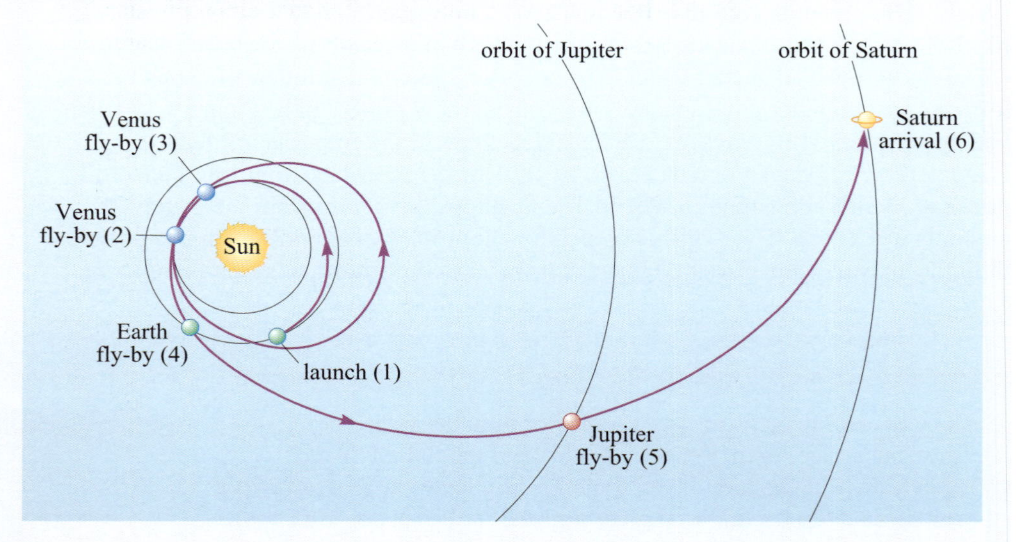

Figure 5.12 The Cassini–Huygens trajectory showing the four gravity assist manoeuvres.

Gravity assists are also known as planetary swing-bys or fly-bys.

Cassini–Huygens represents the second phase of the exploration of Saturn and its system. The first phase was characterized by Pioneer 11 and Voyagers 1 and 2, spacecraft which flew by Saturn giving us a snapshot of that environment. However, the Pioneer 11 and Voyager 2 fly-bys of Titan were relatively distant, limiting the resolution of the observations. In addition, any process or phenomenon that is time varying would probably have been missed. Cassini–Huygens, by remaining in orbit around Saturn and staying in close proximity to its targets will enable series of observations to be made and, based on the results, follow-up observations planned.

The Cassini–Huygens mission has five main scientific study objectives:

- Saturn itself,
- the extensive magnetosphere of Saturn,
- the Saturn ring system,
- the icy satellites,
- Titan.

These are addressed by a combination of the instrument package on the Cassini Orbiter, which comprises 12 scientific instruments, and the Huygens Probe with its payload complement of 6 instruments whose sole function is the investigation of Titan.

The Huygens Probe investigation of Titan has five aims that were agreed between NASA and ESA at the mission's planning stage. These are:

1. Determine the abundance of atmospheric constituents (including any noble gases), establish isotope ratios for abundant elements and constrain scenarios of the formation and evolution of Titan and its atmosphere.
2. Observe the vertical and horizontal distributions of trace gases; search for more complex organic molecules; investigate energy sources for atmospheric chemistry; model the photochemistry of the stratosphere and study the formation and composition of aerosols.
3. Measure the winds and the global temperature; investigate cloud physics, general circulation and seasonal effects in Titan's atmosphere and search for lightning discharges.
4. Determine the physical state, topography and composition of the surface; infer the internal structure of the satellite.
5. Investigate the upper atmosphere, its ionization and its role as a source of neutral and ionized material for the magnetosphere of Saturn.

The 318 kg Huygens Probe is shown in Figure 5.13. The front consists of a shield covered with thermal tiles to protect the probe from the heat generated during the high-speed entry into Titan's atmosphere. The back cover also provides thermal protection as well as housing the parachute compartment.

Figure 5.13 The Huygens Probe during final assembly. The 2.7 m front shield is clearly visible. (ESA)

Both front shield and back cover are jettisoned during the upper part of the descent to leave an inner kernel containing the experiment platform (with an experiment payload mass of 48 kg) to descend to the surface (Table 5.6).

Table 5.6 The Huygens Probe scientific instruments.

Instrument	Acronym	Purpose
Aerosol Collector and Pyrolyser	ACP	Collects aerosol particles by deploying an extendable device into the airflow around the probe at two different altitudes. The collected particles are heated and the products will be passed to the GCMS (see below) for analysis.
Descent Imager and Spectral Radiometer	DISR	A collection of instruments to take both images and spectra of Titan's atmosphere and surface.
Doppler Wind Experiment	DWE	An experiment that uses equipment on both the Huygens Probe and Cassini Orbiter to provide information on the Probe's motion due to wind and turbulence.
Gas Chromatograph/ Mass Spectrometer	GCMS	Measures the chemical composition and determines the isotope ratios of the major gaseous constituents from 170 km altitude down to the surface.
Huygens Atmospheric Structure Instrument	HASI	Measures a wide range of physical properties of the atmosphere, including temperature and pressure profiles; wind speeds and turbulence; atmospheric conductivity; surface permittivity and radar reflectivity. It will also try to detect lightning.
Surface Science Package	SSP	Instruments, designed to study the surface in the region of the Probe landing site. Parameters to be measured include temperature, thermal conductivity, mechanical strength and the speed of sound. In the case of a liquid landing, liquid depth, density and surface wave properties will also be measured.

During Cassini's nominal four-year mission, it will carry out at least 40 fly-bys of Titan, mostly at altitudes of less than 2500 km. During these encounters, many of Cassini's instruments will be directed towards Titan. Of special note are various radar instruments. These include an altimeter capable of mapping the topography to a height precision of 150 m which will cover about 50% of the globe at 25 km spatial resolution by combining information from multiple fly-bys, and *synthetic aperture radar* that will produce radar images of the surface at 350 m to 1.7 km spatial resolution.

An important aspect of the photochemical models of Titan's atmosphere arose when the fate of some of the products of the various chemical pathways was considered. The models suggested that some of the products may well be solids or liquids. This raised the interesting possibility that the surface of Titan would be subjected to a steady rain or snowfall of a variety of organic materials including more complex macromolecular substances. Table 5.3 shows the main products of the photolysis of methane and carbon monoxide and their predicted downward fluxes. Of these, ethane is liquid at the temperature and pressure of Titan's surface while ethyne is solid. Even more intriguing is to consider what would be the accumulated effect of

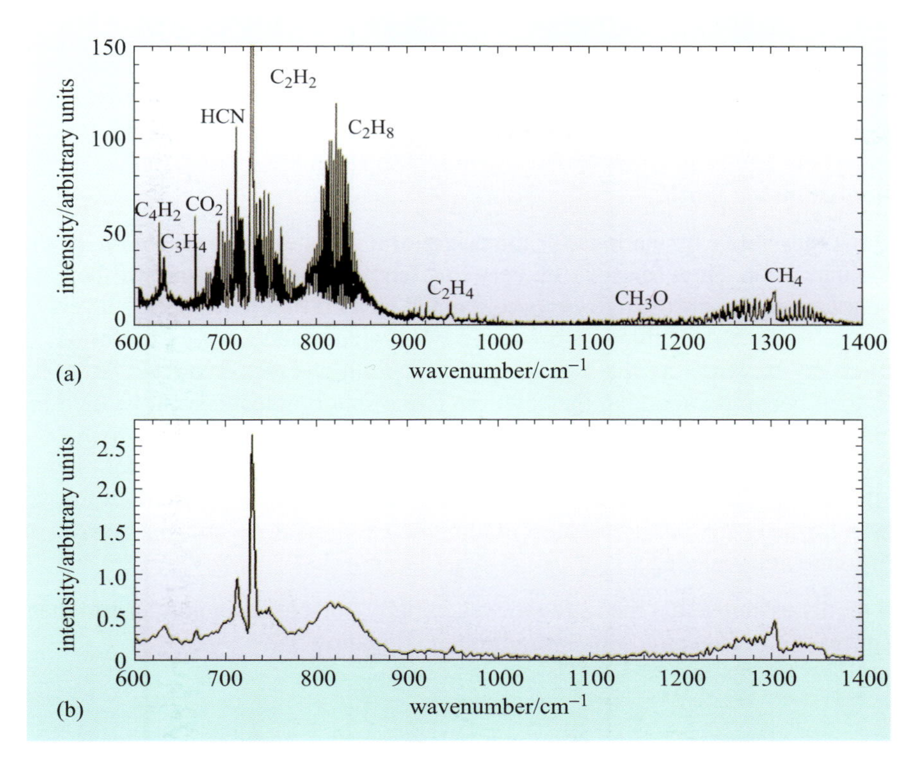

Figure 5.14 The same part of Titan's infrared spectrum observed by (a) ESA's ISO satellite and (b) Voyager infrared spectrometer. The superior spectral resolution of the former is shown by the far greater detail discernible. (Coustenis, A. and Taylor, F. W., 1999)

these fluxes over the age of the Solar System. The last column in Table 5.3 shows the depth of each individual product that would accumulate over the age of the Solar System. This reveals that perhaps 600 m of liquid ethane might have collected on the surface. However, you should note that these depth estimates apply only on a 'billiard ball' type Titan; if there is any topography, the depth would be variable and the entire globe need not be submerged. We will return to these intriguing possibilities later when we look in detail at the nature of Titan's surface.

Table 5.3 Predicted downward fluxes of products of methane and carbon monoxide photolysis and their accumulated depths over the age of the Solar System.

Species	Flux (mol m^{-2} s^{-1})	Depth (km)
C_2H_6	5.8×10^{13}	0.6
C_2H_2	1.2×10^{13}	0.1
C_3H_8	1.4×10^{12}	0.02
CH_3C_2H	5.7×10^{11}	0.006
HCN	2.0×10^{12}	0.02
HC_3N	1.7×10^{11}	0.002
C_2N_2	6.0×10^{10}	0.001
CO_2	3×10^{9}	2×10^{-5}

5.3.2 The origin of nitrogen in Titan's atmosphere

You have already seen that Titan, like the Earth, has an atmosphere whose main constituent is nitrogen as N_2. Titan's size and rocky interior resemble the terrestrial planets more than the giant planets, but the types of molecule (apart from N_2) found in the atmosphere cause the atmospheric chemistry to resemble that on Jupiter rather than that on the Earth.

There are two possible explanations for the origin of a nitrogen atmosphere on Titan. The first is that when Titan formed, the very low temperature of that part of the solar nebula caused the nitrogen to become trapped as a **clathrate** in the icy layers as the satellite formed. A clathrate is produced when a substance, e.g. H_2O, forms with an open crystal structure that can admit small gas molecules, such as CO_2 or N_2, which then become trapped in the cavities. These small molecules can be held in the cavities until the crystalline substance is warmed or perhaps melted at which point they are released, for example by radiogenic or tidal heating. This model suggests that the nitrogen is present as N_2 because it was trapped as N_2 when the satellite was formed. Any ammonia present remained as a solid ice and was never released into the atmosphere.

A second model assumes that the nitrogen was initially present as ammonia, and that subsequent reactions converted this into nitrogen (Equation 5.13).

$$2NH_3 \longrightarrow N_2 + 3H_2 \quad (5.13)$$

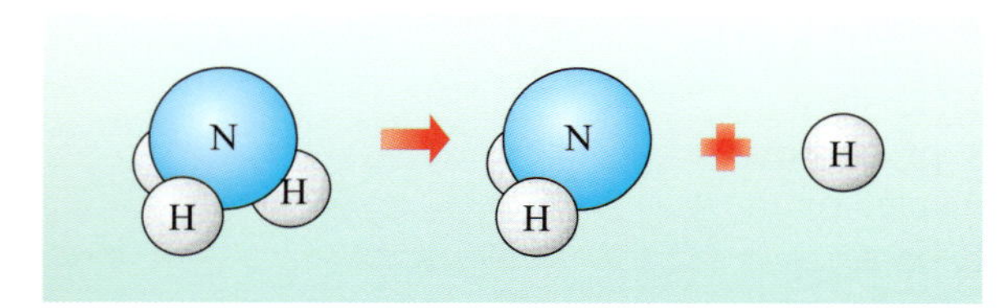

Figure 5.15 Schematic representation of the photodissociation of an ammonia molecule.

However, this reaction would be very slow at the temperatures on Titan and, if the atmosphere were at chemical equilibrium, there should be substantial amounts of ammonia present. One reason for the lack of ammonia in Titan's atmosphere becomes apparent when we consider the photodissociation of ammonia by sunlight. One way in which this can occur is illustrated in Figure 5.15. Here, the energy absorbed is channelled into one of the N—H bonds causing it to break.

The initial products of photodissociation are hydrogen atoms, which escape from Titan, and the radical NH_2, which rapidly reacts with another NH_2 radical to form the molecule **hydrazine**, N_2H_4, as in Equation 5.14:

$$H_2N + NH_2 + M \longrightarrow H_2NNH_2 + M \quad (5.14)$$

in which M represents a molecule that participates in the reaction collision, but emerges unscathed chemically.

Generally when we write chemical formulae, we group together all like atoms as in N_2H_4, but sometimes it can make the structure of the molecule clearer if we write out the formula in its constituent parts. Writing N_2H_4 as H_2NNH_2 tells us for example that hydrazine consists of two NH_2 fragments joined by a bond between the two nitrogens. This sort of representation is particularly used for compounds of carbon.

Similar ammonia photochemistry occurs in the present atmosphere of Jupiter. However, there are two important differences between Jupiter's photochemistry and the proposed route to N_2 on Titan.

First, at the temperatures and pressures in the layers of Jupiter's atmosphere where photodissociation occurs, hydrazine condenses out to form a haze. This haze does not undergo any further chemical reactions. However, on Titan, it has been suggested that the hydrazine remained in the gas phase when Titan's atmosphere was being formed. For this to happen it would be necessary for the surface temperature of Titan to be some 50 K higher than it is at present. Hence the model must include some mechanism for raising the surface temperature.

Second, hydrogen is more readily lost from Titan than from Jupiter owing to the lower escape velocity on Titan. Consequently, hydrogen is more abundant on Jupiter so that the NH_2 recombines with hydrogen, to re-form ammonia. However, on Titan, gaseous hydrazine will undergo a series of photodissociation reactions that eventually leads to the production of N_2H_2, a molecule that will break down to nitrogen and hydrogen:

$$N_2H_2 \longrightarrow N_2 + H_2 \tag{5.15}$$

The hydrogen escapes into space, leaving nitrogen in the atmosphere.

Laboratory experiments to measure the efficiency of trapping gases in clathrates at around 75 K, indicate that argon and nitrogen (in the form of N_2) are trapped with about the same efficiency. So one way in which we might be able to determine the origin of Titan's nitrogen is to measure the abundance of argon. Voyager 1 was unable to detect argon in Titan's atmosphere since argon doesn't have a spectral line in the infrared part of the spectrum covered by the Voyager IRIS instrument. Any argon on Titan almost certainly came from the nebula out of which Saturn and Titan were formed. Thus, if clathrates on Titan trapped both argon and nitrogen from the nebula we would expect a certain ratio of Ar/N_2 (about 0.06), whereas if the nitrogen was produced from the photodissociation of ammonia, we would expect a much lower ratio. The Ar/N_2 ratio is therefore an important measurement for the Cassini–Huygens mission (Box 5.2).

5.3.3 Aerosols

The composition of the particles that make up Titan's haze has been the subject of a series of laboratory-based experiments. In these, a mixture of nitrogen and methane gas in a glass vessel is subjected to an electric discharge in order to simulate the interaction of sunlight, cosmic rays and electrons with Titan's atmosphere. Typically, after a period of several months, a thin layer of brown–orange 'goo' has formed on the inside of the reaction vessel. This so-called **tholin** (from the Greek word 'tholos' meaning mud), when subsequently analysed was found to contain over 75 different constituents, mainly hydrocarbons and nitriles. Although subjecting a gas mixture to an electrical discharge for a few weeks is not the same as irradiating Titan's atmosphere with UV photons and energetic particles for aeons, the results were, at least superficially, very encouraging and gave a qualitative explanation of the visual appearance of Titan (see Figure 5.2). In order to make a more quantitative comparison, it was necessary to compare some of the optical characteristics of the laboratory generated tholins with those of Titan's haze. This was done over wavelengths from ultraviolet through to infrared and showed reasonably good agreement between Titan haze and the laboratory generated particles suggesting that the proposed mechanism, namely the photochemical processing of a methane–nitrogen mix, is indeed the primary source of Titan's aerosols.

5.4 Modelling Titan's surface

Prior to the Voyager fly-bys in the early 1980s, there had not been a great deal of speculation concerning the nature of Titan's surface. However, data from Voyagers 1 and 2 enabled scientists to be far more quantitative about several aspects of Titan's atmosphere, and this had profound effects on our understanding of its surface. As you saw in Section 5.3.1, the abundance of gases in the atmosphere was determined to high degree of precision (Table 5.1). In addition, the radiation environment that Titan was subjected to was also determined quite accurately.

QUESTION 5.4

Assume that 50% of Titan's surface is covered with a methane–ethane–nitrogen ocean of average depth 0.5 km and density $0.66 \times 10^3\ \text{kg m}^{-3}$. The ocean is made up of 70% methane, 25% ethane and 5% nitrogen *by mass*. Assuming that methane is lost from the atmosphere at a rate of $4 \times 10^{-12}\ \text{kg m}^{-2}\ \text{s}^{-1}$, calculate for how long the ocean can resupply the atmosphere. (State any assumptions you make.)

Winds and waves

There is little direct information on winds in Titan's atmosphere but indirect evidence suggests that they should exist. For example, Voyager 1 data showed that there is a temperature difference of some 15 K between the equator and latitudes 60°.

- Why should this fact imply the existence of atmospheric winds?
- When a significant temperature difference exists in a fluid, this tends to result in convection, which is the consequence of mass motion of the fluid, causing winds.

So if a significant body of liquid does exist on the surface of Titan, we can be reasonably confident that waves would be generated just as in terrestrial expanses of water. On Earth, the dominant force that restricts the growth of wind driven waves on an expanse of water is gravity. As Titan's gravity is only 15% of that on Earth, one might expect that waves on Titan's seas should grow to much greater heights under comparable conditions. In what was probably the first example of extraterrestrial oceanography, a group of scientists took the standard mathematical model which described the way in which waves are generated on Earth and changed all the input parameters such as gravity, liquid density and viscosity, to those which would be expected for oceans on Titan. The characteristics of the waves generated are shown in Table 5.5 with those under similar conditions for Earth shown for comparison.

Table 5.5 A comparison of wind-driven ocean waves for a fully developed sea[a] on Earth and Titan assuming a surface wind speed of $5\ \text{m s}^{-1}$ which corresponds to a gentle to moderate breeze.

	Earth	Titan
Significant wave height[b] (m)	0.6	4.5
Wave speed (m s^{-1})	5.5	5.5
Wavelength (m)	11	105
Period (s)	3.5	11.5

[a] For a fully developed sea, there must be a stretch of open water of sufficient size in the direction that the wind is blowing so that the waves can build up to their maximum height. The length of this stretch of open water is called the 'fetch'. Typically on Earth a fully developed sea requires a fetch of 20 km for a wind speed of $1\ \text{m s}^{-1}$ and as much as 200 km or more for a speed of $5\ \text{m s}^{-1}$.

[b] 'Significant wave height' is a term used by oceanographers to get around the fact that waves have a distribution of heights. It is close to the mean of the highest one-third of the waves present in a sea, and approximates visual estimates of wave height.

From Table 5.5, it is apparent that Titan's waves would be higher, more separated, of similar speed and thus less frequent when compared to waves on Earth. It may well transpire that the size of these waves leads to the demise of the Huygens Probe once it lands on the surface of Titan (Box 5.2). The Cassini–Huygens mission will have the capability to probe the surface, both directly and remotely, and will be able to tell us whether features such as wind driven ocean waves are a reality or simply the conjectures of theoretical planetary scientists.

Radar observations of Titan's surface

In 1990, the NASA communications dish at Goldstone, California was used to direct a radio signal at a wavelength of 3.5 cm. towards Titan while the Very Large Array telescope in New Mexico attempted to detect the echo or return signal. Although this was an extremely challenging task, they did manage to detect a reflected signal. This made Titan the most distant object from which a reflected radio signal had been received. Radiation at these wavelengths is relatively unaffected by the presence of haze or clouds in the atmosphere of Titan, yet the strength of the return signal is sensitive to the type of materials present on the surface from which it is reflected. The strength of the detected signal implied that the surface of Titan was reflecting about 10% of the incident radio signal. Note that because of the difficulty of the measurements, there is a fairly large uncertainty in this figure. This value was not consistent with Titan's surface being covered with global ocean of ethane–methane several hundred metres deep as this would be expected to reflect only about 2% of the incident radiation. Although this was not consistent with the idea of a global ocean, it was also at odds with Titan having an icy surface like the largest Jovian satellites, namely Europa, Ganymede and Callisto which reflect between 30% and 90% of the radiation incident upon them at radio wavelengths. Further radar observations have been carried out and these seem to support the earlier observations. However, there is some evidence that different regions of the surface have different radar characteristics, one area in particular appearing to be 'radar bright', implying that the surface of Titan is not uniform but varies in its properties. One interpretation is that the radar-bright region, which seems to cover a surface area approximately equal to that of Australia, is predominantly clean water-ice and the surrounding area is some form of hydrocarbon ocean. Overall, these observations show that Titan doesn't seem to match any other object observed at these wavelengths. In order to circumvent the apparent conflict between these measurements and the hypothesis of a deep global ocean, some theoreticians have tried to see if it is possible to modify this concept to make it consistent with the radar measurements. They have found that if the surface of the ocean were very frothy, perhaps as a result of breaking wind-driven waves as discussed earlier, or was 'dirty' as a result of floating solid products of atmospheric photolysis products, or even from dust from meteoritic impacts, then the radio reflectivity would be significantly higher and therefore perhaps consistent with the observations. This is sometimes referred to as the deep, dirty, frothy ocean model.

Titan's orbital eccentricity

Further insight on the possibility of Titan's oceans came from a theoretical consideration of Titan's orbital eccentricity (see Table 5.2) of 0.0292. A surface layer of liquid on Titan would manifest a tidal bulge just as occurs on Earth, but in the case of Titan this would be generated by Saturn's gravitational pull. This bulge generates tidal currents that act to dissipate Titan's orbital energy, causing the orbit to become circular over time. Calculations in the early 1980s suggested that if there were an ocean of less than 400 m in depth, tidal friction would have caused the orbital eccentricity to have dissipated long ago and the orbit to have become circular. So their argument was that the ocean is either very deep (>400 m) or doesn't exist at all, although this wouldn't prevent the existence of small isolated lakes.

Surface imaging

When Titan is imaged at wavelengths other than those of visible light, in particular the infrared region, some interesting details begin to emerge. Figure 5.16 shows the variation of Titan's albedo with **wavenumber** in the near-infrared part of the electromagnetic spectrum.

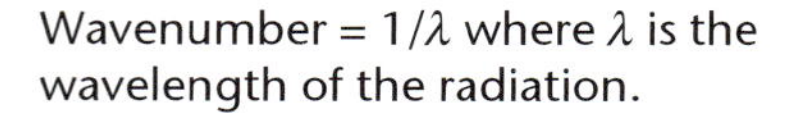

Wavenumber = $1/\lambda$ where λ is the wavelength of the radiation.

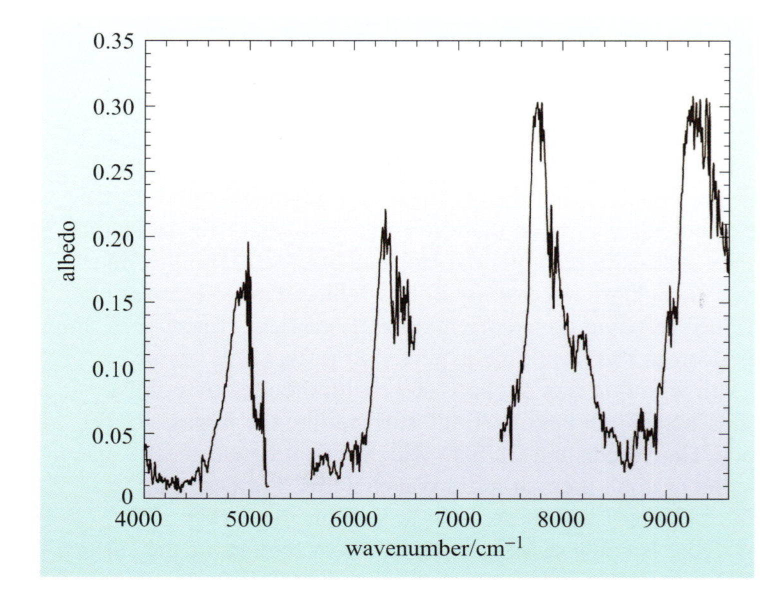

Figure 5.16 The variation of Titan's albedo with respect to wavenumber in the near-infrared part of the spectrum.

The troughs in Figure 5.16 have values close to zero (i.e. values less than 0.05). This means that at these wavelengths almost all the radiation incident on Titan is absorbed in the atmosphere. The peaks, which have values between 0.2 and 0.3, correspond to regions where a good proportion of the incident radiation is reflected back. So in these regions the atmosphere is at least partially transparent. These regions are termed *windows* because they enable us to 'see' through the atmosphere. This region of the spectrum has been investigated using ground-based telescopes and also by the Hubble Space Telescope (HST). The observations are generally in agreement and appear to show variations in the albedo during one axial rotation of Titan (i.e. over a period of 16 days). Significantly, these same variations have been observed over several different axial rotations.

- Why is the consistency of the variations from different rotations significant?

- It implies that the variations are genuinely due to surface property differences because if they were due to cloud or haze variations, they wouldn't be expected to be long-lived (i.e. over many Titan rotations). However, if they were due to real surface variations, the opposite would be the case.

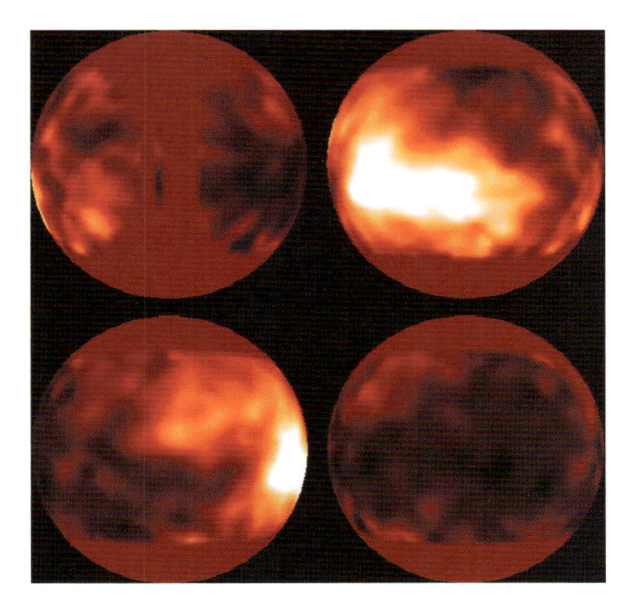

Figure 5.17 Four projections of Titan's surface (see text) determined from observations with the Hubble Space Telescope between 4 and 18 October 1994. The images were taken in one of the postulated atmospheric 'windows' in Titan's atmosphere in the near-infrared part of the spectrum. The top left image is of the area of Titan facing Saturn, and each subsequent image represents a region rotated by 90° from the previous one. The difference between the brightest and darkest regions represents only 10% of the total light collected but appears to be a real effect. (Peter H. Smith and NASA)

These observations have enabled the production of the first crude surface maps of Titan. Figure 5.17 shows four views of Titan from 14 images acquired by the HST in October 1994 in one of these near-infrared atmospheric windows. They represent views of different faces of Titan. The 'false colours' represent different amounts of near-infrared radiation reflected from different parts of the surface. There has been much speculation as to what physical reality they correspond to on the surface. We almost certainly won't know until the arrival of the Cassini–Huygens mission. One intriguing possibility is that the 'bright' regions represent a fairly clean icy surface, while the dominant 'dark' area corresponds to the hydrocarbon seas predicted by some.

5.5 Modelling Titan's interior

Without any direct contact with the surface of Titan, there has been no way of 'sounding' the interior of Titan. Indeed, there is only one strong constraint on the interior structure and composition of Titan.

■ Can you suggest what this constraint is? (*Hint*: look at Table 5.2.)

❑ The mean density (in this case $1.88 \times 10^3\,\mathrm{kg\,m^{-3}}$) constrains the materials that can make up Titan's interior.

From solid bodies whose structure we understand a lot better, we know that most such bodies are made predominantly of icy and rocky material. To get an overall mean density of $1.88 \times 10^3\,\mathrm{kg\,m^{-3}}$, we would expect a dominance of icy rather than rocky material.

■ If Titan is made up of such a mixture, what can you say about how this material would be distributed?

❑ We would expect the denser component to have differentiated to the centre (of perhaps a primitive molten body) with an overlaying layer of ice.

Any further progress, until we have more direct measurements from Titan, must rely on modelling of its structure and, in particular, on possible formation scenarios for Titan. There are several alternatives but the most favoured are a series of evolutionary models in which Titan formed inside a protoSaturn nebula, a flattened disc of gas, rich in methane and ammonia, and dust, encircling the protoplanet during its early stages of contraction. Titan would have been a hot object in the early stages after its formation, and more uniform in its composition. As it cooled, the heavy, rocky material would tend to fall to the centre, leaving the lighter materials, mostly water with an estimated 15% by weight of ammonia in solution, to form the outer core. Irradiation of the atmosphere and surface by solar ultraviolet radiation, cosmic rays and charged particles from Saturn's magnetosphere could have dissociated enough ammonia to form a thick nitrogen atmosphere, a precursor of the present situation.

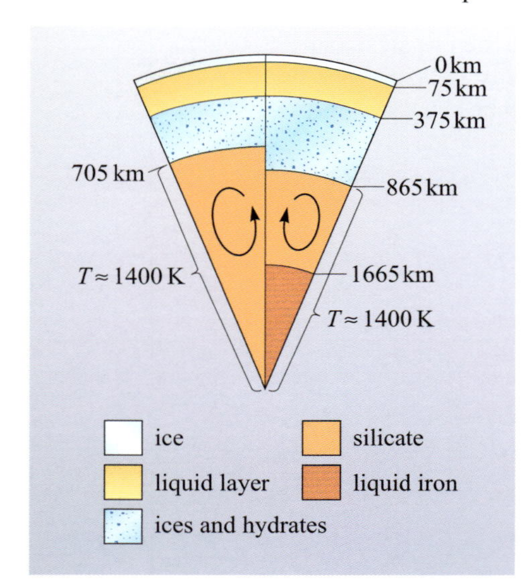

Figure 5.18 Model of a hypothesized Titan interior, with or without a liquid iron core. The location of a possible liquid layer is indicated; the arrows show convection in the icy layers. Distances shown are distances from the surface.

In more detailed interior models, the ice layers outside the rocky core have extents, locations and crystal structures that depend on the distribution of temperature and pressure. They also contain methane and ammonia clathrates, in which these molecules are trapped inside the water ice lattice, and perhaps other ices. The heavier core may itself be differentiated, with, for example, iron and sulfur compounds and other metals at the centre, overlain by the silicates and other rocky material. Figure 5.18 shows two of the possible interior models.

It seems possible, then, though highly speculative, that a liquid layer, perhaps as thick as 350 km, could exist some 75 km below the surface. The temperature in the liquid layer, if it is really present, is probably in the range 220 K to 250 K.

What of the composition of this subsurface ocean? Methane and nitrogen could be present. By analogy with Europa (see Chapter 4), it has been suggested that salty liquid water might be the main component.

- If the latter were the case, how might this be detectable by a fly-by spacecraft?

- Water, especially salty water, is electrically conducting. A conductor in motion in a magnetic field will produce a magnetic field of its own. Therefore, a conducting ocean passing through a giant planet's magnetic field will do just that. This can then be detected by a spacecraft-borne magnetometer, as has possibly occurred with the Galileo spacecraft at Europa (see Section 4.2).

Another possibility is that ammonia is the main constituent, while nitrogen–water, methane–water or ammonia–methane–water mixtures have also all been proposed. Unfortunately, the exact state of affairs is impossible to determine because of the lack of observational data. This is compounded by the lack of data on the properties of suitable high-pressure, low-temperature liquid mixtures. We have already seen how difficult it is to speculate concerning the nature of the possible subsurface liquid on Europa. But at least in the case of Europa, we have excellent and extensive data on the nature of its surface. By comparison, in the case of Titan, we know almost nothing.

QUESTION 5.5

Calculate (a) the closest and (b) the furthest possible distance from Earth to Saturn. How do you explain therefore that the Cassini spacecraft has covered a distance of about 3.2×10^9 km by the time it has reached Saturn?

QUESTION 5.6

Imagine that after the Huygens Probe arrives at Titan in 2005, the following data are obtained. In each case interpret these findings.

(a) The average abundance of argon in the atmosphere is found to be 0.05%.

(b) The Probe lands on a body of liquid in which the significant wave height is measured to be about 3 m.

(c) The strength of sunlight at the surface is found to be about one-thousandth of that at Earth's surface.

5.6 Summary of Chapter 5

- Titan is a planet-sized body with a thick, rich and complex atmosphere within which occurs a wide array of chemical reactions.
- Titan's surface is essentially unexplored by direct means but indirect evidence points to the possibility of extensive reservoirs of liquid methane and ethane, possibly even in the form of hydrocarbon seas or lakes.
- Knowledge of Titan's interior structure is very scant but there exists the possibility of a deep layer of liquid water below an ice crust. There is therefore a remote possibility of life but the lack of oxygen and low temperatures make this seem unlikely.
- Titan will be visited by the Cassini–Huygens space mission which aims to resolve many of the unknown observational features and thus give us an understanding of why Titan, alone among the planetary satellites, possesses a dense atmosphere.

CHAPTER 6
THE DETECTION OF EXOPLANETS

6.1 Introduction

In Chapters 1–5 we examined how life arose on Earth, and whether it might exist, or have existed, on other Solar System bodies. In Chapters 6–9 we will consider how common other planetary systems are and whether they might have life. We will also look at the methods that scientists use to search for such planets and signs of life, signs of intelligent civilizations, and how common intelligent life may be. Our only example of life to date – life on Earth – tells us that life can exist on a planet. So, to understand the search for life we must include the sciences of planets, life and intelligent life.

We will begin the search for life on exoplanets using the framework of the Drake equation and then turn our attention to methods of detecting **exoplanets** (also called extrasolar planets) – planets that are *not* part of our Solar System.

6.1.1 Planets, life and intelligence in the Universe

Is there life elsewhere in the Universe? To many people this is one of the most important questions that scientists and explorers can address. The answer will have profound religious and philosophical implications. Unfortunately, we're not in a position to give you an answer to this question yet, but we can pose, and then answer some related questions: what can science tell us now? How should we direct our future scientific endeavours?

Frank Drake (Figure 6.1), a pioneer in the field of extraterrestrial communication and searching, proposed Equation 6.1 (known as Drake's equation) for the number of civilizations, N, broadcasting their existence in our Galaxy:

$$N = R_b T \qquad (6.1)$$

where R_b is the rate (number per unit time) at which broadcasting civilizations (civilizations that broadcast in radio waves) appear in the Galaxy and T is the time over which they broadcast.

Figure 6.1 Frank Drake, a pioneer in the search for extraterrestrial intelligence and arguably the first human to make a serious attempt to communicate with extraterrestrial intelligence. (© Dr. Seth Shostak/ Science Photo Library)

Drake's equation holds true as long as R_b remains constant, and is approximately true as long as R_b does not vary too much on times comparable with T.

- What does our only example of an intelligent, broadcasting civilization (ourselves) tell us about the three variables in Equation 6.1?
- The most readily observable variable is N, which is at least one (us). Even for our single example, T is unknown – we have no idea how long we may broadcast our existence into space. All we can say is that T must be greater than the 50 years or so that we have been transmitting radio waves. Observations alone cannot tell us much about R_b, though we know it cannot be zero because N is not zero.

R_b can be broken down into specific factors:

$$R_b = R p_p n_E p_l p_i p_c \tag{6.2}$$

where

R = rate at which suitable stars are formed

p_p = probability of planets forming around a suitable star

n_E = average number of suitable planets in habitable zones per planetary system

p_l = probability of life appearing on a suitable planet in a habitable zone

p_i = probability of intelligence developing in life

p_c = probability of intelligent life broadcasting across space.

Equation 6.2 states that the rate of appearance of broadcasting civilizations is equal to the number of suitable stars forming per unit time multiplied by the average number of broadcasting civilizations per star. By 'suitable stars' we mean those that contain enough of the heavier elements to permit the formation of planets and life. The first generations of stars contained only the lightest elements (hydrogen, helium and lithium) but nuclear reactions in their cores manufactured the heavier elements for subsequent generations of stars. The chain of multiplied probabilities following R reflects the fact that a chain of events must occur for broadcasting life to evolve. When probabilities are multiplied together, the result is often much less than 1, even if the individual probabilities are not very small. For example, the probability of rolling a 6 on a die is 1/6. But the probability of rolling four sixes in a row is:

$$\frac{1}{6} \times \frac{1}{6} \times \frac{1}{6} \times \frac{1}{6} = \frac{1}{6^4} = 0.00077$$

It is quite natural for us to interpret a probability in terms of the number of expected events in a given time or, alternatively, the expected time taken for the event in question to be 'realized'. In rolling dice, you would probably have to wait through thousands of throws before seeing four sixes in a row. Likewise, if the chance of broadcasting life evolving in a stellar system is tiny, then we might need to search a great many systems over a long time to find it. We may be encouraged by the fact that there are about 300 billion stars and billions of years of evolution in our Galaxy.

Be wary of Drake's equation tempting you into numerology (this is to arithmetic what astrology is to astronomy). For example, it can be tempting, even subconsciously, to assign values to the factors in the Drake equation with the aim of getting a value for N that you like. Instead, concentrate on the task at hand: assessing science's ability to address the prospect of life elsewhere in the Universe. Science enables us to estimate the variables on the right-hand side of the equation, but with decreasing confidence as you read them from left to right.

■ Given that our Galaxy comprises in the order of 300 billion stars and is approximately 10 billion years old, calculate an order of magnitude estimate (i.e. a rough estimate that is correct to within a factor of 10) for the average rate of star formation R to date. (Assume that all stars are suitable for this calculation.)

❑ Given that all the stars currently in our Galaxy must have formed over the last 10 billion years or so, the average rate of star formation, R = 300 billion stars/10 billion years = 30 stars per year.

The rate of star formation, R, is known to have varied with time, probably reaching a maximum when the Universe was about 4 Ga old. That said, the rate of 30 stars per year is still a valid *average* estimate. For various reasons, the rest of astronomy gives us very good reason to believe that our estimates of both the number of stars in our Galaxy and its age are correct to within 10%. We can therefore be quite confident with the above value of R, certainly as an order of magnitude estimate, averaged over the lifetime of our Galaxy.

Astronomers currently estimate that the Universe is 13–14 Ga old.

Let us now briefly discuss the remaining factors of the Drake equation. If we were writing this course in 1990, we would have said that the quantitative reach of science, certainly observational astronomy, stopped at estimating R. Now, at the time of writing (early 2003), science places real constraints on the value of p_p, the probability of planets forming around a suitable star. For example, the probability of finding Jupiter-like planets orbiting sun-like stars is now known to be less than 1/3. Plans are afoot for space observatories that can begin to measure n_E, the number of 'suitable' (e.g. Earth-like) planets in habitable zones. Worthwhile theoretical studies are being carried out on this, using the latest computers to evolve planets through millions of orbits in hypothetical stellar systems. Astrobiologists are thinking seriously about p_l, though the science is young, and is far from obtaining a reliable estimate of its value. However, the search for life in our own Solar System could well yield essential clues. Lastly, but significantly, there is the issue of intelligence. Currently, we have trouble defining exactly what we mean by intelligence in scientific terms, suggesting that the final two factors in Drake's equation (p_i and p_c) cannot be predicted from any mathematical, scientific theory. Also, a key difference between the search for a planet and the search for intelligence is that intelligent life can *choose* not to signal its presence. This brings psychology and sociology into the definition of p_c. So, perhaps the only way to estimate p_i and p_c is to look for observational signs of intelligence, e.g. radio broadcasts. The Search for Extraterrestrial Intelligence (SETI) is science at its earliest stage – pure observation.

Drake's equation (and Frank Drake has said this himself) may not have the fundamental significance, objectivity or even simplicity, of, say, Newton's laws, but it does provide a framework for science to proceed. Specifically, it neatly summarizes the essence of what has been described using many words in the preceding paragraphs. In your lifetime several of its key variables have come into the observational sights of modern astronomy, namely R, p_p and soon n_E. It is therefore an exciting time to consider the science involved, and to allow yourself to speculate, carefully, about what can be discovered in the coming decades. Such speculation is exactly what professional scientists have to do in planning and funding their often expensive, cutting-edge research programmes decades into the future.

6.1.2 Detecting exoplanets

In general, any type of matter can be detected if it affects passing electromagnetic radiation or other matter nearby.

More specifically, an object can be detected if it:

- reflects radiation, e.g. the Moon
- emits radiation, e.g. the Sun
- absorbs or occults (blocks) radiation, e.g. absorption lines in atmospheres or solar eclipses
- refracts radiation, e.g. a gravitational lens
- affects the motion of nearby matter, e.g. bodies orbiting one another.

All of these methods have been applied to planet hunting, so we will organize our discussion of planet detection in sections corresponding to the above categories. Before proceeding please read Box 6.1, which introduces the mathematical notation we will adopt for Chapters 6 to 9.

BOX 6.1 MATHEMATICAL NOTATION FOR EXOPLANETS

For properties such as mass or radius that are specific to a particular object (star or planet) we use an upper-case letter. For quantities such as distance or the radius of an orbit we use lower-case letters. Do not worry if some of the terms below are not clear at present, as they will be defined in more detail in the subsequent text.

- R radius of a star or planet
- M mass of a star or planet
- L luminosity of (energy radiated per unit time by) a star or planet
- A albedo
- S surface area
- b_{AB} brightness (or flux) of A at B
- a semimajor axis of the orbit of a star or planet (equal to the radius for a circular orbit)
- d_{AB} distance between A and B
- β angular radius or distance
- i_0 inclination of an orbit relative to the line of sight
- P period of an orbit
- t timescale for gravitational lensing
- v_r, $(v_r)_{MAX}$ radial velocity and its maximum value.

Usually, the subscript of a variable will tell you which object is being referred to, though if we are quoting a general rule then the variables may appear without subscripts. The subscripts used in this text are:

$\odot$	the Sun	J	Jupiter
α	alpha Centauri system	E	Earth
$*$	a star	V	Venus
		p	a planet

6.2 Reflected radiation

To our eyes, the Moon, Venus and Jupiter appear to shine more brightly than any of the stars in the night sky. This is because we are seeing them by reflected sunlight, and their proximity makes them appear so much brighter than any of the stars, which are of course intrinsically much brighter. Even during daylight, when planetary bodies face stiff competition from sunlight scattered in the atmosphere, it is easy to see the Moon on a cloudless day, and possible to see the brighter planets.

If this is possible then perhaps, with modern astronomical equipment, it is possible to see exoplanets by their reflected light despite the overwhelming glare from their stars. To do so, such equipment must be able to:

- collect enough light to detect the planet in an image
- resolve the star and its planet as two separate objects in the image.

Both these factors require large collecting apertures (telescope mirrors). Long exposure times (sometimes called integration times) are needed to collect enough light to detect the planet. From ground-based telescopes, good optics and stable atmospheric effects (known to astronomers as good **seeing**) are required to resolve the star and planet as two separate objects.

We can begin to assess the proposition of detecting extrasolar planets directly from their reflected radiation by calculating their **brightness**. Brightness, also referred to as **flux**, and sometimes **flux density**, can be defined as: the energy carried by radiation across an area of 1 m^2 placed at right angles to its path. The brightness of the Sun at the Earth (above its atmosphere) is about 1360 W m^{-2}, i.e. the radiation delivers 1360 J of energy per second to a 1 m^2 area facing the Sun. From this we can work out how much energy the Sun must be emitting per second, which astronomers call the luminosity. For present purposes we can regard luminosity as the total radiation emitted (or, for planets, scattered) in all directions whether the emission is isotropic (equal in all directions) or not. Since isotropic radiation spreads out into space equally in all directions, at any distance, d, from the Sun, the radiation it emitted at some time in the past (d/c, where c is the speed of light) has been spread across the surface of a sphere of radius d. This is depicted in Figure 6.2. If you denote the luminosity by L, and brightness at distance d by b, and the brightness is not changing with time (so the instant of emission doesn't matter) then:

$$b = \frac{L}{4\pi d^2} \tag{6.3}$$

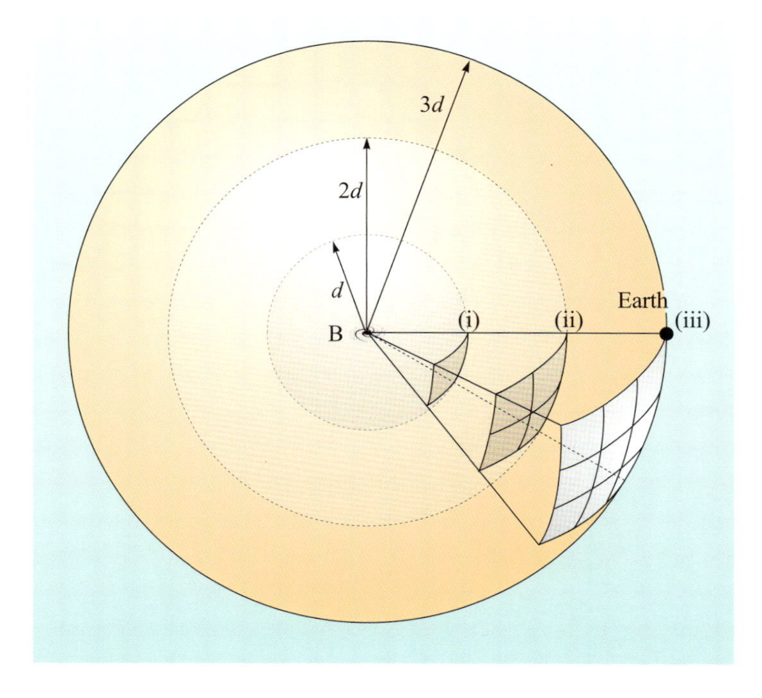

Figure 6.2 Radiation from a star is spread out across a sphere of radius d. We can use this figure to understand why the energy falling per unit area on a surface facing the star at distance d, known as the brightness, is $b = L/4\pi d^2$. The inverse square law can be interpreted as follows: (i) at radius d, a given amount of radiated energy crosses one unit of area; (ii) at radius $2d$ the same amount of energy is spread across 2^2, which is 4 units of the area; (iii) at radius $3d$ the same amount is spread across 3^2, which is 9 units of area.

This equation states that the brightness equals the emitted rate of energy, the luminosity L, divided by the surface area of a sphere, $4\pi d^2$. The fact that brightness decreases with the square of distance from the source means that this relationship is called an inverse square law. The brightness of radiation b from sources that don't emit equally in all directions will not obey Equation 6.3, but it turns out that the decrease of brightness in a particular direction for such sources will still obey an inverse square law. That is, they obey an equation like Equation 6.3 where $b \propto 1/d^2$, but with a different constant of proportionality.

The luminosity of the Sun, $L_\odot$, is 3.84×10^{26} W and it is at a distance of 1 AU from the Earth. 1 AU = 1.50×10^{11} m. With this information we can calculate the brightness of the Sun at the Earth.

a_E and $b_{\odot E}$ are explained in Box 6.1.

$L_\odot = 3.84 \times 10^{26}$ W and $a_E = 1.50 \times 10^{11}$ m, so:

$$b_{\odot E} = \frac{L_\odot}{4\pi (a_E)^2} = \frac{3.84 \times 10^{26}\ \text{W}}{4\pi(1.50 \times 10^{11}\ \text{m})^2} = 1.36 \times 10^3\ \text{W m}^{-2}$$

Light-year is abbreviated to ly.

■ Calculate the Sun's brightness at alpha Centauri, 4.3 light-years distant (1 light year is the distance travelled by light in one year at speed $c = 3.00 \times 10^8\ \text{m s}^{-1}$).

❑ The distance to alpha Centauri ($d_{\odot\alpha}$) can be calculated as:

$$d_{\odot\alpha} = 4.3 \times 3.00 \times 10^8\ \text{m s}^{-1} \times 365 \times 24 \times 3600\ \text{s} = 4.07 \times 10^{16}\ \text{m}$$

Replacing a_E with $d_{\odot\alpha}$ in the above equation for $b_{\odot E}$ we find $b_{\odot\alpha} = 1.84 \times 10^{-8}\ \text{W m}^{-2}$. Another way to do this is to multiply the brightness of the Sun from the Earth $b_{\odot E}$ by $(a_E / d_{\odot\alpha})^2$.

We will now calculate Jupiter's luminosity. The first step is the one we have just practiced on the Earth and alpha Centauri: calculating the brightness of the Sun at Jupiter. Jupiter's distance from the Sun, a_J, is 5.2 AU. The brightness of the Sun at Jupiter is therefore a factor of $(5.2\ \text{AU}/1\ \text{AU})^2 = 27.0$ times less than at the Earth, i.e. $b_{\odot J} = 50.4\ \text{W m}^{-2}$. Jupiter's radius, R_J, is approximately 70 000 km (actually it's slightly bigger across the equator than it is pole to pole, but we'll assume it's spherical for present purposes), so it presents an area, $S_J = \pi(7 \times 10^7\ \text{m})^2 = 1.54 \times 10^{16}\ \text{m}^2$ to the incoming solar radiation, as shown in Figure 6.3. This means that the total amount of solar radiation energy falling on Jupiter per unit time is $b_{\odot J} S_J$. Not all of this radiation is reflected back into space (some is absorbed).

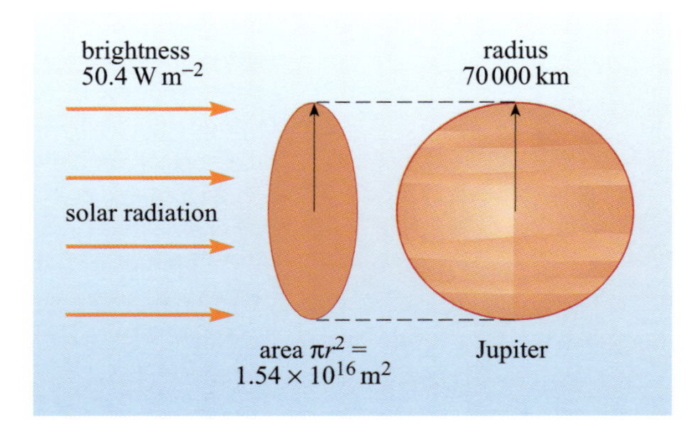

Figure 6.3 The rate of solar radiation intercepted by Jupiter can be calculated by imagining the rate of energy flow across a disc of the same radius placed just in front of the planet. The rate of intercepted radiation is simply the brightness, or flux, multiplied by the area of the disc.

The fraction reflected back into space is called the albedo, which we'll call A_J, and is known to be about 0.7. So, we can write down an expression for the luminosity of *reflected* radiation from Jupiter:

$$L_J = A_J b_{\odot J} S_J \qquad (6.4)$$

Remember that, unlike the Sun, only half of Jupiter is lit at any given time, but nonetheless, we can still meaningfully call L_J the luminosity – the rate at which reflected radiative energy is leaving the surface.

However, the fact that only one half of Jupiter is lit and so reflecting radiation does make a difference to how we calculate the brightness. Now, instead of the luminosity being spread over an entire sphere of area $4\pi d^2$, it is spread over a hemisphere that has area $2\pi d^2$. So, the equivalent of Equation 6.3 for a planet, which will always be half lit, is:

$$b = \frac{L}{2\pi d^2} \qquad (6.5)$$

QUESTION 6.1

Calculate Jupiter's maximum possible brightness, b_{JE}, as viewed from the Earth, neglecting any absorption effects of the atmosphere. Make sure to state the physical conditions for maximum brightness. Comment on Jupiter's prominence in the night sky by comparing your brightness value for Jupiter with the value for alpha Centauri, which at a magnitude (see Box 6.2) of (almost) 0 is one of the brightest stars in the night sky.

BOX 6.2 ASTRONOMICAL MAGNITUDE

In ancient times, the brightest stars visible to the naked eye were placed in one of six classes: from the handful of bright magnitude 1 stars to the thousands of barely visible magnitude 6 stars. Modern astronomy kept this system and defined it in relation to brightness. The magnitude, m, of a star of brightness, b, (in SI units of W m^{-2}) is:

$$m = -2.5 \log_{10} \frac{b}{b_0}$$

where $b_0 = 2.29 \times 10^{-8}\ \text{W m}^{-2}$ and defined as the brightness corresponding to a magnitude of zero. The magnitude system is designed so that a *difference* of 5 magnitudes is equal to a brightness change of a *factor* of 100. So, a star of magnitude 1 is in fact one hundred times brighter than a star of magnitude 6.

Modern telescopes can reach down to about magnitude 30.

A negative magnitude means that the brightness is greater than the reference brightness, b_0. In practice, a negative magnitude means that the object is very bright; only the Sun, Moon and a few planets and stars have negative magnitudes.

The '$\log_{10}$' function simply tells you what power of ten a number is, for example $\log_{10} 1000 = 3$ because $1000 = 10^3$. If the number is not an integer power of ten, then the same applies, but you will need a calculator to work it out. Make sure you do not use the 'ln' function, which is another kind of log. For example, $\log_{10} 1360 = 3.13$ because $1360 = 10^{3.13}$.

So, the separation of star and planet, β_J, is larger than the telescope's resolution by a factor of about 300. At face value, it appears that modern telescopes have sufficient resolution to do the job of detecting a Jupiter orbiting a nearby star. Even accounting for seeing, a resolution of 4″ (arcseconds) is well within the reach of professional ground-based observatories. However, you must also acknowledge the fact that even a large planet like Jupiter will be many times dimmer than a Sun-like star that it orbits. This is the problem. It is easy to see a firefly in an otherwise dark night, but could you see one next to car headlights? In fact, the formula for spatial resolution, Equation 6.7, is specifically for two sources of *similar* brightness. The star and planet are of very different brightness, and so it turns out that they cannot be resolved with modern telescopes.

- Calculate the ratio of the Sun's brightness to Jupiter's brightness as they would be measured from another star, i.e. at a distance much greater than Jupiter's orbital radius.

- From such a distance we can regard both Jupiter and the Sun as being, to a good approximation, at the same distance from the star. Therefore the ratio of their brightnesses will simply be equal to that of their luminosities (see Equation 6.3). $L_{\odot} = 3.84 \times 10^{26}$ W and from the answer to Question 6.1 you know that $L_J = 5.42 \times 10^{17}$ W, giving us the ratio $L_{\odot}/L_J = 7.08 \times 10^8$, which is of the order of a billion to one.

This large ratio is a problem for resolution because of unavoidable optical effects and seeing that cause a star to appear as a fuzzy disc rather than a point on any astronomical image. The radius of this disc will be much larger than the actual disc of the star in the image (which is beyond the resolution of even today's conventional telescopes). Figure 6.5 shows an image of a bright star in a field of dimmer stars. The relative sizes of the stars' discs depend only on the brightness, with brighter stars having larger discs. The problem for planet hunting is that the dim planet will be lost in the outskirts of the bright star's extended image. The solution to this problem is to reduce the enormous star to planet brightness ratio – one method of doing this is discussed in the next section.

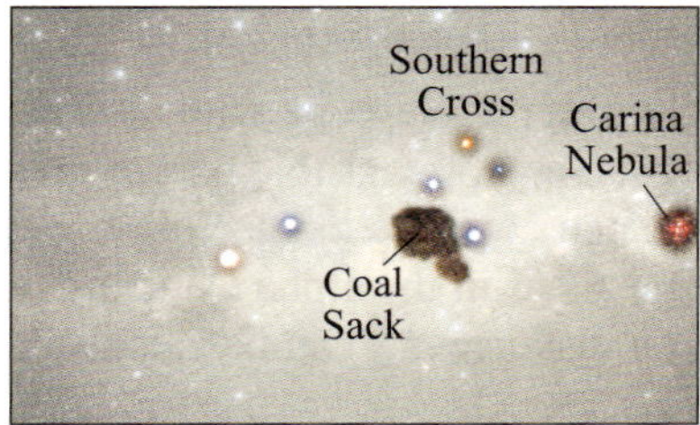

Figure 6.5 An image showing many bright and faint stars in the south polar region of the night sky – the Southern Cross can be seen to the right of centre. A bright star in this image has an apparent size that is much larger than that of the star's actual size. Fainter stars or planets close to it will be lost in its glare. (Steve Mandel/Galaxy Images)

Despite the great technical challenges involved, there are serious investigations being undertaken of direct space imaging in the visible part of the spectrum with space telescopes equipped with 4 m to 10 m mirrors (Hubble's mirror is 2.4 m in diameter). These mirrors require a great surface smoothness and must work in a telescope equipped with devices known as coronagraphs, which physically block out the light of the star but not that of the planet.

Efforts are also being made to disentangle the planet light from the star light without actually spatially resolving the two. From the ground this is being attempted by searching in the spectrum of the star for the very much fainter, but moving, spectrum of the reflected light. With a certain prospect for success, the space missions, starting with MOST (Canadian) in 2003, and then COROT (French), Kepler (NASA) and Eddington (ESA), aim to see the rise and fall of the reflected light from close-in Jupiters superimposed on the constant star light.

6.3 Emitted radiation

Every object in the Universe emits some form of radiation. For most objects, including stars, the luminosity and overall spectrum of the emitted radiation is almost entirely determined by the temperature. For example, the surface of the Sun is at a temperature of 5770 K, and so it emits radiation in the visible part of the spectrum. Planets, such as the Earth, at a typical temperature of no more than a few hundred K, emit radiation most strongly in the infrared. You cannot see this radiation, but you can feel it warming your hand if you place it near a (dark-coloured) surface that has been warmed by direct exposure to sunlight. Figure 6.6 compares the spectrum of the Sun with that of the Earth. It is clear that the Sun–Earth brightness ratio must be much less in the infrared band, where the Earth's spectrum peaks, as compared to the visible band, where the Sun's spectrum peaks. It turns out that the billion to one ratio in visible wavelengths that we calculated for a Jupiter orbiting a nearby star in Section 6.2 can be brought down to a million to one or better in the infrared.

- There is a drawback in working at the longer wavelengths of infrared, over working in the visible. What do you think it is?

- Box 6.2 tells us that the resolution of a telescope depends on wavelength. In fact the minimum separation angle that can be resolved, α, is proportional to wavelength. This means that the resolution of the same telescope will be poorer – that is the minimum separated angle will be larger – in the infrared than in the visible.

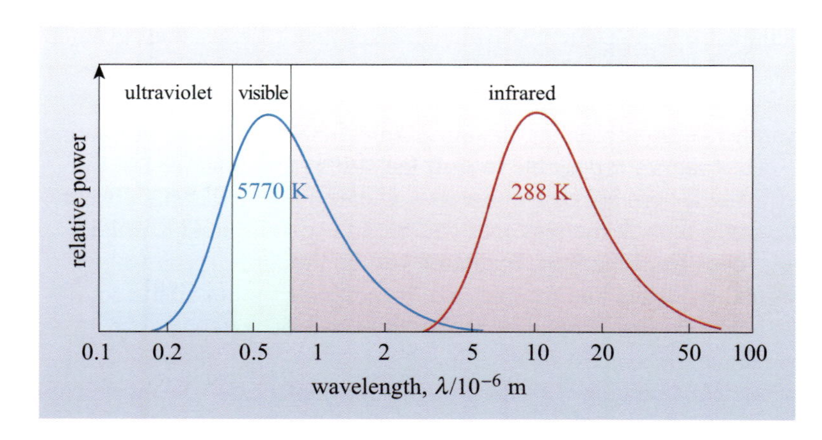

Figure 6.6 A schematic of the Sun's spectrum and the Earth's spectrum. They peak in the visible and infrared, respectively, because they are at different temperatures. The spectra have been scaled to have the same peak value. (Real spectra contain features due to spectral lines of gases which are not shown here.)

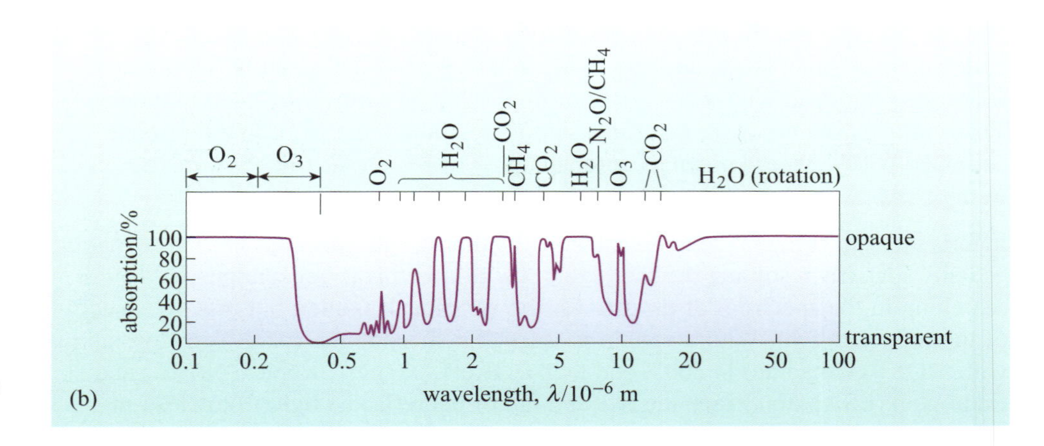

Figure 6.7 The absorption of the Earth's atmosphere.

In addition to resolution being intrinsically poorer in the infrared, the Earth's atmosphere strongly absorbs at many infrared wavelengths, as shown in Figure 6.7. As indicated in the figure, this is due to the presence of many absorption lines due to certain gases, notably carbon dioxide (CO_2), water (H_2O) and ozone (O_3). This is a particular problem for ground-based observations because it precludes the detection of these same gases in Earth-like planets (see Chapter 8), which would give us vital clues to the existence of life.

Probably the best way to see planets directly in their own infrared emitted light is to take large telescopes into space. This is the conclusion that ESA and NASA have arrived at in planning missions to make such observations in the next decade or two. NASA's mission is called the Terrestrial Planet Finder (TPF), and ESA's is called Darwin. Both missions are very similar in concept, and will most likely merge to become a joint ESA–NASA effort. For this reason, and because their details will probably change between now and launch, we will just describe the common concept behind them.

Figure 6.8 Artist's conception of ESA's Darwin mission. (ESA)

The missions will work in the infrared and consist of several separate spacecraft, each carrying a large mirror (larger than 1 m in diameter), plus one or more spacecraft to collect and relay the data back to the Earth (Figure 6.8). The distances between these spacecraft are planned to be between 100 m and 1000 m, and must be maintained to an accuracy of less than a centimetre. This configuration will allow the whole ensemble of spacecraft to act as an **interferometer**. The method of interferometry was developed in radio astronomy (where wavelengths are much longer than infrared wavelengths) to combine several radio telescopes that had low resolution to form one large, effectively high-resolution telescope. Another reason for using the interferometer set-up in planet hunting, is that it makes it possible to **null** the star's radiation. We use the term 'null' here because an interferometer does not simply obscure the star's light in the way that you might use your hand to shield your eyes from the Sun. Radiation can be thought of as a wave, formed by crests and troughs. So, if radiation from a single source – the star – can be brought together from two of the spacecraft, with one signal delayed in time with respect to the other, then the crests of one can coincide with the troughs of another and so cancel each other out. In this way, it is possible to null the starlight and leave only the signal from orbiting planets. Once this is done it will be possible to observe spectra of Earth-like planets orbiting the nearest few hundred stars. Of prime interest are the carbon dioxide, water vapour, methane and ozone absorption lines in the planets' atmospheres, as they give vital clues of any extant life.

A project is currently being considered that could contribute greatly to the study of exoplanets that involves OWL – the OverWhelmingly Large telescope. OWL is a ground-based telescope with a segmented 100 m mirror that could well herald revolutions in all branches of astronomy. With its tiny diffraction limit and its advanced optics to negate seeing problems, OWL could directly observe a Jupiter or a Saturn orbiting a nearby Sun-like star. It will also be able to perform spectroscopic studies of planetary atmospheres.

6.4 Absorbed or occulted radiation

In astronomy, the term 'absorption' refers to radiation having its energy absorbed by some matter. Strong absorption at certain wavelengths allows astronomers to detect the presence of gases in interstellar space, or in atmospheres of stars and planets. As mentioned in Section 6.3, once a planet has been discovered, these absorption lines can give us important information on what gases are present. However, this does not offer an effective means of planet hunting because exposure times of several days may be required to observe absorption lines of a star's atmosphere.

Occultation, at least as used in astronomy, has nothing to do with magic or witchcraft. It simply means that one object blocks out some light from another object. A solar eclipse is an example of an occultation, as are planetary **transits** in front of the Sun (Figure 6.9).

The passage of one object across another of larger apparent diameter, e.g. Mercury or Venus in front of the Sun, is known as a transit.

Let us now consider the drop in solar radiation incident on the Earth that would result if Venus passed directly between the Earth and the Sun. We will work this out mathematically, and only insert figures at the end. This will allow us to re-apply the theory to planets around other stars. The simplest way to proceed is to work out the

Figure 6.9 (a) A solar eclipse, where the Moon covers the bright solar disc. (b) A transit of Mercury in front of the Sun. ((a) NASA; (b) Jim Ferreira)

fraction of the solar disc that Venus obscures, as this will equal the fraction by which the brightness is reduced (assuming the Sun is uniformly bright over its disc which is not strictly true, but an acceptable approximation for present purposes). Firstly, we can calculate the angular radius of Venus, β_V, as $\beta_V = R_V/d_{EV}$, where R_V is the radius of Venus and d_{EV} is the Earth–Venus distance. Similarly, the angular size of the Sun, $\beta_\odot = R_\odot / a_E$. In the same way we talk of angular radius, we can define an angular area on the sky, and this area is proportional to the angular radius squared (you don't need to know the constant of proportionality). This means that Venus obscures a fraction f_V of the area of the solar disc:

$$f_V = \left(\frac{\beta_V}{\beta_\odot}\right)^2 = \left(\frac{R_V / d_{EV}}{R_\odot / a_E}\right)^2 \tag{6.8}$$

■ Assuming that the solar disc is uniformly bright, use Equation 6.8 to estimate the fraction of solar radiation that is occulted when Venus transits the solar disc. Venus's radius and orbital radius are $R_V = 6200$ km and $a_V = 0.72$ AU, and the Sun's radius, $R_\odot = 6.96 \times 10^5$ km.

❑ The Earth–Venus distance, $d_{EV} = a_E - a_V$ when Venus is directly between the Sun and the Earth. Rearranging Equation 6.8 and inserting the values for the radii we find:

$$f_V = \left(\frac{R_V / d_{EV}}{R_\odot / a_E}\right)^2 = \left(\frac{R_V}{R_\odot}\right)^2 \left(\frac{a_E}{a_E - a_V}\right)^2$$

$$= \left(\frac{6.2 \times 10^6 \text{ m}}{6.96 \times 10^8 \text{ m}}\right)^2 \left(\frac{1\,\text{AU}}{0.28\,\text{AU}}\right)^2 = 1.01 \times 10^{-3}$$

So an occultation by Venus would produce a 0.1% ($1.01 \times 10^{-3} \approx 0.1\%$) drop in the apparent brightness of the Sun. Such a small drop may not be noticeable with our eyes, but it is easily detectable with the right equipment. This formula can be extended to exoplanets by replacing the 'Sun' with the star and 'Venus' with the extrasolar planet. Notice that for an extrasolar planet the distance between it and the Earth, d_{Ep}, will virtually equal the distance between the star and the Earth a_E because the Earth–planet distance will be so much greater than the planet–star distance.

The fraction, f_p, of a star's light that is blocked out by an extrasolar planet of radius R_p is therefore:

$$f_p = \left(\frac{R_p}{R_*}\right)^2 \tag{6.9}$$

In actual fact this formula is not exact because the Sun's (or indeed any star's) disc is not uniformly bright, but brighter in the centre than at the limb (the edge of the disc). However, our formula will give an estimate of the fractional decrease that occurs when the planet is some way between the centre and the limb of the star's disc.

For a Jupiter-sized planet around a Sun-like star, we should expect a fraction of $(7 \times 10^7\,\text{m}/6.96 \times 10^8\,\text{m})^2 = 0.01$ of the star's light to be blocked out. Notice that this fraction is independent of the distance to the planetary system.

7×10^7 m is the radius of Jupiter and 6.96×10^8 m is the radius of the Sun.

To date (early 2003), only one planet (OGLE-TR-56) has been discovered by occultation, and one planet has been confirmed by this method. This is in spite of many searches with ground-based telescopes and a determined attempt to look for planet transits in a star cluster (called 47 Tucanae) by the Hubble Space Telescope (HST). Figure 6.10 shows the **light-curve** (an astronomer's term for a plot showing how brightness varies with time) of the planet whose existence was confirmed by occultation, around the star HD 209458 (HD stands for Henry Draper, the creator of a widely used star catalogue). The size of the dip allows an estimate of the radius of the planet. If the mass of the planet can be estimated from some other method (see Section 6.6) then knowledge of radius from the occultation method allows us to estimate the average density of the planet.

OGLE refers to the optical gravitational lensing experiment that discovered the planet.

During an occultation, a planet's atmosphere will also absorb some radiation. This will have a negligible effect on the observed dip in radiation measured over all wavelengths, but absorption lines from gases in the atmosphere may be observable. This has been seen in an HST spectrum of HD 209458 where absorption lines of sulfur were present.

■ From Figure 6.10, estimate the radius of the planet as a fraction of the star's radius.

❑ The dip is about 1.6%, i.e. $f_p = 0.016$, so the radius of the planet divided by the radius of the star will be given by Equation 6.9 as:

$$\left(\frac{R_\text{p}}{R_*}\right) = \sqrt{f_\text{p}} = \sqrt{0.016} = 0.13$$

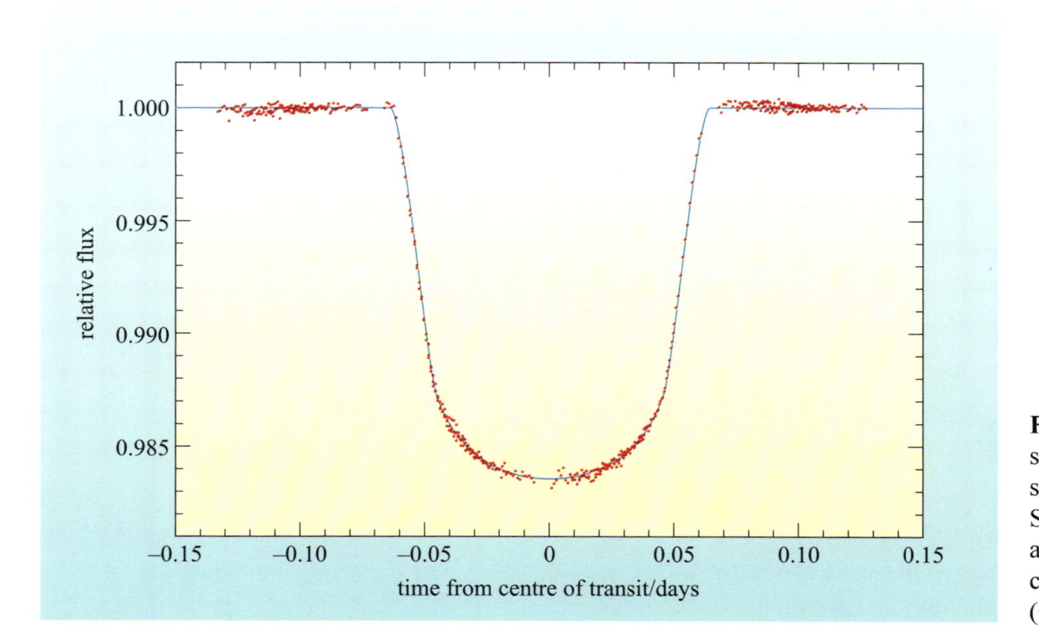

Figure 6.10 Light-curves showing a planetary transit for star HD 209458 from the Hubble Space Telescope. The red points are observed data, and the blue curve is a theoretical fit to the data. (Observatoire de Paris)

Of course, the occultation method (also known as the transit method) can only work if the planet's orbit is such that it takes the planet between the star and the Earth. Assuming that planetary orbits are oriented randomly about stars, this equates to about a tenth of stars with short period planets like HD 209458, and less for longer period planets. Future space missions, mentioned in Section 6.2, i.e. MOST, COROT, Kepler, and Eddington, will all be able to detect dips due to occultation of stars by planets.

6.5 Refracted radiation

Refraction simply refers to the bending of light. Inside some telescopes, and other optical devices, this is done using glass or plastic lenses. On astronomical scales, the path of radiation can be noticeably altered when it passes near a mass such as a star or even a planet. This effect is commonly called **gravitational lensing**, because in some senses, a mass's gravity is refracting the light (actually, Newton's law of gravity isn't appropriate, instead Einstein's general relativity is required – it is more correct to say that light is traversing a curved region of space-time). Just like normal lenses, gravitational lenses can distort the appearance of what is behind them, and amplify brightness by a focusing effect. When the distorted image due to lensing cannot be resolved, but the presence of a lens is known due to its brightness amplification, astronomers call the phenomenon **micro-lensing**.

Figure 6.11 illustrates some gravitational lensing geometries. In Figure 6.11a, where there is no lens, the radiation from a star travels in a straight line. In Figure 6.11b, the nearer star (the lens) is directly between the Earth and a far away star (the source). In the lensed case, the radiation does not travel directly towards us, but follows a curved path, having its path altered in the vicinity of the lens. It turns out that there are two routes that the radiation can take around either side of the star. This means that we will see an image of the source on either side. The case where the observer, lens and source all lie in a line is shown in Figure 6.11c. However, to appreciate the symmetry of this case properly we must imagine it in three dimensions, as shown in Figure 6.11d. For a source, lens and observer all in perfect alignment, the image on the sky is that of a circle, often called the **Einstein ring**.

Whether the alignment is perfect or not, gravitational lensing effectively focuses more rays onto the observer. In the case of micro-lensing, where no image can be resolved, it will appear as if the background star undergoes a temporary brightening. Even in non-aligned cases of micro-lensing, the concept of the Einstein ring is still important, as it defines an important spatial scale to the problem. For any gravitational lens of given mass, an Einstein ring around it can be imagined on the sky (as if there were a source directly behind it). If the lens's motion against background stars brings any background stars close to, or inside, its Einstein ring, then significant amplification can result. The angular radius (in radians) of this ring on the sky, commonly called the angular **Einstein radius**, for a lens of mass M at distance d lensing a background object at distance d_b is given by:

$$\beta_e = \sqrt{\frac{2R_s}{d}\left(\frac{d_b - d}{d_b}\right)} \tag{6.10}$$

where the **Schwarzchild radius**, $R_s = 2GM/c^2$. The constants are Newton's gravitational constant, G, which is $6.672 \times 10^{-11}\,\text{N m}^2\,\text{kg}^{-2}$ and the speed of light, $c = 3.00 \times 10^8\,\text{m s}^{-1}$. The important fact to remember from Equation 6.10 is that the

Einstein radius is proportional to the square root of the mass of the lens. For a black hole, the Schwarzchild radius gives the radius of the event horizon, hinting that Einstein's general relativity lies behind the derivation of Equation 6.10. Also note that for a very distant background object ($d_b \gg d$) Equation 6.10 becomes:

$$\beta_e = \sqrt{\frac{2R_s}{d}}$$

so that in this particular case the Einstein radius only depends on the mass of, and distance to, the lens and not any quantity relating to the background object.

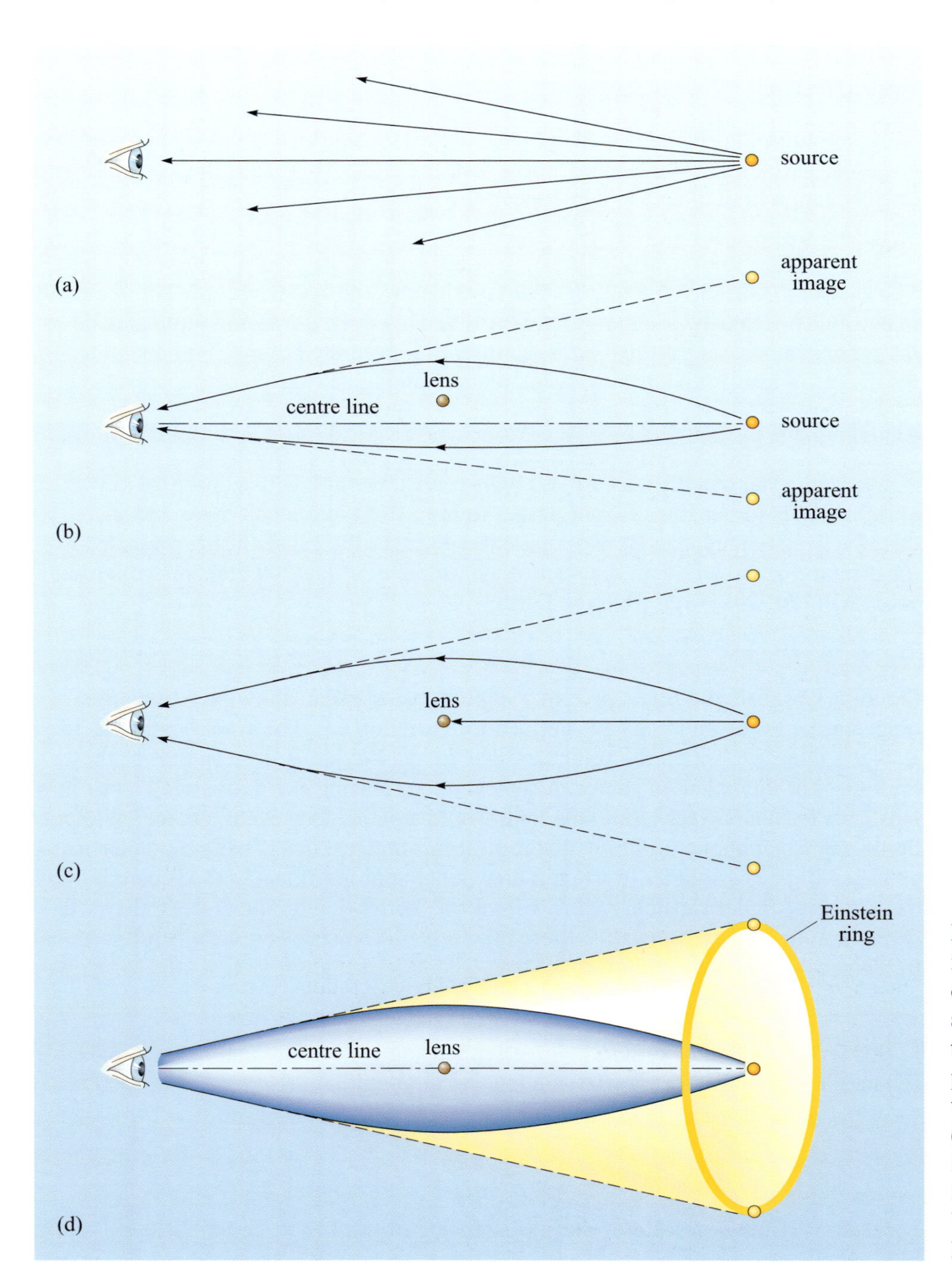

Figure 6.11 (a) Path of radiation in the absence of a lens. (b) In two dimensions light can be bent around either side of the lens. In this case the alignment is not perfect (the common case) so there will be two images of the source. (c) Here we see the symmetric case in two dimensions. (d) The case as in (c) but now in three dimensions. We can see that there is an Einstein ring.

■ What is the Schwarzchild radius of the Sun ($M_\odot = 1.99 \times 10^{30}$ kg) and how does it compare with the Sun's radius?

□ Inserting values we find

$$R_s = \frac{2GM_\odot}{c^2} = \frac{2 \times 6.672 \times 10^{-11}\,\text{N m}^2\,\text{kg}^{-2} \times 1.99 \times 10^{30}\,\text{kg}}{(3.00 \times 10^8\,\text{m s}^{-1})^2} = 2950\,\text{m}$$

The Schwarzchild radius is very small compared to the Sun's radius, $R_\odot$, which is = 6.96×10^8 m.

Question 6.3 represents an extreme (and unlikely case) of directly imaging an Einstein ring around our nearest star. Objects at greater distances will have smaller angular Einstein radii so will be more difficult to resolve.

QUESTION 6.3

What would be the angular Einstein radius of the star alpha Centauri A (approximately one solar mass) if it were lensing a very distant object? How does this compare with the diffraction limit of an 8 m telescope in the visible?

The answer to Question 6.3 tells us that, in principle, the best telescopes could directly image an Einstein ring around our nearest star. However, the seeing (atmospheric distortion) imposes a practical limit on observing from the ground. Without resolvable images of the lensing event, planet hunting has to rely on micro-lensing.

QUESTION 6.4

A solar mass star at a distance of 100 light-years travelling with a speed $20\,\text{km s}^{-1}$ at right angles to our line of sight passes right in front of a much more distant, effectively stationary star. By working out the size of the Einstein ring projected to the distance of the lens star, obtain a rough estimate of how long this micro-lensing event will last, as viewed from the Earth.

Figure 6.12 shows the light curve of a micro-lensing event, showing a timescale quite similar to the one you have derived for Question 6.4. The amplification factor, μ, is the factor by which the brightness has increased; $\mu = 1$ corresponds to the brightness of the unlensed source. A typical micro-lensing event involving a star as a lens lasts for many weeks, but how long would a planet lens event last for? Without doing any calculations we know that it must be shorter. Planets have a smaller mass, so they will have a smaller Einstein radius, it being proportional to the square root of mass. Their smaller Einstein radius will take a shorter time to traverse, so they will cause more rapid lensing events. More precisely, the lensing timescale is proportional to the square root of the mass of the lens.

So, for a star–planet system involved in lensing we can say that a planet of mass M_p will have a characteristic lensing timescale of:

$$t_p = t_* \sqrt{\frac{M_p}{M_*}} \tag{6.11}$$

where t_* is the timescale of the star's lensing event.

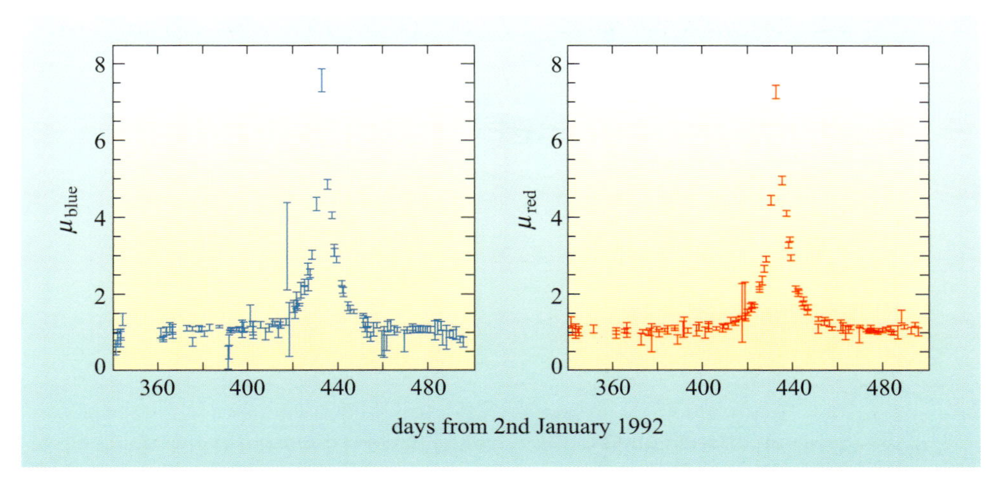

Figure 6.12 Light curve of a micro-lensed star. The brightness variation is shown here for two wavelength bands, one in the red and one in the blue part of the spectrum. The fact that they are so similar confirms this as a micro-lensing event, as gravitational lensing affects all wavelengths in the same way. (The Macho Project)

- For the same star as in Question 6.4, calculate the lensing timescales if it had a planet of mass: (a) 1.90×10^{27} kg (Jupiter); (b) 5.98×10^{24} kg (Earth).

- By multiplying 43.2 (the star's lensing timescale from the answer to Question 6.4) by $\sqrt{\dfrac{M_p}{M_*}}$ in each case: (a) Jupiter mass, $t_p = 1.33$ days; (b) Earth mass, $t_p = 0.075$ days $= 1.80$ hours

If a planet is involved in the lensing of a stellar system it can show up as a very obvious spike in the time profile of the background star's brightness. To understand more we need to consider the geometry of the situation in some detail (even though we cannot resolve it with present telescopes). The properties of a gravitational double lens, such as a star and planet system, are complicated. However, we can appreciate the important points by considering a particular example. Figure 6.13a shows the situation for a lens with a background star passing just above it. (Alternatively, we could imagine that the source is fixed and the lens is moving.) As the source position moves from left to right, the upper image moves in the same direction but the lower image moves in the opposite direction. Both images are most distorted when the lens and source are closest together at time t_2. Figure 6.13b shows the amplification (the factor by which the unresolved star has brightened – we'll discuss this further below) as the event progresses. Also shown in Figure 6.13b, is the spike at t_3 caused by the presence of a planet at position P. If the planet was at position Q, there would be no spike at all. This example illustrates that it is quite likely for a star to be involved in micro-lensing whilst its planet is not. So, in practice, the lensing of the star serves to attract our attention, alerting us to the possibility of a planet being lensed too. However, if no spike is seen, we cannot conclude that there is no planet because it is quite possible that a planet may have been there, but in the wrong position to produce a spike.

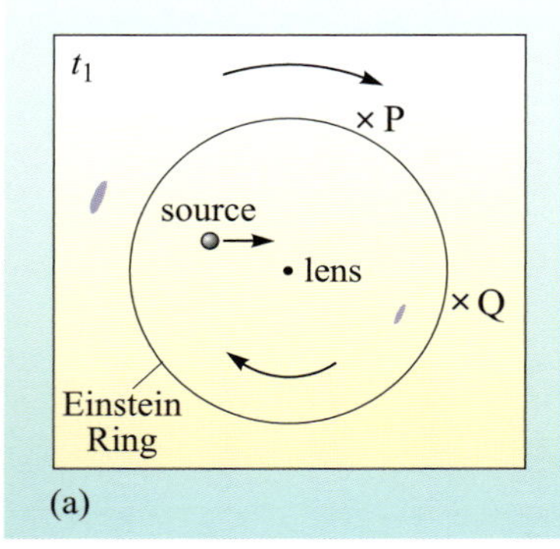

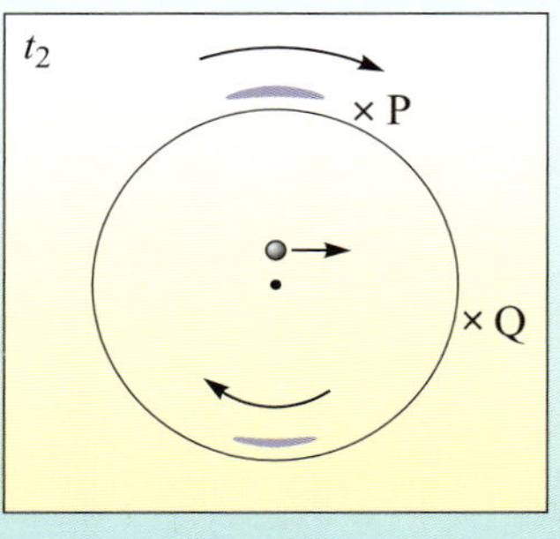

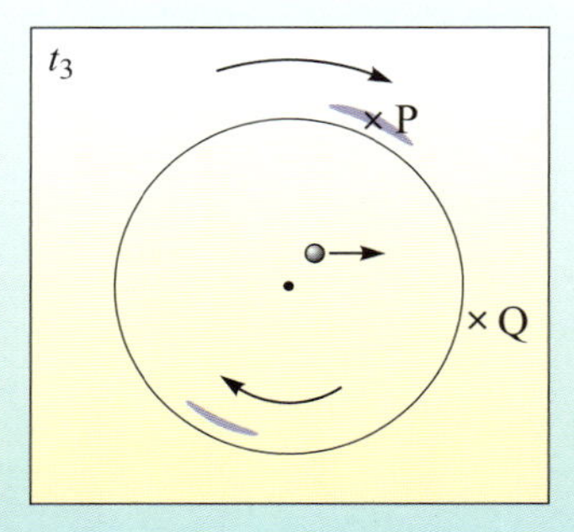

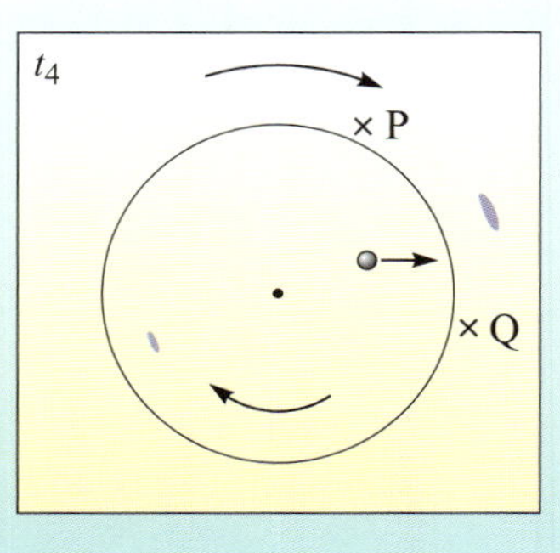

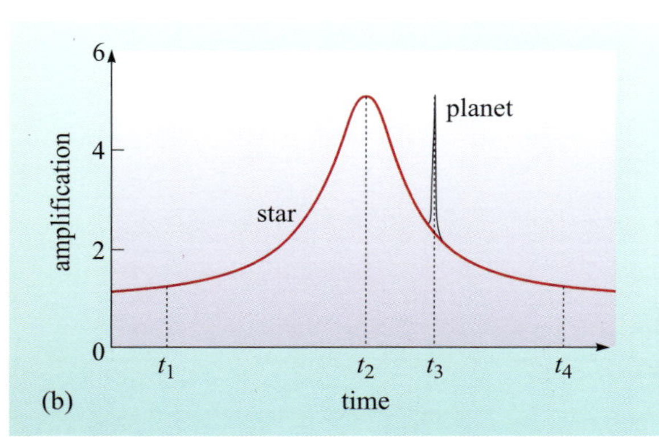

Figure 6.13 (a) These images show what would be seen in the sky at different times (t_1, t_2, t_3 and t_4) if a micro-lensing event could be observed by a telescope with a spatial resolution well beyond what our current technology permits. The dot in the centre is the lensing star and the circle is the Einstein ring. As the source moves from left to right the images (shown in blue) above and below move in the directions indicated by the arrows. The distorted images of the circular source can be seen becoming progressively more flattened as they approach the middle. (b) The light curve of the event depicted in (a). If the planet is in position P then the black spike is produced. If the planet is in position Q then there is no spike.

The remaining issue is that of amplification. It turns out that if a lens is an angular distance β from the source (a background star) on the sky, then the amplification factor (lensed brightness/normal unlensed brightness) μ is given by:

$$\mu = \frac{(\beta / \beta_e)^2 + 2}{(\beta / \beta_e) / \sqrt{(\beta / \beta_e)^2 + 4}}$$

(This formula breaks down for the case of perfect alignment, $\beta = 0$, but is a good approximation for present purposes.) This form of the magnification factor is hard to interpret, but it can be simplified for highly amplified cases ($\beta < \beta_e$), which are likely to be observed:

$$\mu = \frac{\beta_e}{\beta} \tag{6.12}$$

Equation 6.12 shows that the amplification is proportional to the angular radius of the Einstein ring, β_e. In turn, Equation 6.10 shows that β_e is proportional to the square root of the mass (i.e. $M^{1/2}$). This means that the amplification, μ, will also be proportional to the square root of the mass.

■ Will the planet or the star be more amplified in brightness, and by what factor, if the star and the planet happen to be at the same distance from the background star (and they have the same β)?

❑ The star, having the greater mass, will be more greatly amplified. Since the amplification factor is proportional to the square root of the mass of the lens, the ratio of amplification factors is $(M_* / M_p)^{1/2}$.

We can now conclude that the star will, in general, be more strongly amplified than the planet. However, it is possible, though unlikely, to have certain configurations of star, planet and source in which the planet is the more greatly amplified. So, the hope is that extrasolar planets can be detected in geometries such that they are much more greatly amplified than the star that they are orbiting, appearing as large, short-duration peaks on the time-profile of the star's lensing time profile (as in Figure 6.13b). Searching for planets by micro-lensing involves long-term monitoring of many distant stars, and waiting for some invisible foreground object to pass in front. To date, after less than a decade of monitoring, no planets have been found by micro-lensing. Although this may seem like a negative result, it has been used to make the following statement: less than a third of all solar type stars (G and K type stars, see Box 7.1) have a Jupiter-sized planet, in a Jupiter-sized orbit. This result is given more weight because astronomers are successfully using micro-lensing to detect binary stars at a rate consistent with the fraction of binary stars in our Galaxy (about half of stellar systems are binary). The snag with micro-lensing is that the event is a 'one-off' so that other methods must be employed for confirmation of the planet's existence and any other follow-up studies.

6.6 Movement of the star

In Sections 6.2–6.5 we have considered how a planet may be detected by its effect on radiation. We now consider the possibility of finding a planet by its effect on nearby matter, i.e. by the effect of the planet, which we cannot see, on the motion of the star, which we can see.

6.6.1 Astrometry

If two stars of equal mass orbit one another, the symmetry of the situation alone demands that they must each be orbiting the midpoint of the line that joins them. If one star had a greater mass than the other, then the centre of the orbits would be between the midpoint and the more massive star. Figure 6.14 illustrates these examples.

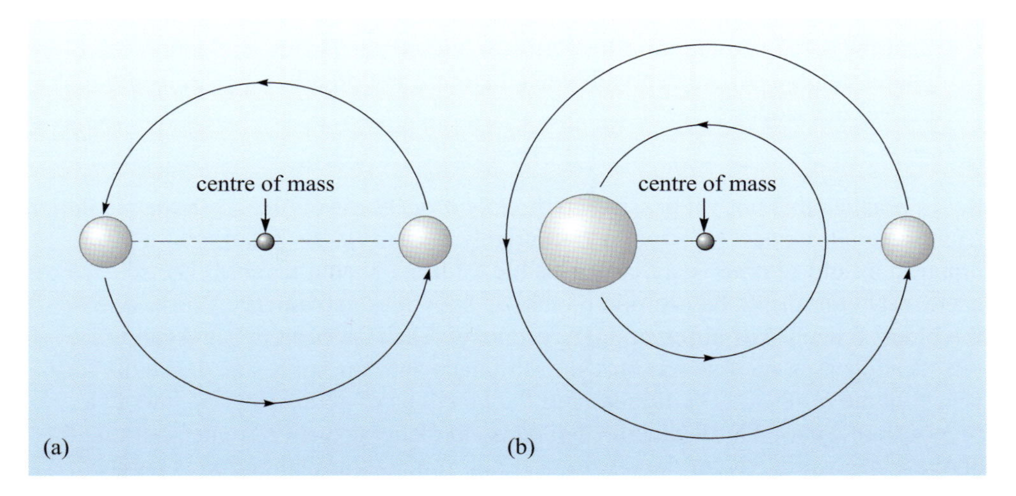

Figure 6.14 The orbits of a two-body system, showing the centre of mass in each case, for (a) two equal masses; (b) one mass greater than the other.

Consider two masses labelled A and B. The distances, or radii, a_A and a_B to the common centre of their orbits, called the **centre of mass**, are given by:

$$M_A a_A = M_B a_B \qquad (6.13)$$

where M_A and M_B are the masses. Even for different masses, the orbital periods of both masses about the centre of mass are the same (a little thought should reveal that this must be true by symmetry).

- If the Solar System only contained the Sun and Jupiter (the most massive planet), where would the centre of mass be? (Jupiter's mass is 1.90×10^{27} kg.)
- Re-arranging Equation 6.13 above we find:

$$a_\odot = \frac{M_J}{M_\odot} a_J = \frac{1.90 \times 10^{27}\ \text{kg}}{1.99 \times 10^{30}\ \text{kg}} \times 5.2 \times 1.5 \times 10^{11}\ \text{m} = 7.45 \times 10^8\ \text{m}$$

where $a_\odot$ is the distance of the Sun from the centre of mass and a_J is the distance of the planet from the centre of mass.

The centre of mass of the Solar System is not at the centre of the Sun, but happens to be not far from its surface. This means that all the planets and the Sun itself are orbiting that point. In effect, it appears that the Sun is wobbling back and forth as it orbits a point near its surface. So, when we look at other planetary systems, even if we cannot directly see the planets, can we see the wobble of the star?

- Would the largest modern telescopes be able to detect a solar mass star's wobble due to the presence of a Jupiter at the alpha Centauri distance?
- The angle of the wobble on the sky would be (see Box 6.3):

$$\beta_\odot = \frac{a_\odot}{d_{\alpha E}} = \frac{7.45 \times 10^8\ \text{m}}{4.07 \times 10^{16}\ \text{m}} = 1.83 \times 10^{-8}$$

$$= 1.83 \times 10^{-8} \times 2.06 \times 10^{5\prime\prime} = 3.77 \times 10^{-3\prime\prime} = 3.77\ \text{mas}$$

where 'mas' stands for milli-arcseconds. The angle of wobble, β, is much less than the 0.13″ resolution of modern 8 m telescopes derived in Section 6.2. Direct detection of such a wobble is therefore not possible with conventional, optical telescopes.

So, in practice it is not yet possible to directly observe a wobble, given the resolution of modern telescopes. However, by accurately measuring the position of a star amongst a field of other stars, it is possible to detect such a wobble. The science of accurately measuring the position of stars is known as **astrometry**, and is one of the oldest branches of observational astronomy. The European Space Agency's (ESA) Hipparcos satellite has brought astrometry into the space age during its four year lifetime (1989–1993). It measured the positions of about 100 000 stars to a precision of 2 mas. This is tantalizingly close to planet detection (remember that the above 4 mas wobble is for the nearest star and Jupiter's orbital period is nearly twelve years). Hipparcos data has confirmed the presence of one planet by its wobble – the same one as mentioned in Section 6.4 around the star HD 209458, that was first discovered by the Doppler spectroscopy method (Section 6.6.2).

NASA's forthcoming contribution to astrometry is the SIM satellite, planned for launch in 2009. Its goal is to measure the positions of stars in all visible regions of our Galaxy with a resolution of 4 μas (4 micro-arcseconds). ESA's next astrometry mission is called Gaia (pronounced Guy-yah), scheduled for 2012, promising a precision of 10 μas for a billion stars in our Galaxy.

■ Will SIM or Gaia be able to detect the wobble caused by Earth around the Sun at the alpha Centauri distance? (The Earth's mass is 5.98×10^{24} kg.)

❑ The wobble caused by the Earth will be:

$$a_{\odot} = \frac{M_{\mathrm{E}}}{M_{\odot}} a_{\mathrm{E}} = \frac{5.98 \times 10^{24}\ \mathrm{kg}}{1.99 \times 10^{30}\ \mathrm{kg}} \times 1.5 \times 10^{11}\ \mathrm{m} = 4.51 \times 10^{5}\ \mathrm{m}$$

The angle of the wobble will therefore be:

$$\beta_{\odot} = \frac{a_{\odot}}{d_{\alpha \mathrm{E}}} = \frac{4.51 \times 10^{5}\ \mathrm{m}}{4.07 \times 10^{16}\ \mathrm{m}} = 1.11 \times 10^{-11} = 2.29 \times 10^{-6}\,'' = 2.29\ \mu\mathrm{as}$$

This is below SIM's detection limit and also below that of Gaia.

6.6.2 Doppler spectroscopy

Spectroscopy, the analysis of radiation split up into its spectrum of wavelengths, is another old branch of astronomy. In fact the entrance of spectroscopy into astronomy in the late nineteenth century brought the science to a new level, initiating what we now call astrophysics. Without spectroscopy, stars would be little more than dots of light on the sky. At the end of the twentieth century, spectroscopy provided the most successful method for detecting planets, namely **Doppler spectroscopy**, also called the **radial velocity method**. Perhaps the biggest surprise, as you will see, is that it allowed the discovery of planets that scientists did not think should exist. Given Doppler spectroscopy's success in planet detection, we have intentionally made this section more rigorous than previous sections. If you find it hard going, bear in mind that once you have worked through it you will have attained a solid, quantitative understanding of how humans first detected planets outside their own Solar System.

The fact that moving objects emit light at apparently shifted frequencies – the **Doppler effect** – has allowed astronomers to deduce valuable information on the speed at which astronomical objects are moving:

- if an object is moving towards you, the shift will be to shorter wavelengths (a blue-shift);
- if an object is moving away from you, the shift will be to longer wavelengths (a red-shift).

The size of the shift is proportional to the speed at which the object is moving either towards or away from you.

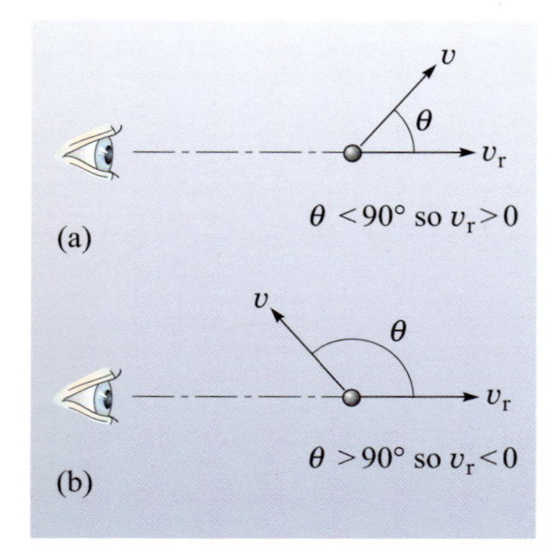

Figure 6.15 The relation of the radial velocity to an object's true velocity. (a) An observer sees an object moving away. The angle θ between the object's velocity and the line of sight to the observer is less than 90°, so the radial velocity is positive. (b) An observer sees an object approaching, θ is more than 90°, so the radial velocity is negative.

To be quantitative, the change in wavelength $\Delta\lambda$ caused by an object moving *directly towards* or *away from* you at velocity v_r (this is not a speed because v_r can be positive or negative) will be:

$$\frac{\Delta\lambda}{\lambda} = \frac{v_r}{c} \tag{6.14}$$

where c is the speed of light. For an object moving away from you, v_r will be positive and so $\Delta\lambda$ will be positive. Likewise, v_r and $\Delta\lambda$ will be negative for an object moving towards you. Notice that the Doppler shift does *not* depend on the distance to the object being observed. Usually the measurement is performed by comparing the wavelength of a spectral line from a particular element (e.g. hydrogen, helium, oxygen) from the astronomical object with that of a stationary source on Earth. The difference between the two will therefore be $\Delta\lambda$, the shift in wavelength, where λ is the wavelength of the spectral line measured from the stationary source in a laboratory.

Remember that v_r is not simply the speed of the object, but the component of velocity of the object along the observer's line of sight, often called the **radial velocity**. Figure 6.15 illustrates how the radial velocity is related to the true velocity of the object. Remember that velocity refers to both the speed and direction in which an object is moving, i.e. to specify a velocity you must indicate the direction in which an object is travelling (an arrow will do) as well as its speed in that direction. If the angle between the line of sight and the velocity of the object, θ, is known, then the radial velocity is related to the object's true speed v:

$$v_r = v\cos\theta \tag{6.15}$$

You can work out the value of $\cos\theta$ using a calculator. Figure 6.16 shows how the radial velocity changes with angle θ.

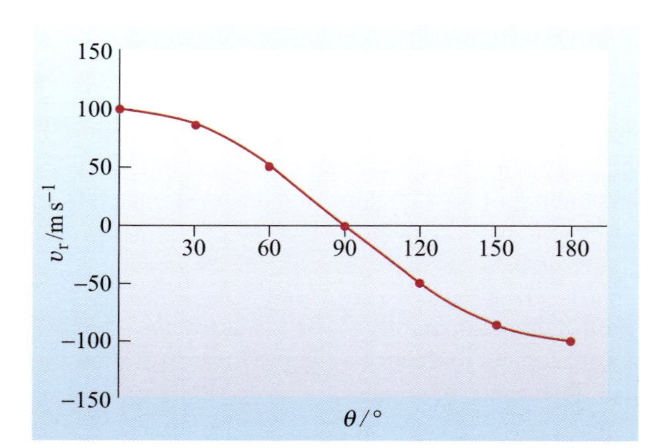

Figure 6.16 A plot of radial velocity against line of sight at angle θ for a speed of 100 m s^{-1}.

QUESTION 6.5

An object is known to move with speed 100 m s^{-1}. What is its radial velocity if the angle between its velocity and the line of sight is: (a) 0° (b) 30° (c) 60° (d) 90° (e) 120° (f) 150° (g) 180°? Sketch a graph of radial velocity against θ. What do negative values of velocity indicate?

The small orbit of the star about the centre of mass, the one we assessed with astrometry in Section 6.6.1, should also cause Doppler shifts in spectral lines emitted by the star. Whilst the star is moving away from us, spectral lines will move to slightly longer wavelengths, and then to shorter wavelengths when the star moves towards us. What's more, this will be periodic (i.e. repeat itself exactly with a fixed period) because the star is in an orbit that is periodic. From hereon we will only consider a star's orbital motion and not its motion through interstellar space which can be 'subtracted' in real observations. So when we refer to radial velocity it should be understood that we mean its orbital radial velocity.

■ A star, 51 Pegasi, is observed to show periodic shifts in wavelength. From the data (and this is *real* data) shown in Figure 6.17, make a rough estimate of the star's maximum radial velocity due to its orbit and the period of the suspected planet.

❑ The maximum radial velocity is somewhere between $50\,\mathrm{m\,s^{-1}}$ and $60\,\mathrm{m\,s^{-1}}$. The period of the suspected planet will be equal to one complete period of the star's radial velocity. This can be most easily measured as the time between maxima or minima, which in this case is about 4 days.

Using the radial velocity of the star's motion together with the orbital period it is possible to work out further properties of the putative planetary system. Before doing this, we must take account of the fact that we are only measuring the radial component of velocity. In itself this is not a problem for a *circular orbit* that is edge-on to our line of sight (Figure 6.18a). In this case, the maximum and minimum radial velocities correspond to times when the star is moving directly away from and towards us, respectively. For this reason the maximum radial velocity will be equal

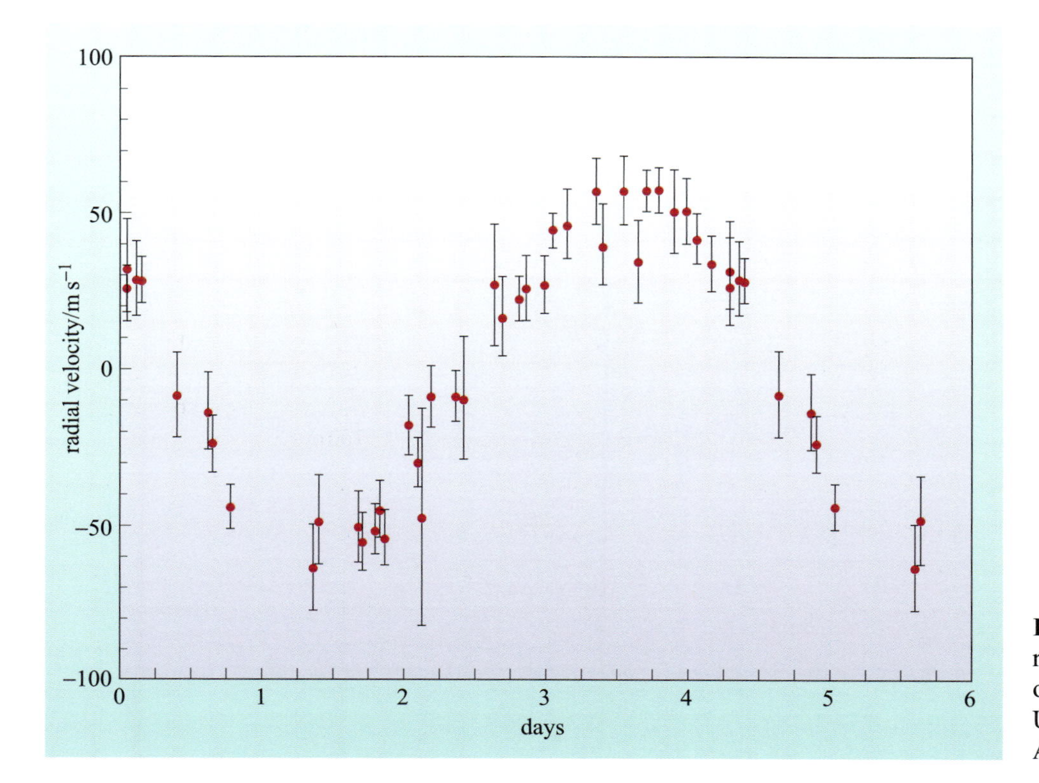

Figure 6.17 Measurements of radial velocity for 51 Pegasi over one orbit. (S. Korzennik, Harvard University, Smithsonian Centre for Astrophysics)

to the star's orbital speed. Now consider the situation shown in Figure 6.18b, where the stellar system is tilted at some angle. Even at times when the star's radial velocity is at a maximum, the direction of its velocity does not exactly point directly away from us (and similarly for times of minimum velocity). Since the cosine function, i.e. $\cos\theta$ in Equation 6.15, can never exceed a value of 1, we can see that the radial velocity v_r must be less than or equal to the star's orbital speed v_*. By convention, the tilt of a stellar system, as shown in Figure 6.18b, is measured by an angle between the line of sight and the line perpendicular to the plane of the orbit called the normal. This angle is called the inclination, i_0. Clearly for $i_0 = 90°$ we have the most favourable situation, and for $i_0 = 0°$ (Figure 6.18c) we have the worst situation where Doppler spectroscopy is useless because the radial velocity is always zero.

We can now work out a relationship between θ, the angle between the star's velocity and the line of sight, and i_0, the inclination of the orbit to the line of sight. Figure 6.19 shows the planet's orbit and an enlarged view of the star's orbit. Notice that the star is always on the opposite side of its orbit to the planet (this has to be so otherwise the centre of mass would not be at the centre). From the position at time 1 we can deduce that the inclination angle i_0 and the angle between the line of sight and the star's velocity, θ_1, are related by:

$$\theta_1 = 90° - i_0$$

In practice, almost all astronomers work with i_0, so it is worth re-writing Equation 6.15 for the time of maximum radial velocity (i.e. position S1) in the following form:

$$(v_r)_{MAX} = v_*\cos\theta_1 = v_*\cos(90° - i_0) = v_*\sin i_0$$

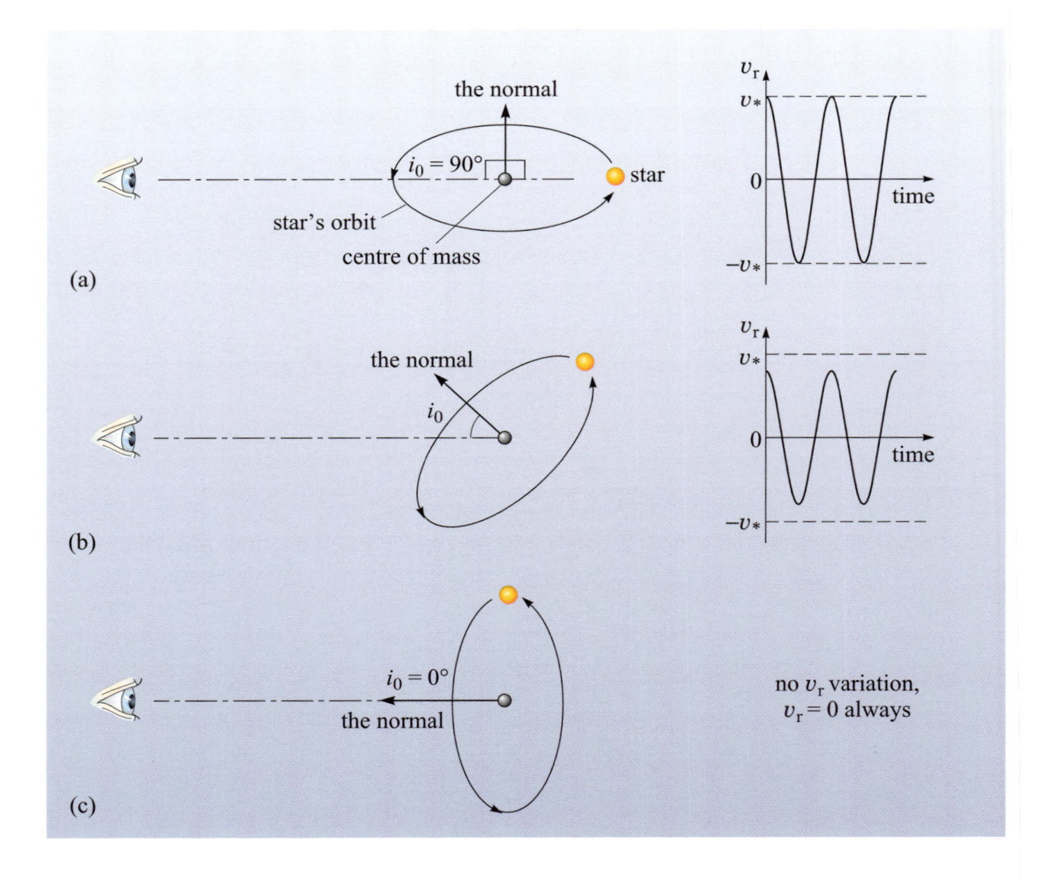

Figure 6.18 The inclination of a star's circular orbit. The dot represents the centre of mass of the planetary system. (The planet, which is much further out, is not shown.) (a) $i_0 = 90°$, the most favourable case with the largest radial velocities; (b) $0° < i_0 < 90°$ (c) $i_0 = 0°$, the worst case where the radial velocity is always zero.

(The cosine and the sine, i.e. cos and sin, are closely related, as their names suggest. To verify the conversion from cosine to sine in the above formula, pick any angle, work out the sine and note down the result. Then, subtract the angle from 90° and take the cosine. The result will be the same; if it isn't, make sure your calculator is in 'degrees or 'deg' mode.)

- Explain why the radial velocity must be zero when the star is at position 2 in Figure 6.19.

- The star's velocity is at 90° to the line of sight at position 2, which means that the radial velocity will be zero (see also Figure 6.15). Alternatively you could have shown this mathematically as:

$$(v_r)_2 = v_* \cos \theta_2 = v_* \cos 90° = 0$$

You should now understand that the star's maximum radial velocity of $v_* \sin i_0$ occurs at position 1. The radial velocity then falls to zero at position 2 and will decrease to $-v_* \sin i_0$ when the star and planet are opposite position 1 and then return to zero when they are opposite position 2 at position 4. This explains the profile and amplitude of the radial velocity cosine curves that you saw in Figure 6.16 and that were plotted in the answer to Question 6.5.

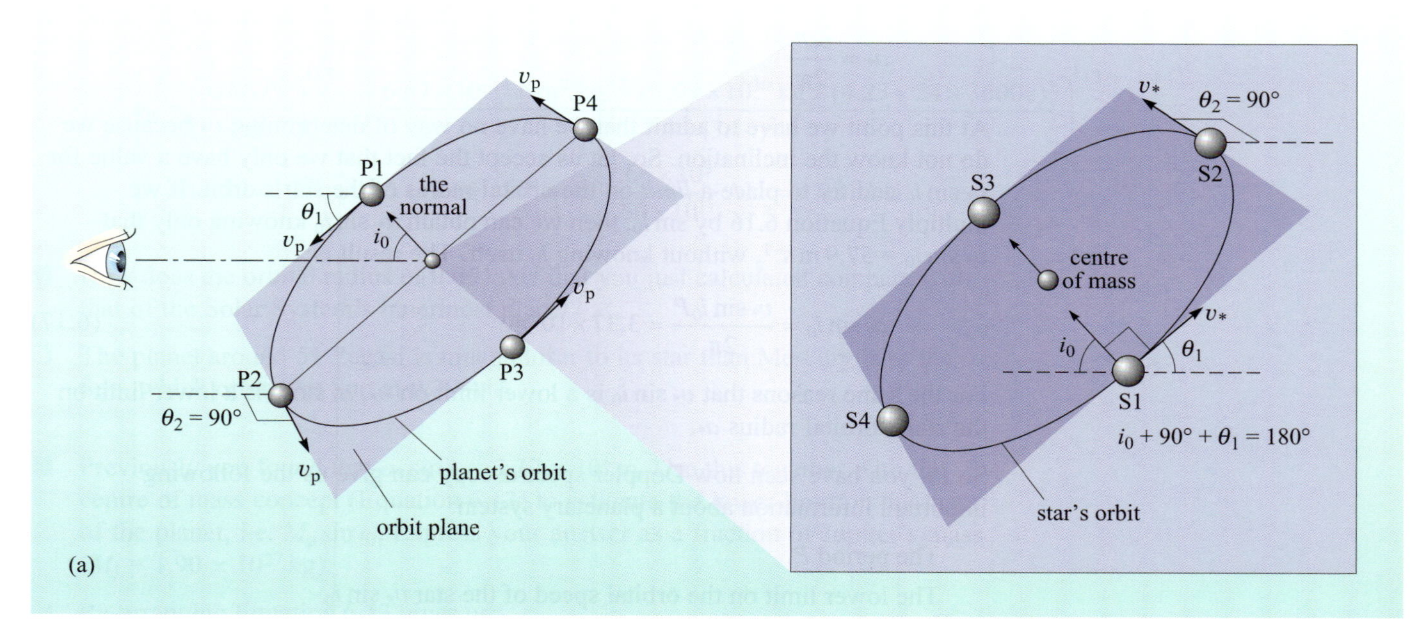

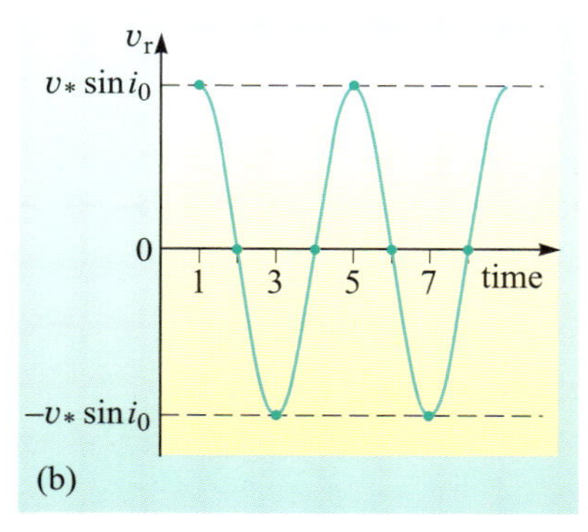

Figure 6.19 (a) The positions of the star (S) and planet (P) at four equally spaced times: 1, 2, 3 and 4. We can deduce the relationship between θ and i_0 at the time when the planet and star are at positions P1 and S1, when the radial velocity is maximum. The enlargement of the star's orbit (the magnification factor is M_* / M_p), shows that the star's orbit is the same as the planet's orbit, except that it is much smaller and that the star is always opposite the planet. According to their definitions θ and i_0 and a right angle must add up to 180° to form a straight line when the planet and star are in position 1. When the planet and star are in position 2 the angle between the line of sight and the star's velocity is $\theta_2 = 90°$. (All the horizontal dashed lines represent the line of sight.) (b) The graph of the star's radial velocity corresponding to the marked positions in (a).

Now you have discovered the big surprise that awaited the planet hunters using Doppler spectroscopy. The planet orbiting so close to the star 51 Pegasi is not a small planet like Mercury, but a giant planet comparable in mass with Jupiter. Although this is a surprise, a little bit more mathematics reveals that it is exactly these kinds of planets – known as **hot Jupiters** – that Doppler spectroscopy is most likely to detect. The reason is simply that observations will most easily pick up stars exhibiting large radial velocity variations, which is of course closely related to the orbital speed of the star. From Equation 6.16 we saw that the orbital speed of the star v_* is:

$$v_* = \frac{2\pi a_*}{P} \tag{6.23}$$

substituting for P using Equation 6.19 and a_* using Equation 6.21 you will find:

$$v_* = 2\pi \frac{M_p a_p}{M_*} \left(\frac{GM_*}{4\pi^2 a_p^3} \right)^{1/2} = \left(\frac{G}{M_*} \right)^{1/2} \frac{M_p}{a_p^{\frac{1}{2}}} \tag{6.24}$$

The important thing to realize is that the star's orbital speed is greatest for massive planets (large M_p) that are orbiting close to the star (small a_p). In addition to this fact, such planets will have short orbital periods of order a few days, so will not require many years of observations as would, say, a real Jupiter with an orbital period of about 12 years. For these reasons Doppler spectroscopy is most suited to finding massive planets that have a small orbital radius. This is a classic example of a **selection effect** – an effect whereby our discoveries are biased towards the most easily detectable examples. It is essential to recognize selection effects if one is interested in looking at the statistical distribution of the properties of discovered exoplanets, such as their mass or orbital periods.

■ What are the problems of observing a periodic phenomenon involving a faint astronomical object at short timescales (seconds) and long timescales (centuries)?

□ On the timescale of seconds you may not have enough time to collect enough incident radiation to build up an image or spectrum of an object. For very faint objects you may not even be able to detect it at all on the timescale of a few seconds. On the timescale of centuries you have to rely on scientists embarking on projects that span human generations. To ensure consistency you would have to ensure careful calibration of any new equipment and/or the continued use of old technology in the future. (Perhaps funding such long-term endeavours is more problematic than ensuring your descendants are interested in continuing the work.)

The vast majority of planets discovered orbiting normal stars to date (early 2003) have been found using Doppler spectroscopy, with other methods confirming their existence. With improving techniques and technology, together with growing datasets, Doppler spectroscopy can now reach to larger orbital radii and smaller masses (although it can only place a lower limit on the mass). Currently, finding a Jupiter in a Jupiter orbit is a realizable goal, but it seems likely that finding Earth-like planets may be achieved first by space missions such as Darwin or the Terrestrial Planet Finder. In any case, since Doppler spectroscopy does not directly observe the planet, we must rely on the space missions for direct analysis of an extrasolar planet's physical nature (such as composition, spin rate, axis inclination).

You may have noticed that we have only used the maximum values of the spectral shift plots, such as in Figure 6.16; we have ignored the shape of the curve traced out by the plotted data. The shape does tell us some useful information about the shape of the star's orbit, and therefore of the planet's orbits. For simplicity, we have only considered circular orbits. However we know that planets do have elliptical orbits. In elliptical orbits the orbital speed is not constant, as it is in circular orbits. This fact means that the time profile of the radial velocity, as shown by the spectral shift plots, will be more complicated. Figure 6.20 illustrates how elliptical orbits will affect observed time profiles of radial velocity. From these plots it is possible to deduce what is known as the eccentricity of an orbit, e. An eccentricity of zero corresponds to a circle, whereas larger values indicate more elliptical orbits.

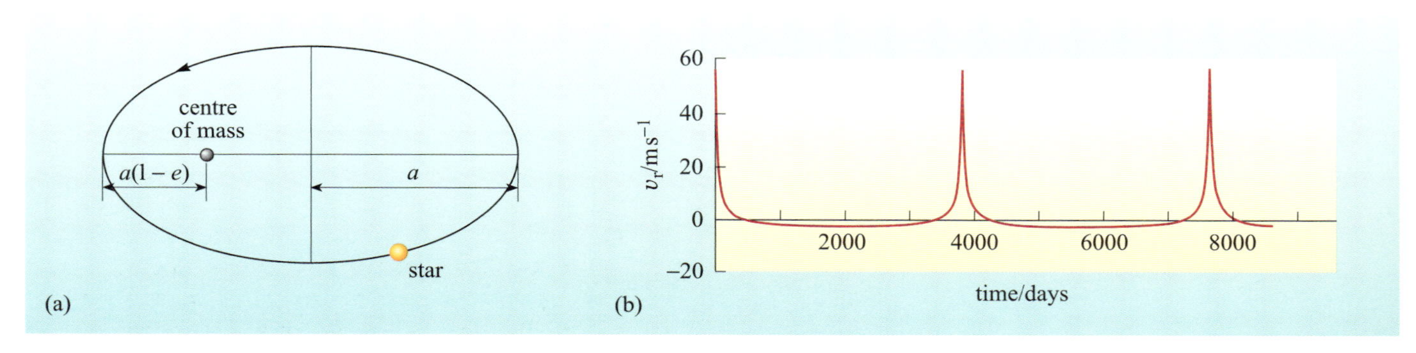

Figure 6.20 (a) An elliptical orbit. The semimajor axis a is half the widest distance across the ellipse, and the star is located at $a(1-e)$ along the major axis, where e is the eccentricity. For $e = 0$, the orbit is circular and a equals the radius of the orbit. (b) The radial velocity profile for an elliptical orbit having an inclination of 90° and oriented so that its major axis lies in the plane of the sky. The large, narrow, positive peaks correspond to the times when orbital speeds are greatest (at closest approach).

In the general case of an elliptical orbit, and also for a circular orbit, the difference between $(v_r)_{MAX}$ and $(v_r)_{MIN}$ is often written as $2v_{rA}$, where v_{rA} is called the observed radial velocity amplitude.

6.7 Observables and important properties

Any science, but especially one that uses mathematics in theoretical modelling, needs to relate quantities that can be observed – known as **observables** – to the important properties involved in the theory. Equations 6.25–6.30 have important physical properties concerning exoplanets on the left-hand side and important observables on the right-hand side. Also on the right-hand side are physical constants and the properties of the star (radius and mass) that can be accurately determined from standard observations. All of these equations follow from equations presented earlier in this chapter, and their derivation is outlined in the answer to Question 6.6. Make sure you attempt Questions 6.6 and 6.7, as these will help consolidate your understanding of the various methods of observing exoplanets presented in this chapter, especially if you found the mathematics hard-going.

$$a_p = \beta_p d_{*E} \tag{6.25}$$

$$a_p = \frac{M_*}{M_p} \beta_* d_{*E} \tag{6.26}$$

$$a_p = \left(\frac{GM_* P^2}{4\pi^2} \right)^{\frac{1}{3}} \tag{6.27}$$

$$R_p = R_* \sqrt{f_p} \tag{6.28}$$

$$M_p = M_* \left(\frac{t_p}{t_*} \right)^2 \tag{6.29}$$

$$M_p \sin i_0 = \left(\frac{M_*^2 P}{2\pi G} \right)^{\frac{1}{3}} (v_r)_{MAX} \tag{6.30}$$

(and we can write v_{rA} in place of $(v_r)_{MAX}$)

QUESTION 6.6

This question concerns Equations 6.25 to 6.30. (a) Write definitions for each variable appearing in the equations. (b) For each equation, state the method of detection it refers to and identify the numbers of the equations in Chapter 6 that are used in deriving it.

QUESTION 6.7

Consider the following methods for detecting exoplanets:

Direct; occultation; gravitational lensing; astrometric; Doppler spectroscopy.

(a) For a planet of given mass, which methods are favoured for smaller orbital radii? Briefly explain the reason for your answer. (b) For a planet of given orbital radius, which methods favour larger planets (either in mass or radius)? (c) Aside from the obvious observational difficulties involving smaller brightnesses at greater distances, which methods work independently of distance?

In answering questions (a–c), give references to equations where necessary to back up the point that you are making.

6.8 Summary of Chapter 6

- The Drake equation describes the expected number of broadcasting civilizations in the Galaxy and provides a framework for science to follow in searching the Galaxy for planets and life.
- Planets can in principle be detected by radiation reflected from their star, but current technology cannot resolve the star and planet because their difference in brightness is so large.
- Infrared radiation emitted by the planet itself is more promising than the reflected radiation because the brightness ratio between planet and star is smaller. Infrared lines can tell us about the presence of certain gases in the planet's atmosphere, though observation of these lines is beyond present technology.
- A planet can be detected by occultation, where the star's light is apparently dimmed as the planet passes in front of the star. This will be possible for only a small fraction of planets with orbits in planes perpendicular to the sky.
- Gravitational micro-lensing causes a background star to apparently brighten and then fade over many weeks when a foreground star passes in front of it. The presence of a planet can in principle be detected as a several-day duration peak on top of the star's micro-lensing time-profile.
- The presence of a planet will cause the star to execute an orbit that can be detected either directly, by way of astrometry (although the technology is not yet available), or by Doppler spectroscopy, which measures the periodic variation of the star's velocity along the line of sight.
- Only the Doppler spectroscopy method has found planets, though the presence of one was confirmed using the occultation method. The micro-lensing method has not yet found any planets (early 2003) but has placed an upper limit of one in three on the proportion of solar-like stars with Jupiter-like planets in Jupiter-like orbits.
- Selection effects mean that each method is biased towards discovering particular types of planets in particular orbits. Importantly, Doppler spectroscopy is biased towards detecting more massive planets orbiting close to their stars.

CHAPTER 7
THE NATURE OF EXOPLANETARY SYSTEMS

Now that the various methods of detecting exoplanets have been described, we turn to the results of the various searches. What types of planet have been detected? What types of planet await detection? Have we already discovered potential planetary habitats, or is this a discovery still to be made? In terms of the Drake equation (Section 6.1.1), what have we learned about the probability, p_p, of planets forming around a suitable star, and the average number, n_E, of suitable planets in a habitable zone per planetary system?

7.1 The discovery of exoplanetary systems

Ever since it has been known that the Sun is a star, people have longed to know whether there are planets around other stars. But it is only since the 1930s that astronomers have believed they might be able to find out. Since then, astrometry has been applied to nearby stars, notably by the Dutch astronomer Peter van de Kamp working at the Sproul Observatory, USA. From as early as 1937 he strove to detect the tiny stellar motions that would reveal the presence of companions of planetary mass. He even thought that he had detected some planets, but later work by others showed that he had not – the apparent stellar orbits were the result of changes made to his telescope in 1949 and 1957. Nevertheless, van de Kamp went to his grave believing that he had discovered at least one planet, around Barnard's star, just 5.9 light-years from the Sun – not much further than the closest star Proxima Centauri, at 4.2 light-years.

Then, in 1992, after decades of disappointment, the first exoplanets were confirmed. The US astronomers Alex Wolszczan and Dale Frail announced that they had detected two planets in orbit around a star of a rare type called a **pulsar**. Each planet had a mass just a few times that of the Earth. The claim that a pulsar had planets was greeted with considerable surprise by other astronomers, but this claim has withstood further investigation. The surprise stems from the way that a pulsar is formed, as the remnant of the explosion of a massive star at the end of its life – a **supernova** explosion. It had not been anticipated that planets could survive such a catastrophe. Perhaps they didn't. Perhaps the planets formed from debris left by the explosion, or perhaps they were captured from a companion star as the pulsar travelled near it. More important for us, life could not have survived the explosion, and even if the planets formed or arrived afterwards, they would be uninhabitable so close to the pulsar, with its deadly radiation. We will therefore discount these and other pulsar planets, and confine our attention to planets around stars that are more like the Sun.

Figure 7.1 Michel Mayor and Didier Queloz, discoverers of the first exoplanet around a Sun-like star. (Geneva Observatory)

In October 1995, two Swiss astronomers, Michel Mayor and Didier Queloz, of Geneva Observatory, announced the discovery of the first non-pulsar planet, in orbit around the Sun-like star 51 Pegasi (star number 51 in a catalogue of the constellation of Pegasus). The result was soon confirmed by others, and by early 1996 astronomers knew that the long drought in exoplanetary discoveries had ended. A steady stream of discoveries has

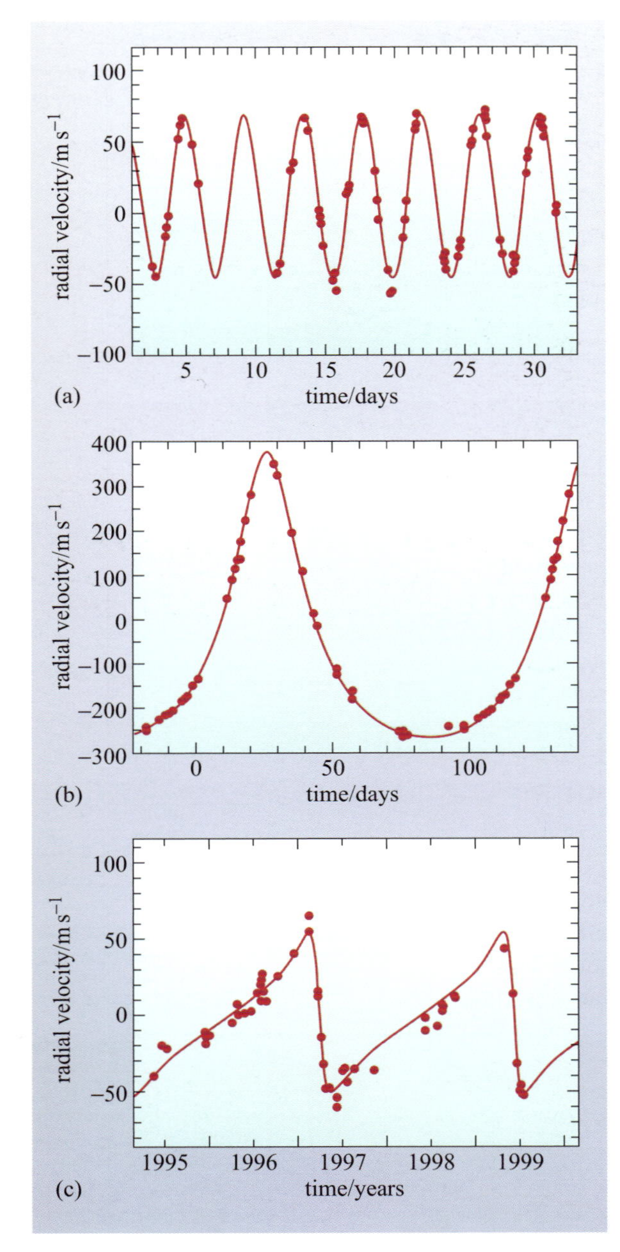

Figure 7.2 The variation of the radial velocity of some stars with planets, as obtained from the Doppler shifts in the stellar spectral lines.
(a) 51 Pegasi (its planet is in a low-eccentricity orbit – data from late 1995); (b) 70 Virginis (its planet is in an eccentric orbit); (c) 16 Cygni B (its planet is in an eccentric orbit), where 'B' denotes star B in a multiple-star system, in this case a binary system, so the only other star is 16 Cygni A. ((a) Marcy & Butler)

followed, and still continues today. At early 2003, 105 planets (non-pulsar) had been confirmed, in 91 planetary systems. Of these, 12 are known to be multiple-planet systems, but it is likely that all the others are also multiple-planet systems – the formation or survival of a single planet would be rare. The number of discovered planets is growing fairly rapidly.

Exoplanets are named after their stars. The letter 'b' is used to denote the first planet discovered, 'c' the next, and so on ('a' is not used). Thus the only known planet of 51 Pegasi is named 51 Pegasi b. If several planets in a system are discovered at the same time, then the lettering starts with the innermost and works outwards, as in Upsilon Andromedae b, c, and d.

All but one of the 105 exoplanets have been discovered using the Doppler spectroscopy method. This relies on the detection of the cyclic Doppler shifts in the stellar spectral lines induced by the radial velocity of the star (Section 6.6.2). Figure 7.2 shows data for three stars, along with best-fit curves. From the sinusoidal fit for 51 Pegasi in Figure 7.2a, it can be deduced that the orbit has a low eccentricity. The different-shaped curves in the other two cases indicate more eccentric orbits. The eccentricity can be deduced, though the details will not concern us.

All 90 stars so far known to have planets through Doppler spectroscopy are nearby – the most distant is at 401 light-years (HD47536b), which is only about 0.4% of the diameter of the disc of the Galaxy in which we live. This proximity is because the closer a star the greater the flux of radiation we receive from it. The spectral lines are thus clearer, and the Doppler shifts are consequently easier to measure. (Remember that the Doppler shifts themselves are *independent* of the distance to the star – see Section 6.6.2.) Subsequently, one exoplanet discovered by the radial velocity method, around star HD209458 (HD denotes a catalogue), has also been observed by the occultation (or transit) method (see Section 6.4), i.e. observed in transit across the face of its star (Figure 7.3). To observe a transit the planet's orbit has to be presented to us very nearly edge-on. Exoplanet orbits are orientated randomly on the sky. Calculations then show that, on average, out of 100 exoplanets we can only expect to observe transits for about one or two. Another planet, Gliese 876b (another catalogue) has since been detected astrometrically.

The only confirmed discovery of a planet by a method other than Doppler spectroscopy is of a planet orbiting the star OGLE-TR-56. This star was unnamed before the discovery, and its name carries that of the survey – the Optical Gravitational Lensing Experiment, which is carried out at Las Campanas Observatory in Chile. In spite of the survey name, the discovery was made by the transit method. This is because a survey looking for the changes in apparent stellar brightness due to gravitational micro-lensing (see Section 6.5) could also detect the changes during a transit. The transit and lensing methods have greater range than Doppler

Figure 7.3 An artist's impression of the transit of HD209458b across its star, seen from Earth. See Figure 6.10 for the light curve. (Copyright © Lynette R. Cook)

spectroscopy, and this is exemplified by the large distance of OGLE-TR-56, about 5000 light-years away. In spite of this distance, the mass of the planet has been obtained by Doppler spectroscopy, aided by the very short 1.2 day orbital period. The mass is $0.9M_J$, where M_J is the mass of Jupiter (318 times the mass of the Earth, M_E). Many more transit discoveries are thought to be imminent. There are several unconfirmed candidates from micro-lensing.

We will now look at the known exoplanetary systems in more detail. In Section 7.2 the properties of the host stars are outlined, and then the properties of the exoplanets themselves. You will see that, as yet, largely because of observational selection effects, no Earth-mass planets have been discovered, but only giant planets of the sort unlikely to support life. Nevertheless, the discoveries that have been made have important consequences for the possible existence of Earth-mass planets, as discussed in Section 7.3. In Section 7.4 we summarize the properties of the planets so far discovered, and consider the prospects for future discoveries, including Earth-mass planets.

QUESTION 7.1

Using the graphs in Figure 7.2, measure the radial velocity amplitudes in Figure 7.2a–c. Express your results in (i) $\mathrm{m\,s^{-1}}$, and (ii) as a multiple of the speed limit for cars in a residential district (30 miles per hour in the UK, which is $13.4\,\mathrm{m\,s^{-1}}$).

7.2 Properties of exoplanetary systems

7.2.1 The stars in the known exoplanetary systems

Most of the stars presently known to have planets are similar to the Sun – they are Sun-like stars. Stars and stellar classification are outlined in Box 7.1.

BOX 7.1 STARS AND STELLAR CLASSIFICATION

Stars, at birth, are made mainly from the two lightest elements, hydrogen (H) and helium (He). These elements dominate the Universe, and the Sun when it was young would have consisted by mass of about 73% hydrogen, 25% helium and 2% everything else. In astronomy elements other than H and He are called heavy elements, which astronomers also call metals. The proportion of heavy elements is called the **metallicity** of the star. Note that over 99% of the Earth's mass is heavy elements, but that the giant planet Jupiter is much more like the young Sun, though almost certainly with a greater proportion of heavy elements, perhaps as much as 5% to 10%.

All stars we see today, including the Sun, had approximately the same composition when they were born, the most important variation being in metallicity, the range 0.05% to 3% covering most stars at birth. Their evolution depends mainly on their mass (unless there is interference by a close companion star). If the mass is greater than 0.08 solar masses ($0.08M_{\odot}$), then, at some point, the central core of the star becomes hot enough for nuclear fusion to be sustained, in which hydrogen is converted to helium. This releases energy that keeps the star in a fairly stable state. The star is then said to be a **main sequence star**, and the onset of this phase of a star's life marks its zero-age. During the main sequence phase the star roughly triples its luminosity (power output), the rate of increase getting steadily greater through this phase. The surface temperature (which is much lower than the core temperature, where nuclear fusion is occurring) changes much less – by the order of 10%. Table 7.1 shows how the luminosity and surface temperature depend on mass in typical cases, roughly half-way through the main sequence age.

■ Referring to Table 7.1, by what factors are the main sequence lifetime, luminosity and surface temperature of a star of mass $1.6M_{\odot}$ different from those of a star of mass $0.79M_{\odot}$?

❑ Main sequence lifetime is less by a factor of 5; luminosity is greater by a factor of 12; surface temperature is greater by a factor of 1.4.

The fourth column in Table 7.1 is the spectral class of the star. This is essentially a way of specifying the surface temperature. It is widely used in preference to surface temperature because it relates to observed properties of stellar spectra, from which the temperature is then deduced. These properties will not concern us, and neither will the reasons for the seemingly arbitrary association of letters with temperature – these are strictly historical. For our purposes we can take the spectral class as a surrogate for main-sequence surface temperature. The temperature decreases along the sequence O, B, A, F, G, K, M. Each class is subdivided 0, 1, … 9, with temperature decreasing as the numbers increase.

■ Place the following spectral classes in order of decreasing temperature: M5, A9, G2, B0, G5.

❑ B0, A9, G2, G5, M5.

A common mnemonic for the spectral class sequence is 'Oh Be A Fine Guy/Girl, Kiss Me'.

The proportion of stars born with a particular mass increases greatly as mass decreases. For example, there are about 25 times more stars of $0.5M_{\odot}$ than

Table 7.1 Properties of main sequence stars of different mass.

Mass/$M_{\odot}$	Luminosity/$L_{\odot}$	Surface temperature/K	Spectral class	Main sequence lifetime/Ma
23	87 000	35 000	O8	3.0
17.5	40 000	30 000	B0	15
2.9	50	9800	A0	500
1.6	5.7	7300	F0	3000
1.05	1.35	5940	G0	10 000
0.79	0.46	5150	K0	15 000
0.51	0.070	3840	M0	200 000
0.21	0.0066	3170	M5	very long

$2M_{\odot}$. The abundant main sequence M stars are called **M dwarfs**, because they are quite small.

The duration of the main sequence lifetime is very sensitive to the mass of the star, decreasing as mass increases, as shown in Table 7.1. The main sequence lifetime of a star ends when there is no longer enough hydrogen in its core to sustain sufficient energy output through fusion to helium. Subsequent events are dramatic and take place on timescales considerably shorter than the main sequence lifetime. For stars up to about $8M_{\odot}$ the star swells hugely and cools to become a **red giant**. It then flings off a large proportion of its mass, leaving a very hot, Earth-sized though massive remnant, which cools very slowly. This remnant is called a **white dwarf**. For a star of greater mass, the end is even more dramatic. Such a star swells to become a supergiant, it then undergoes a (Type II) supernova explosion in which nearly all of its mass is flung into space, leaving an exotic remnant, either a very dense object a few tens of kilometres in diameter, called a **neutron star**, or a **black hole**. Note that a pulsar is a particular kind of neutron star.

By 'Sun-like stars' we mean main sequence stars of spectral class F, G, or K, the Sun being G2, with properties not very different from those of the G0 star in Table 7.1. Their predominance in the known exoplanetary systems is because such stars attracted most of the initial search effort. The reason for this is that they suit the widely used Doppler spectroscopy method by having plenty of sharp spectral lines, good surface stability, and because there are plenty of bright examples. They also have sufficiently long main sequence lifetimes for any biosphere to become detectable from afar.

■ Why can't we include the lifetime after the main sequence phase?

❑ Any life that might have existed on a planet as its star ended its main sequence phase would not survive the post-main sequence events.

Recall from Chapter 2 that the presence of banded iron formations and carbon isotope evidence suggest that life may have emerged on Earth 600 to 700 Ma after the Earth was formed. But it would have been extremely difficult to detect this life from far away. It became much easier when life had greatly modified the atmosphere (Section 2.4.4). This had happened by the time the Earth was about 2000 Ma old. By then oxygenic photosynthesis had caused a considerable increase in atmospheric oxygen, and you will see in Chapter 8 how this can be detected from afar. Sun-like stars (F, G, and K) have sufficiently long main sequence lifetimes for life as we know it to be well established on any habitable planets.

You might wonder why M dwarfs were not initially favoured in the exoplanet searches – the proportion of stars increases greatly from O to M, and M dwarfs are by far the most common. M dwarfs also have huge main sequence lifetimes (Table 7.1). The reason for their neglect stems partly from their more complex spectral lines and their low luminosities. There was also a belief that the proximity of the habitable zone to the star was problematical. Recall (Section 2.3) that the habitable zone is that range of distances from a star in which the stellar radiation on an Earth-like planet would sustain water as a liquid on at least parts of the surface. Figure 7.4 shows the habitable zones for stars like the young Sun (G2) and for a young M0 star. A young M0 star has such a low luminosity that the habitable zone is close to the star, even closer for M1–M9 stars. For any planet in the habitable zone this results in a tidal interaction between the star and the planet strong enough to establish synchronous rotation in the planet, i.e. it keeps one

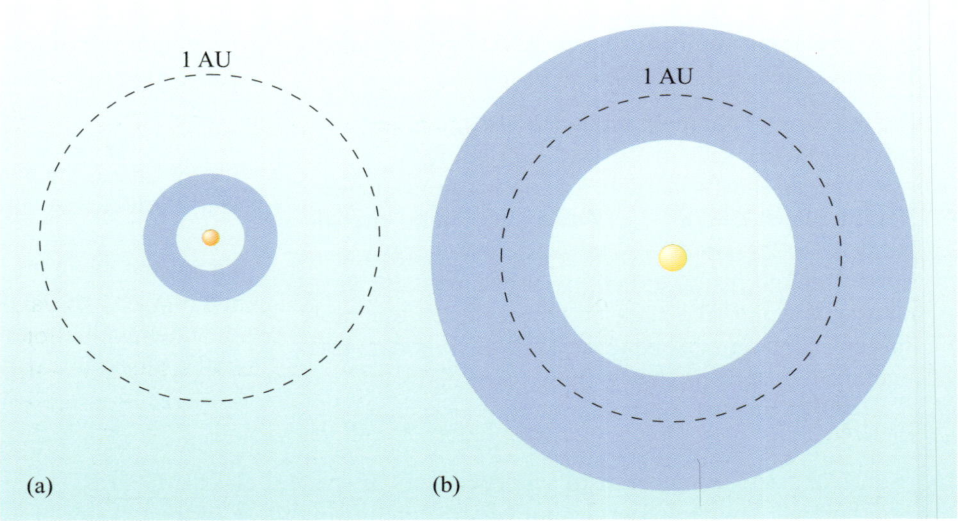

Figure 7.4 Habitable zones (blue-shading) around (a) an M0 main sequence star and (b) a G2 main sequence star (like the Sun), when the stars are young. Intermediate criteria have been adopted for the habitable zone boundaries. (1 AU is the radius of the Earth's orbit.)

hemisphere facing the star, just as the Moon does to the Earth. Tidal interactions were discussed in Box 4.2, where their heating effect was of primary concern, though synchronous rotation was also mentioned. When synchronous rotation is established the surface of the outward-facing hemisphere of the planet can get extremely cold, and it has been argued than this could cause the whole atmosphere to condense on this hemisphere. This would prevent life developing on the surface. However, if the atmosphere were sufficiently massive, then its circulation could carry heat from the star-facing hemisphere to the other hemisphere at a sufficient rate to prevent extensive condensation. This is a good reason to survey M dwarfs, and more of them are now being scrutinized.

Doppler spectroscopy has (by early 2003) discovered planets orbiting 8% or so of the nearby Sun-like stars, mostly with the comparatively high metallicities of the Sun, perhaps because this favours planet formation (Box 7.2). This fraction with planets can only grow as the more difficult discoveries are made. To these must be added planets that will be discovered around other types of star, particularly the abundant M dwarfs. Whether the majority of nearby stars have planets is unknown, but astronomers are hopeful that the proportion is substantial. Transit and gravitational micro-lensing surveys reach further out, and the tiny yield (one planet so far) indicates that fewer than about one-third of stars have Jupiter-mass planets within Jupiter-like distances (5 AU) of their stars. In some surveys this might be because of the metal-poor stars that have been examined, but in others this upper limit might apply also to higher metallicities – this subject is very young, so these conclusions are very uncertain.

Of the 91 stars known to have planets, five of the stars are members of multiple-star systems, i.e. systems in which two or more stars orbit each other. In all five cases the planet(s) orbits one of two stars in a binary system (two stars), the other star being further away. For example, in Gamma Cephei, a giant planet is in an orbit with a semimajor axis of 2.15 AU. The second star is at 21.4 AU. It might have been the case that a star so close would have prevented planetary formation, or ruled out stable orbits. This is not so. About one-half of the 'stars' in the sky are in fact multiple, mainly binary systems. The fact that multiple systems can have planets increases considerably the potential number of stars with planets.

QUESTION 7.2

Redraw Figure 7.4 to show where the habitable zones would lie when the stars are about 5000 Ma old. Justify your estimates. Only an indicative shift of the zone is needed, not its actual position. *Hint*: think what happens to the luminosity of a main sequence star as it ages.

7.2.2 Exoplanet masses

Figure 7.5 shows the number of measured values of $M_p \sin i_0$ in the mass intervals 0 to $1M_J$, 1 to $2M_J$ etc. M_p is the actual mass of the planet and i_0 is the inclination of its orbit with respect to the plane of the sky – remember that the radial velocity method gives $M_p \sin i_0$ not M_p (Section 6.6.2). The least massive planet definitely discovered at the time of writing (April 2003) is $0.12M_J$, which is about $38M_E$, a bit less than one-half the mass of Saturn ($95M_E$). At the other extreme, we are not concerned with bodies of greater mass than $13M_J$, because they are better classified as stars than as planets. Such massive bodies will be of roughly stellar composition, made largely of hydrogen and helium. For such a composition $13M_J$ is the approximate threshold above which the deep interior becomes hot enough for thermonuclear fusion to occur, involving the rare isotope of hydrogen, called deuterium (^{2}H). This fusion process can last the order of 1000 Ma, because even though deuterium is a scarce isotope, it is consumed at a low rate at the modest temperatures in these comparatively low-mass objects.

At masses greater than about $80M_J$ ($0.08M_\odot$) the interior temperatures are high enough to initiate the sustained fusion of the far more abundant isotope ^{1}H, a copious source of power, and so the star then enters on its protracted main

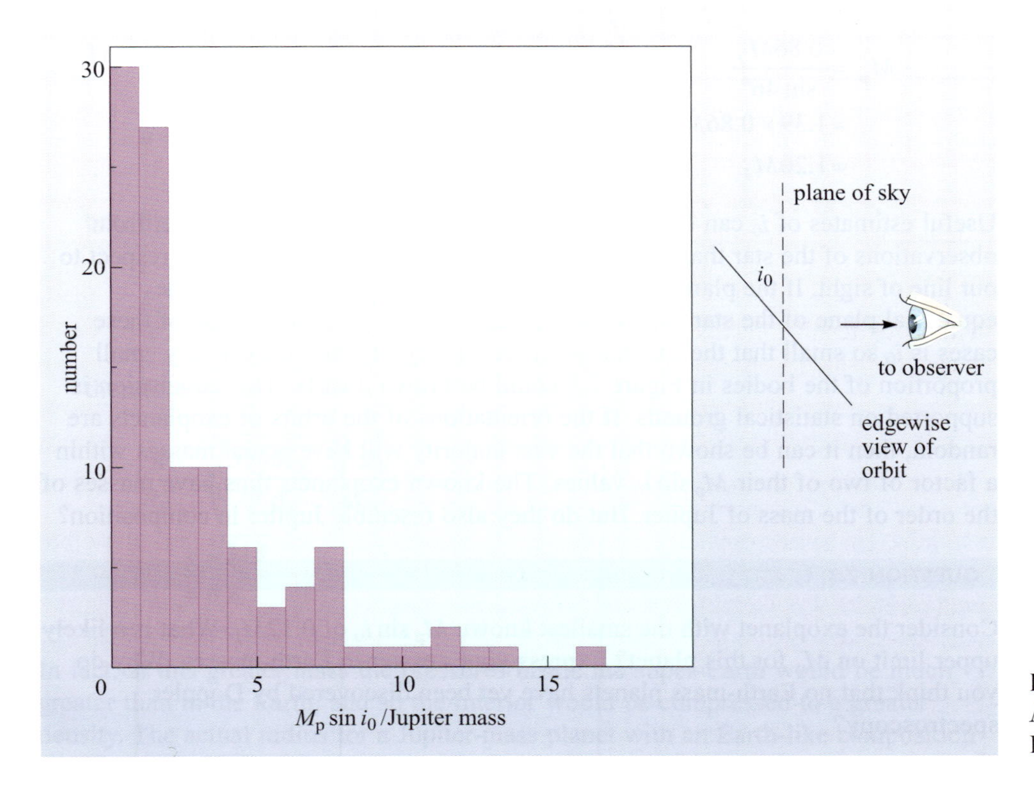

Figure 7.5 The distribution of $M_p \sin i_0$ for the known exoplanets. Inset: illustration of i_0.

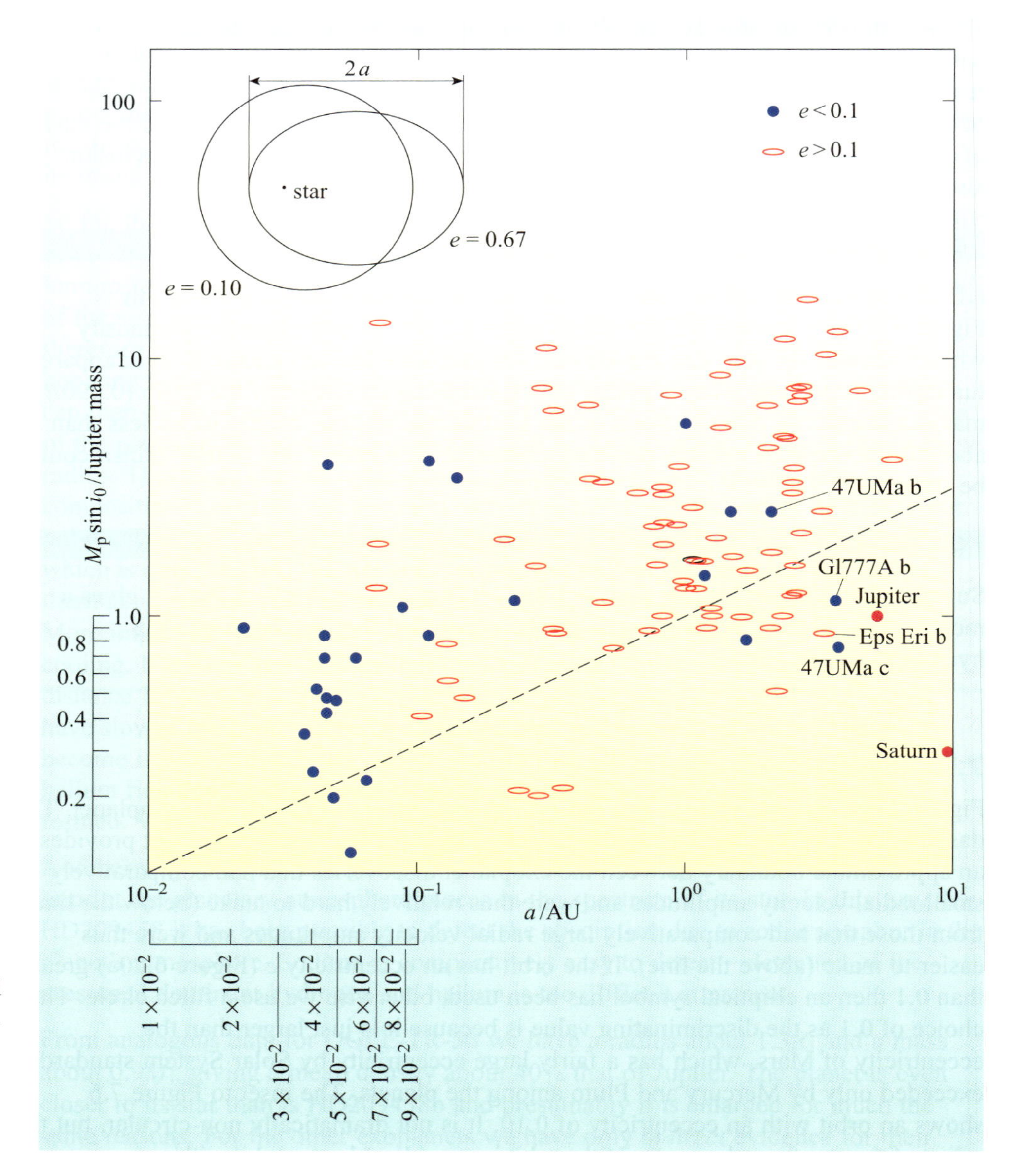

Figure 7.6 $M_p \sin i_0$ versus orbital semimajor axis a for the exoplanets. See text for explanation. Inset (top left): orbits with eccentricities of 0.10 and 0.67. Both have the same semimajor axis a.

This proximity would be unremarkable if the exoplanets were low-mass iron–silicate bodies, as in the Solar System, but they are hydrogen–helium giants. To see why proximity is then remarkable you need to know something about the two current models for the formation of giant planets. A sufficient outline is provided in Box 7.2.

BOX 7.2 THE FORMATION OF PLANETS

Almost all astronomers adhere to the view that planets form from a circumstellar disc of material left over from the formation of the central star. Earlier, all this material was a slowly rotating fragment of an interstellar cloud, mainly of hydrogen and helium, that began to contract under its own gravity. As it contracted its rotation rate increased, for the same reason that the rotation rate of an ice-skater with extended arms increases as the arms are pulled in. The underlying reason is the conservation of a quantity called angular momentum. The net outcome of gravitational contraction, increasing rotation, and collisions of molecules and dust grains within the fragment is not at all obvious, but is as shown in Figure 7.7. Most of the mass is in a central condensation that becomes the star, and a few percent is in a flattened disc orbiting the star.

Though the star ends up with nearly all the mass, star and disc have the same elemental composition. In each of them about three-quarters of the mass is hydrogen, about one-quarter is helium, and the proportion of heavy elements – the metallicity – can be anything up to a few percent. In the disc a proportion of the heavy elements is in compounds, mainly with the abundant hydrogen (e.g. CH_4, NH_3 and H_2O, which are common icy materials) but also with each other (e.g. $FeMgSi_2O_6$, which is a silicate). Compounded or otherwise, nearly all of the heavy elements are initially in dust grains spread throughout the disc. The main exceptions are the chemically unreactive ones, such as neon and argon – these remain as gases. Low metallicities might not lead to planet formation, because little dust then forms.

When the star forms it heats the disc, and this evaporates some of the dust to give a radial variation in dust composition. Very close to the star there is no dust at all, just gas. Then there is a zone where dust consists of only the most condensable compounds (the most refractory), such as iron, and silicates. Further out a very important boundary is encountered. This is where water can condense, to form an icy component of the grains.

- Given that oxygen is the most abundant of the heavy elements, what is the relative abundance of water?

- With hydrogen even more abundant, water must be a particularly abundant compound.

Figure 7.7 An artist's impression (Julian Baum) of a young star and circumstellar disc from which planets will form. Note that this is an oblique view – the disc is circular. (Julian Baum/Take 27 Ltd)

The water is in fact about 20 times more abundant than anything else (except hydrogen and helium). The boundary where water condenses is called the **ice-line** (sometimes the snow-line). In the Solar System it was at about 4 AU, and at this boundary, because water and therefore water-ice is so abundant, there is a considerable increase in the mass in the dust grains.

In the inner disc the dust coagulates to form kilometre-sized planetesimals, and these eventually build up to form **planetary embryos** with masses of order $0.1M_E$, which then combine to form a few iron–silicate bodies with masses up to about that of the Earth. Beyond the ice-line, because of the abundance of water-ice, a few bodies build to the order of $10M_E$, mainly composed of water. At this mass the **kernels**, as they are called, gravitationally capture gas from the disc.

- What is this gas made of?
- It is predominantly hydrogen and helium.

Note the distinction between 'inside the habitable zone' which is between the inner and outer boundaries of the habitable zone, and 'interior to the habitable zone' which is the region closer to the star than the inner boundary of the habitable zone.

Thus, giant planets rich in hydrogen and helium are formed. Giants do not form interior to the ice-line because there is no water-ice to build the massive kernels. Gas capture stops when the circumstellar disc is dissipated, aided by convulsions in the young star that cause it to emit a violent wind – the so-called T-Tauri phase. Throughout gas capture, and perhaps for some time afterwards, icy–rocky planetesimals are captured, giving further heavy element enrichment, though the composition remains dominated by hydrogen and helium. In the Solar System, Jupiter has a mass of $318M_E$, of which, in this model, 5 to 10% would be heavy elements. Saturn, with a mass $95M_E$, if it has a similar total mass of heavy elements, would be proportionally more enriched. Saturn is less massive than Jupiter because it formed more slowly and therefore had captured less gas when the T-Tauri phase started. Uranus and Neptune, at about $15M_E$, must have formed even more slowly, and water is their main component.

That was the standard view until the late 1990s. Then, a theory that had been around in earlier decades and had never quite vanished, had a resurgence. This is the formation of giants in *one* stage, rather than the two-stage kernel-gas capture process. Models showed that at a sufficient distance from the star, roughly beyond the ice-line, the disc might become gravitationally unstable in such a way that giant planets contract directly from it. In this case the giants will initially have the same proportion of heavy elements as the star, up to about 3%, but the subsequent capture of icy–rocky planetesimals could raise this, even to 5 to 10%. Both theories are currently viable, and perhaps both processes operate, one predominating in some circumstellar discs, the other in the rest.

In either the two-stage or one-stage process of building giant planets, they form around or beyond the ice-line, about 4 AU in the case of Sun-like stars. Therefore, if giant planets do form in one of these two ways then those giant exoplanets that are now well within the ice-line of their star must have subsequently moved inwards. It was not long after the discovery of the first of these close-in giants, 51 Pegasi b in 1995, that theoreticians found migration mechanisms that could produce the **hot Jupiters**, as they are called. That these mechanisms are fairly convincing adds weight to the view that the close-in giants are hydrogen–helium in composition. The mechanisms are outlined in the next section.

QUESTION 7.6

A few of the points in Figure 7.6 are labelled with the name of the exoplanetary system. A system more like the Solar System than nearly any other system is 47 Ursae Majoris (47 UMa).

(a) Make a table of the values of the minimum masses of its planets and the semimajor axes of their orbits.

(b) Draw the orbits of its giants with the orbits of Jupiter and the Earth to the same scale. You can approximate all the orbits as circles with the star at the centre. Label each giant orbit with the minimum mass of its giant in units of M_J. Include on your drawing the present-day habitable zone boundaries for 47 UMa, which extends from about 1.0 AU to 1.9 AU. (47 UMa is similar to the Sun in mass, though at 7000 Ma it is a bit older and thus its habitable zone is further out – see the answer to Question 7.2.).

7.3 Migration of exoplanets within exoplanetary systems

The discovery in the mid 1990s of 'hot Jupiters', much closer to their stars than the models of their formation predicted, led within months to plausible mechanisms by which planets could migrate inwards. Of course, if the 'hot Jupiters' could, somehow, have formed near to where we find them, then migration is unnecessary, but we don't understand how they could have formed so close in. By contrast, migration, extensive or limited, now seems almost inevitable. Theoreticians were not entirely being wise after the event, because migration had been predicted over a decade earlier, but largely overlooked. Migration mechanisms are outlined in Section 7.3.1. Then, in Section 7.3.2, we examine the implications of giant-planet migration for the formation and survival of Earth-mass planets.

7.3.1 Migration mechanisms, and consequences for giants

The key to migration is the gravitational effect of the giant planet on the circumstellar disc of gas and dust. The details are complex so only a qualitative outline is given here. At first the disc is symmetrical around an axis through its centre, as shown in Figure 7.7. But as the mass of the giant planet grows, its gravitational field produces spiral structures in the disc that destroy this symmetry, as in Figure 7.8. The disc mass considerably exceeds the mass of the planets that will form from it, so there's plenty of disc left. Consider the two-stage formation of a giant at the point where its icy–rocky kernel has a mass less than M_E. The spiral structure in the disc interior to the kernel has a gravitational influence on the kernel's orbit that tends to push it outwards, whereas the spiral structure exterior to the disc exerts a gravitational influence that tends to push it inwards. For any plausible disc model, the inwards push is the greater and so the

Figure 7.8 A computer model of a circumstellar disc with a spiral structure created by a planetary kernel. The disc is not modelled close to the star, hence the hole, which is an artefact. (Pawel Artymowicz)

Figure 7.9 A massive kernel opens up a gap in its circumstellar disc. (Pawel Artymowicz)

net effect is inward migration. The rate of migration is proportional to the mass of the disc and also to the mass of the kernel, so as it grows it migrates inwards ever more rapidly. This is **Type I migration**. Note that the disc itself is also migrating inwards, but always more slowly than the kernel.

Type I migration continues until the kernel has grown to sufficient mass to open up a gap in the disc, as illustrated in Figure 7.9. The gap causes a major change in the migration. It now slows dramatically, by a factor of between 10 and 100, until the kernel and the disc are migrating inwards at the same rate. This is **Type II migration**. The kernel mass at which the transition takes place depends on various properties of the disc (its density, thickness, viscosity, temperature, and so on), and on the distance of the kernel from the star. An approximate range is $10M_E$ to $100M_E$, and so it is likely that there is a fully fledged giant kernel plus some captured gas at this transition. Gap formation reduces the rate at which the kernel acquires mass from the disc, but does not halt it, and so planets up to several M_J can form.

- If giant planets form in one stage, how is the story modified?
- There is no kernel build-up, and so it is likely that there will be no Type I migration – the giant enters the story at the point of a fully fledged Type II migration.

Migration needs to be halted, otherwise all the giant planets will end up in the star. Either the disc must be removed, or there must be counteracting effects. The disc will be removed partly through gradual infall to the star and partly through bursts of activity that young stars are observed to undergo, when outflowing stellar winds and intense UV radiation push the disc away (T-Tauri phase). Observations of young stars suggest that the disc lasts 1 to 10 Ma. This will be too long in some cases for the giant planet to survive, because the Type II migration time, depending on the disc properties and other parameters, can be less than this. There are however several ways in which opposing effects can appear, including tidal interactions between the planet and the star, magnetic interactions between the star and the disc, and evaporation by stellar radiation of a narrow zone in the disc a few AU from the star, creating a migration barrier. These are rather subtle effects and are peripheral to the story, so they are not detailed here, but note that they could save otherwise doomed planets.

So far we have considered a sole giant planet in a disc, yet the Solar System and at least a few of the known exosystems have more than one giant, and it is presumed that most or all of the others do too. Computer models of discs with two giants have shown that interactions between the giants can slow and even reverse Type II migration, and this is one way in which we could have ended up with a Solar System in which Jupiter and Saturn are still beyond the ice-line. It is also possible under special choices of circumstellar disc parameters to end up with limited migration without invoking giant–giant interactions.

The multiple-giant case can explain giants in highly eccentric orbits. Figure 7.10 shows the eccentricities e versus semimajor axis a, with Jupiter and Saturn for comparison. Some of these eccentricities are very high, particularly at larger a. It is difficult to produce such large eccentricities by formation from the circumstellar disc, without something extra. One way to get high e is when the orbital periods of the two giants are in a simple ratio, such as 1 : 2 or 1 : 3 – these are called orbital resonances (Box 4.2). In such a case the gravitational interaction between the giants can disturb the orbits in a cumulative way, resulting in large orbital changes. An everyday analogue is the pushing of a child on a swing – if you time your pushes correctly a large amplitude builds up.

Another way to achieve large eccentricities is through close encounters between giants. Simulations show that in each case a common outcome is that one giant is flung into the cold of interstellar space and the other is retained in a high e orbit. These scenarios are also the basis for an additional explanation of giants close to the star. If the surviving giant is in a high e orbit with a small **periastron distance** (the closest distance to its star), then tidal interactions with the star, perhaps aided by residual disc gas, will reduce e and we can end up with a giant in a small, low e, orbit.

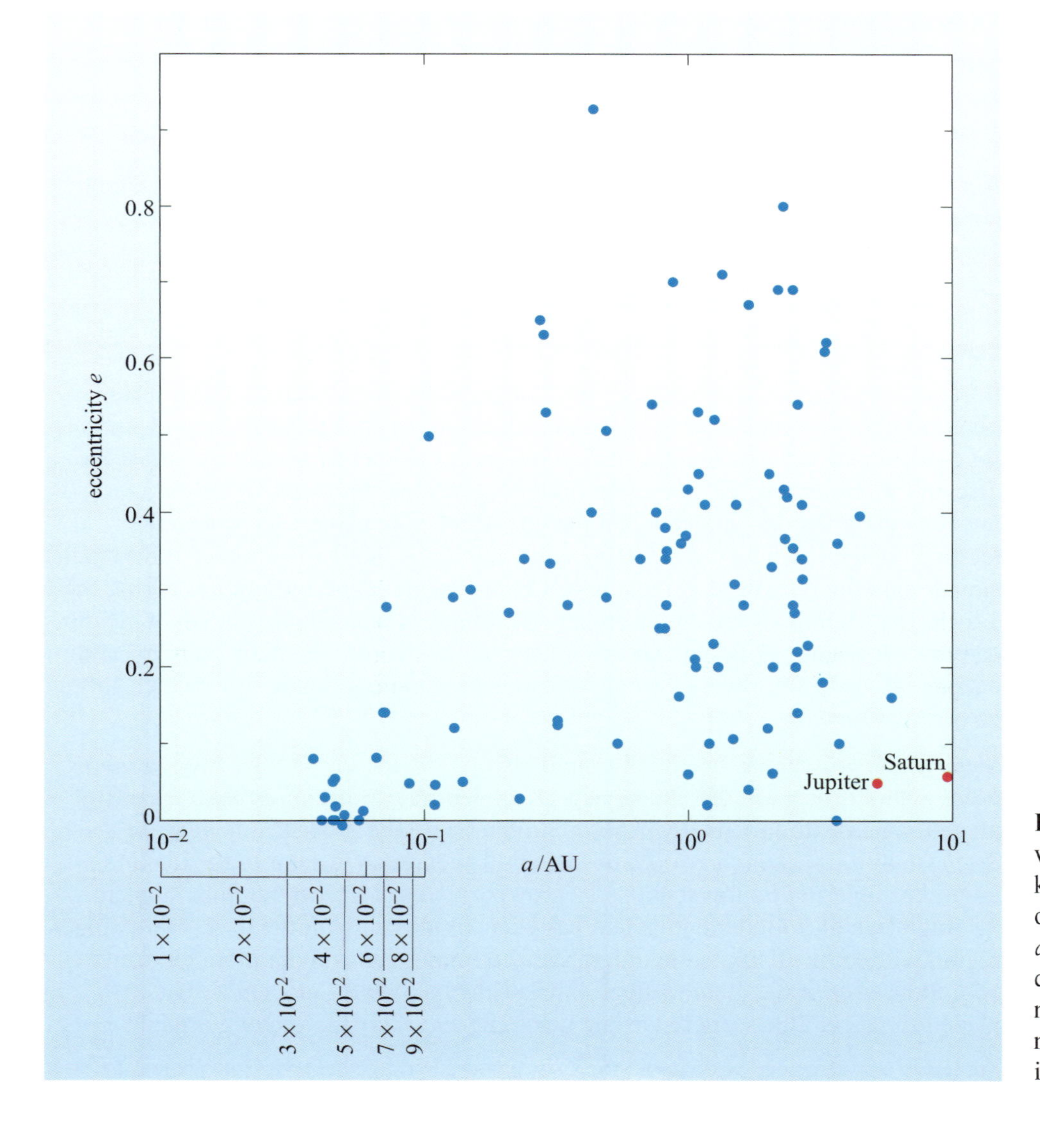

Figure 7.10 Eccentricities e versus semimajor axis a for the known exoplanets. The clustering of several $e = 0$ values around $a = 5 \times 10^{-2}$ AU has forced us to draw one of them as if e were negative. This is not the case; negative eccentricities are impossible.

In conclusion, the system parameters specifying the star, disc, and the giant planets, are sufficient in number and sufficiently variable, that giant planets with a great range of masses, orbital semimajor axes, and eccentricities can result. We certainly have plausible explanations of the observed exoplanetary systems. The giant planets in our Solar System emerge as just one of many possible types of outcome. However, plausibility does not mean that it actually happened that way. There is no guarantee that the exosystems came into existence in the manner described above. The best current explanations have been given.

QUESTION 7.7

Take a look at Table 7.2 which shows some properties of two exoplanetary systems. For each one outline the sequence of events that could have led from each giant being an Earth-mass kernel beyond the ice-line to the systems we have today. (We have ignored an unconfirmed distant giant planet of Epsilon Eridani.)

Table 7.2 Properties of two exoplanetary systems.

System	$M_p \sin i_0/M_J$	Semimajor axis a/AU	Eccentricity
Epsilon Eridani	0.86	3.3	0.608
HD168746	0.23	0.065	0.081

But what about Earth-mass planets? Could these form and survive in the habitable zones of the known exoplanetary systems? If not then the chances of finding potential habitats there are much reduced.

7.3.2 Implications of giant-planet migration for the formation and survival of Earth-mass planets

A major problem with the speedy Type I migration is that it carries Earth-mass planets rapidly into the star. Only by careful choice of disc parameters is it possible for planets with a mass of about $1M_E$ to migrate sufficiently slowly to outlast the disc and thus survive. This can seem rather contrived. Fortunately for our understanding, models indicate that the growth of $1M_E$ planets interior to the ice-line, which is where the habitable zone will lie, is likely to be rather slow, much slower than the growth of kernels beyond the ice-line. Type I migration during the 1 to 10 Ma lifetime of the circumstellar disc of gas is then slight, and when the disc disperses it is quite possible that this inner region will contain many embryos with masses up to $0.1M_E$, plus a swarm of lower-mass planetesimals from which further growth can occur.

Meanwhile, growth has been more rapid beyond the ice-line. In the two-stage process this is because the abundance of condensable water provides a swarm of planetesimals with low relative speeds. In the one-stage process it is because the disc instabilities occur early on. Subsequently, the region interior to the ice-line might or might not be traversed by a (growing) giant. If, as in the Solar System, the migration of a Jupiter equivalent has been, at most, slight, the terrestrial-planet region will not be disrupted by migration. In other systems, with more extensive migration, even if the giant(s) stops short of the habitable zone, embryos and planetesimals will be scattered as the giant moves inwards, sweeping orbital resonances across the habitable zone. If the habitable zone is traversed by a migrating giant then embryos and planetesimals will certainly be scattered. So, could Earth-mass planets form after migration has ceased, and could any such planets survive in the habitable zones today? Let's consider their formation first.

The few theoretical studies have shown that it is possible for Earth-mass planets to form in the habitable zone after giant migration is complete. In cases where the giant comes to reside beyond the habitable zone, sufficient embryos and planetesimals are left provided that the giant planet never comes close to the outer boundary of the habitable zone. In cases where the giant traverses the habitable zone and ends up parked near the star, well away from the interior boundary of the habitable zone, there might be so little material left that the formation of Earth-mass planets from planetesimals and embryos is very unlikely. If such 'hot Jupiters' are ruled out, then in only about 10% of the exoplanetary systems known at the time of writing (early 2003) could Earth-mass planets have formed in the habitable zone. But should they be ruled out? There could be sufficient planetesimals left in some cases, and additionally there could be sufficient dust in the disc to form a new generation of planetesimals that lead to embryos and to an Earth-mass planet. If 'hot Jupiters' are not ruled out then perhaps as many as one-half of the known exoplanetary systems could have Earth-mass planets in the habitable zone.

If, in any system, an Earth-mass planet somehow formed in the habitable zone, the next question is whether it could have *survived* there for the present age of the star. This is not the same question as that of formation, which takes less than 100 Ma, perhaps a lot less. By contrast, gravitational buffeting by the giant(s) militates against survival for the age of the star, which is typically billions of years. Computer studies have been made of representative systems. On the basis of these it can be concluded that in perhaps one-third of the known exoplanetary systems, Earth-mass planets, provided that they can form, will today still be in the habitable zone. If an Earth-mass planet does not remain confined to the habitable zone its usual fate is to be flung into the cold of interstellar space, rather than have a collision with the giant or the star.

Figure 7.11 shows the sort of regions in the habitable zone of typical systems where Earth-mass planets could survive. Figure 7.11a shows a giant planet in a low-eccentricity orbit near its star, very much interior to the habitable zone, and Figure 7.11b shows a giant in a more eccentric orbit not far outside the habitable zone. The systems awaiting discovery (Section 7.4) are likely to include ones that more resemble the Solar System, where the giants, Jupiter and Saturn, are well beyond the habitable zone (Figure 7.11c). In such systems the whole habitable zone, like ours, would be a safe harbour for Earth-mass planets.

QUESTION 7.8

Suppose that a giant planet was in an orbit that lay entirely inside the habitable zone (inside the zone, not interior to the zone).

(a) Where might an Earth-mass planet be found today in this system?

(b) Where might a planetary body be found that is actually habitable? (Remember how narrow is the definition of 'habitable zone'.)

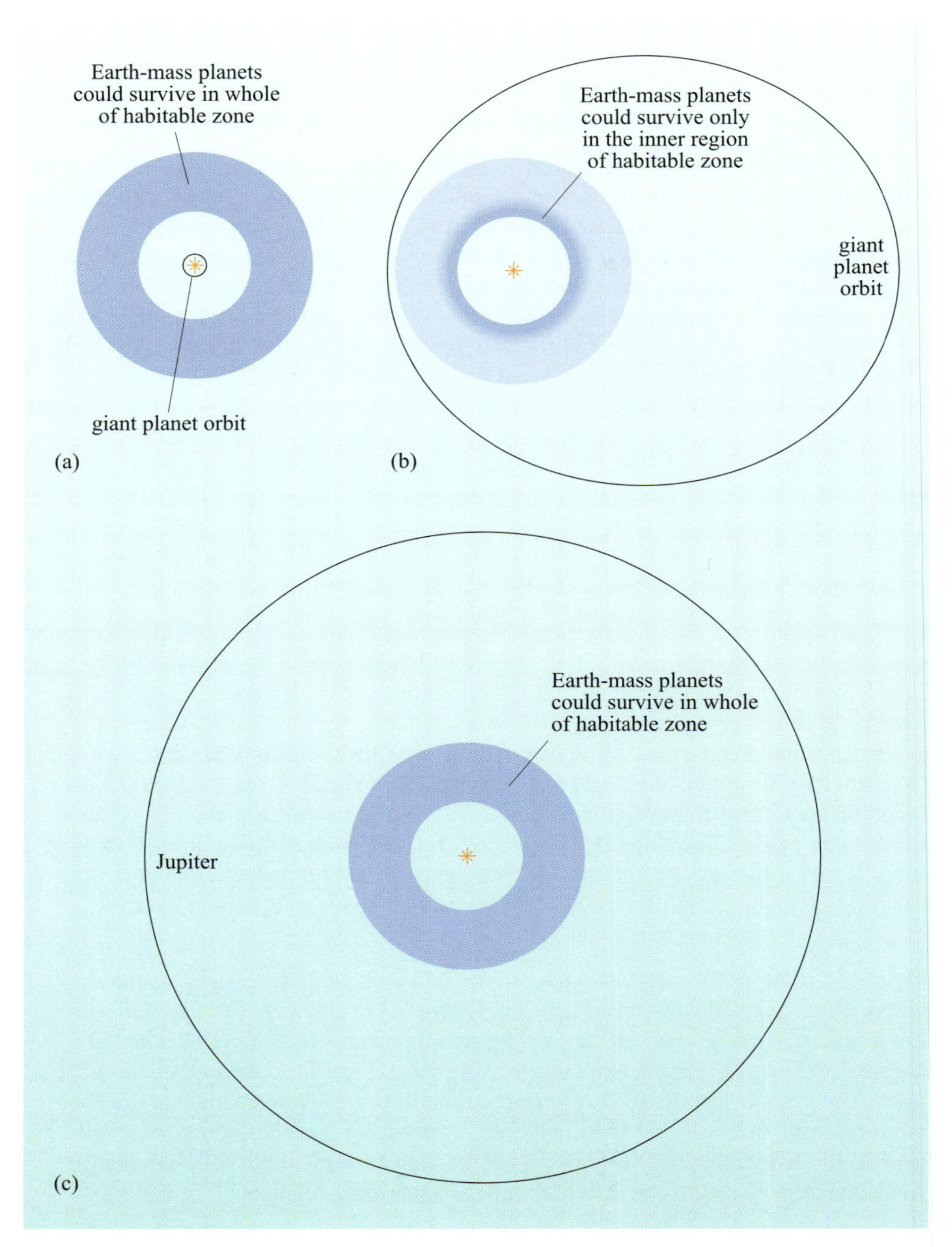

Figure 7.11 Survivable orbits for Earth-mass planets in the habitable zone where (a) the giant is very much interior to the habitable zone, (b) the giant is not far outside the habitable zone, (c) the giant is well outside the habitable zone as is the case with Jupiter in the Solar System.

7.4 The undiscovered exoplanets

7.4.1 The known exoplanetary systems – a summary

So far (early 2003), Doppler spectroscopy has discovered all but one of the 105 confirmed discoveries of planets, in 91 systems, 12 of which are known to be multiple-planet systems. The one exception, OGLE-TR-56b, was discovered in transit, and later confirmed by Doppler spectroscopy. Subsequent to their discovery, HD209458b has been observed in transit, and Gliese 876b has been detected astrometrically.

The properties of the known exoplanetary systems can be summarized as follows (pulsar planets are excluded).

- The furthest planet discovered by Doppler spectroscopy is around the star HD47536, at 401 light-years. OGLE-TR-56 is at about 5000 light-years.
- About 8% of nearby solar-type stars have giant planets within 4 AU of the star.
- Stars with high metallicity seem to be favoured for having planetary systems.
- The planets range in minimum mass ($M_p \sin(i_0)$) from $0.12M_J$ (a bit less than one-half the mass of Saturn) to the brown-dwarf limit, $13M_J$, with a preponderance at the lower end of the mass range.
- The planets are thought to be rich in hydrogen and helium, though for only HD209458b and OGLE-TR-56b is there direct observational evidence for this, from the mean densities.
- Nearly one-half of the planets orbit closer to their star than Mercury does to the Sun, and of the remainder only two planets (perhaps three) are beyond 4 AU, in (different) multiple systems.
- Three-quarters of the planets have orbital eccentricities greater than 0.1.
- Five stars with planets are in binary stellar systems, the planet(s) orbiting just one of the two stars.

Transit and gravitational micro-lensing surveys have shown that further afield, fewer than one-third of stars have Jupiter-mass planets out to Jupiter-like distances (5 AU) from the stars.

7.4.2 What planets await discovery?

Exoplanets continue to be discovered, even within a few hundred light-years of the Sun. Moreover, we have not yet acquired a representative sample. In particular, observational selection effects are playing a major role. You met such effects in Chapter 6. Here we consider them further, first, those associated with the Doppler spectroscopy method that has been so fruitful.

The first selection effect arises from the need to collect enough light to display the spectral lines sufficiently clearly to measure their wavelengths accurately. This discriminates against large distances and explains why all the exoplanetary systems discovered by Doppler spectroscopy are nearby (though it can reach further, as exemplified by the post-discovery detection of OGLE-TR-56b). Another selection effect has been the preference given until recently to solar-type stars – this is because they have an abundance of sharp spectral lines, and are bright (Section 7.2.1).

Other selection effects can be seen in Figure 7.12, which is derived from the equation for the observed radial velocity amplitude v_{rA} of the star, to which the Doppler shift is proportional. This equation was developed in Section 6.6.2. In Figure 7.12 v_{rA} is shown plotted against the orbital period P. Two lines are given, one for a $1M_{\odot}$ star, the other for a star with $0.5M_{\odot}$. The value used for $M_p \sin(i_0)$ is M_J in both cases – for other values note that v_{rA} is proportional to $M_p \sin(i_0)$.

- Referring to Figure 7.12, what happens to v_{rA} if (i) $M_p \sin(i_0)$ is halved, (ii) P is decreased by a factor of 8?

- (i) v_{rA} is also halved (ii) v_{rA} is doubled.

So, the smaller the value of $M_p \sin(i_0)$ the smaller the Doppler-shift and the harder the planet is to detect, which is why Earth-mass exoplanets have not yet been detected by Doppler spectroscopy. Conversely, the smaller the value of P the greater the Doppler shift and the easier the planet is to detect. Overall, massive planets in short-period orbits are favoured by this technique. Short-period orbits have small semimajor axes, and so the Doppler spectroscopy technique favours planets close to their stars. The upper scale in Figure 7.12 gives values of the semimajor axis a for the planet for each of the two stellar masses.

Another advantage of small P is that the radial velocity goes through its cycle in a shorter time. About one orbital period of data is needed to identify a planet with reasonable certainty, and so this can be achieved more quickly the shorter the period. For a planet like Jupiter with an orbital period of 12 years, about 12 years of high quality data would be required.

For nearby solar-type stars, Doppler spectroscopy has discovered giant planets within 4 AU of the stars for about 8% of them. A further small proportion might also have such planets, undetected because the orbit is nearly face-on to us, thus reducing the Doppler signal. This technique has been in action long enough to be on the threshold of detecting Jupiter-mass planets at the Jupiter distance of 5 AU. Radial velocities can currently be measured with a detection limit approaching 1 m s^{-1}, and this is sufficient for any Earth-mass planets around the low-mass M dwarfs, particularly in the habitable zones, which are close to such stars.

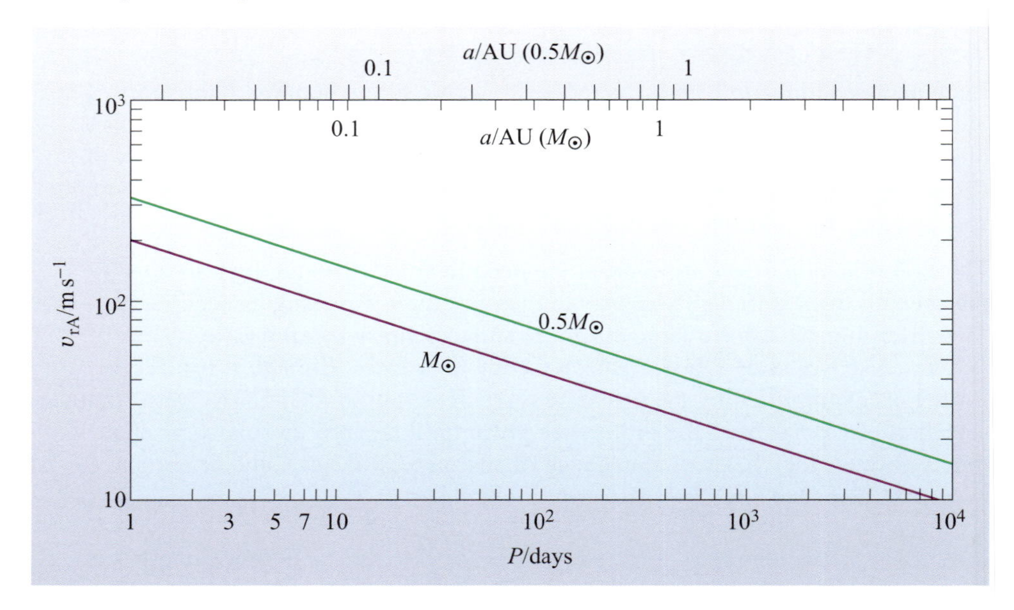

Figure 7.12 How the radial velocity amplitude v_{rA} varies with P (and a) for a solar-mass star and for a star of mass $0.5M_{\odot}$, using a value of $M_p \sin(i_0)$ of $1M_J$. Note that a is the semimajor axis of the planet's orbit.

In the astrometric technique it is the periodic shift of the star's position that reveals the planet (Section 6.6.1). Figure 7.13 shows βd versus P where d is the distance to the star and β is the angular movement of the star in the sky as seen from a distance d. We give βd because then you can calculate β for any value of d. The planet has a mass M_J, and βd is proportional to this mass. One line in Figure 7.13 is for a solar-mass star and the other for a star of one-half this mass. The upper scale gives values of a for the planet. It is now the case that larger P is favoured, corresponding to larger a, though this means that a long series of observations is required.

- For a given semimajor axis of the *star's* orbit, what happens to β if (i) d is doubled (ii) the inclination i_0 is halved?
- (i) β is halved (so, distant stars are disfavoured) (ii) β is unchanged (this technique gives us the mass, not the minimum mass).

As yet astrometry has detected only one planet – Gliese 876b, some years after its discovery. This detection was facilitated by the low stellar mass of $0.32M_\odot$, and the large mass of the giant, $3.3M_J$. Astrometry is ideal for detecting planets in large orbits, and will surely do so when it acquires the required sensitivity, and as soon as observations have been made for long enough. The space telescopes SIM and GAIA (Section 6.6.1), due about 2010, will have the giant planets in Solar System analogues within their capabilities out to the order of 1000 light-years. Earth-mass planets might be detectable around M dwarfs within a few tens of light-years.

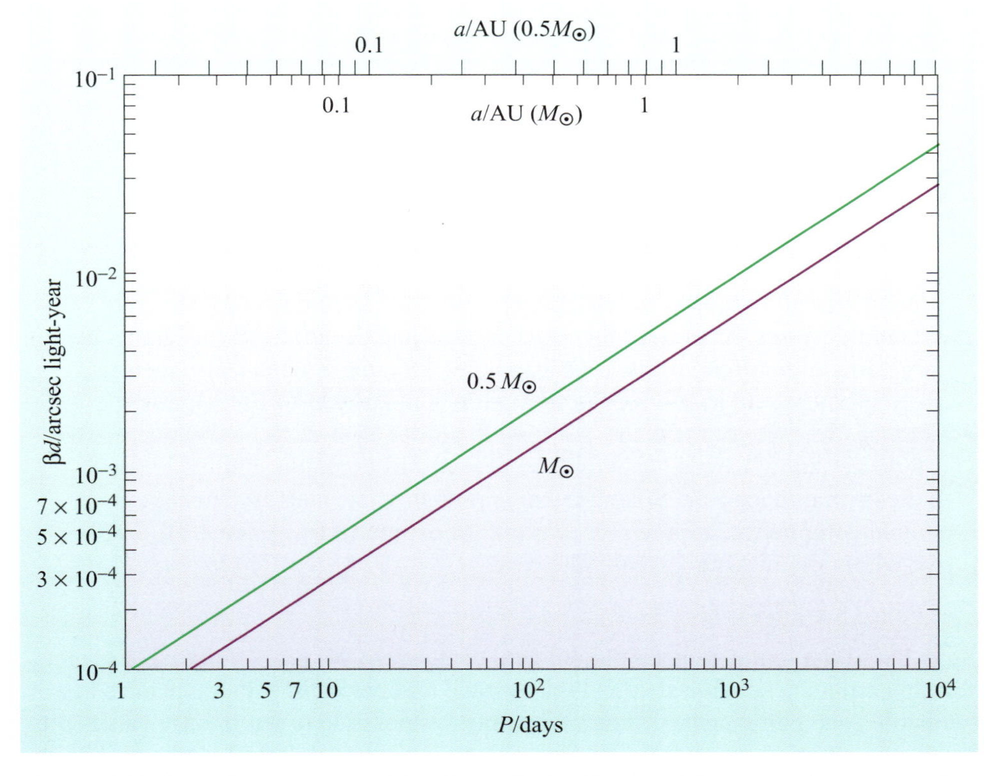

Figure 7.13 How, in astrometry, βd varies with P and a for a $1M_\odot$ star and for a star of mass $0.5M_\odot$, for a planet with a mass $1M_J$. Note that a is the semimajor axis of the planet's orbit.

Within the next few years other techniques are likely to yield further discoveries, such as transit photometry, and gravitational micro-lensing outlined in Section 6.5. It is even possible that on this timescale giant planets at wide separations from their star will be directly imaged with ground-based telescopes. Different selection effects are associated with these techniques. For example, transit photometry can only detect planets for which the orbital inclination i_0 is within a few degrees of 90°, so that the orbit is nearly edge-on to us. So far, just one planet (OGLE-TR-56) has been discovered in this way. It is also necessary for the drop in apparent luminosity of the star to exceed 0.1% (see Question 7.4). On the other hand it is possible to detect Earth-sized planets around M dwarfs by this technique right now, and gravitational micro-lensing could detect an Earth-mass planet around any star. Both techniques can reach out to thousands of light-years, so either there will be a cornucopia of discoveries, or we will discover that planetary systems are rare beyond the solar neighbourhood.

Though we have an increasing capability to detect Earth-mass planets, we will, of course, only detect them if they are there. Recall that theoretical studies have shown the following.

- Migration of giant planets makes it less likely that Earth-mass planets could form in the habitable zone of the star, though 10 to 50% of the known exoplanetary systems could have such planets.
- Perhaps in one-third of the known exoplanetary systems, Earth-mass planets, provided that they can form, might today still be in the habitable zones.

We can thus expect to find Earth-mass planets, even in habitable zones, though perhaps in only a small proportion of systems where the giant planets are within the ice-line.

To summarize, we expect the proportion of stars known to have planets to rise, perhaps considerably. Among these new discoveries there should be lower mass planets, and planets in larger orbits. By the middle of the next decade we will have space telescopes such as Darwin and TPF (Section 6.3) capable of seeing the planets themselves, even Earth-mass planets. Huge ground-based telescopes such as OWL might also have this capability in favourable cases. At present the Solar System stands alone with its giants in large, low eccentricity orbits, and its whole habitable zone is consequently a particularly secure abode for the formation and survival of Earth-mass planets. We expect systems resembling the Solar System to be discovered, but we do not know whether they will be revealed as common. Simulations indicate that though the Solar System is probably unusual, systems like it should comprise an appreciable minority of exoplanetary systems. By 2020 we will know if this is actually the case.

So, what of the two parameters in the Drake equation: the probability p_p of planets forming around a suitable star, and the average number n_E of suitable planets in a habitable zone per planetary system? It is not yet possible to put precise values on these parameters, but we do know that p_p is not vanishingly small, and could be as high as several tens of percent. More speculatively, it is also likely that n_E is not vanishingly small, and could be at least one-third, i.e. one per three systems, on average.

In the next chapter we anticipate the discovery of suitable planets and ask how we could determine if they are habitable, even inhabited.

QUESTION 7.9

Consider the case of an Earth-mass planet 1 AU from a solar-type star.

(a) In a few sentences, discuss whether the planet is in the habitable zone of its star.

(b) With the help of Figure 7.12 (which is for a $1M_J$ planet), calculate the radial-velocity amplitude v_{rA} of the star due to the Earth-mass planet ($M_J = 318M_E$). Hence decide whether this planet could be detected. Assume a detection limit on v_{rA} of 1 m s^{-1}.

(c) With the help of Figure 7.13 (which is also for a $1M_J$ planet), calculate βd due to the Earth-mass planet. Calculate the stellar distance d (in light-years) at which the star's motion could be detected, assuming an astrometric detection limit of 0.001 arcsec (which is the width of a finger from a distance of about 3000 km). Are there any stars closer to us than this?

7.4.3 A note on evidence from circumstellar discs

Among the planets awaiting discovery, it is expected that some will be around those stars known to be surrounded by circumstellar gas and dust. Many young stars are known to have discs of gas plus some dust, and a few older stars are known to have discs or rings predominantly of dust, most or all of the gas having been dissipated by stellar activity, such as during the T-Tauri phase. As noted earlier, Epsilon Eridani is known to have a dust ring *and* a planet. The planet is interior to the ring and might be too far from it to cause gravitational distortions in the ring. The ring however is non-uniform, and the cause of the distortion might be planets near its inner boundary. Distortion is also present in the dust disc around Beta Pictoris (Figure 7.14). The disc is presented to us nearly edge-on, and it is clearly warped. This warp could well be due to the gravitational influence of a giant planet. Moreover, the very existence of dusty discs or rings around older main sequence

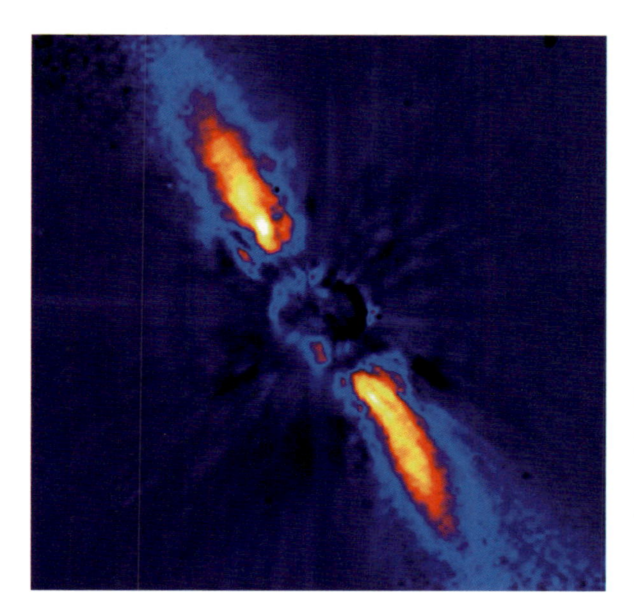

Figure 7.14 The dust disc around the star Beta Pictoris. The dark patch in the middle is part of the imaging system. The inner boundary of the disc is at about 20 AU. (European Southern Observatory (ESO))

stars calls for a mechanism to replenish the dust that otherwise would have been long gone. A likely mechanism is collisions between asteroids and perhaps comets. If there are such bodies, then there could be planets too.

QUESTION 7.10

Table 7.3 contains data on *undiscovered* planets of two main sequence stars in circular orbits. Consider the data carefully and then answer parts (i)–(v).

Table 7.3 For use with Question 7.10.

	Mass of planet M_p/M_J	Orbital inclination i_0/degrees	Orbital period of planet/days	Mass of star $M_*/M_\odot$	Distance of star/light-years
A	0.0030	89.9	200	1.0	25
B	2.0	30	2000	0.50	250

(i) Obtain the semimajor axis of the planets' orbits, expressing your answers in AU.

(ii) With aid of Figure 7.12 obtain the radial velocity amplitudes v_{rA} that would be observed in each case, expressing your answer in metres per second.

(iii) With the aid of Figure 7.13, obtain the angular movement in the sky β of the stars' positions that would be observed, expressing your answer in arc seconds.

(iv) State whether each planet is likely to be discovered by Doppler spectroscopy, astrometry, or transit photometry, or by more than one technique, giving reasons. You can assume that the detection limits for each technique are, respectively, $1\ \mathrm{m\,s^{-1}}$, 0.001 arcsec, 0.1% apparent luminosity change.

(v) Discuss whether either planet is likely to be habitable, assuming that the star has an age of 5000 Ma in each case, and that residence in the habitable zone is a requirement.

7.5 Summary of Chapter 7

Much of the data here applies to early 2003, and is subject to change, even in the short-term.

- Most of the Sun-like stars within a few hundred light-years of the Sun have been observed by Doppler spectroscopy long enough to reveal any planets of the order of Jupiter's mass in orbits within about 4 AU of the star. A small proportion of stars less like the Sun have also been observed, particularly the abundant M dwarfs. Doppler spectroscopy has yielded 104 planets in 90 systems, 12 of which are known to be multiple-planet systems.
- About 8% of the nearby Sun-like stars have giant planets within 4 AU of the star. A further small proportion might also have such planets, undetected, because the orbit is nearly face-on to us, thus reducing the Doppler signal.
- Stars with high metallicity seem to be favoured for having planets.
- The planets range in minimum mass ($M_p \sin i_0$) from $0.12M_J$ (a bit less than one-half the mass of Saturn) to the brown dwarf limit, $13M_J$, with a preponderance at the lower end of the mass range.

- Nearly one-half of the planets orbit closer to the star than Mercury does to the Sun, and of the remainder only two planets (perhaps three) are beyond 4 AU, in (different) multiple systems.
- Three-quarters of the planets have orbital eccentricities greater than 0.1.
- Five stars with planets are in binary stellar systems, the planet(s) orbiting just one of the two stars.
- In addition, there have been transit and gravitational micro-lensing surveys that so far (by transit) have discovered one planet in one new system. Several planets await confirmation. These surveys have established that, at greater ranges than are presently accessible to Doppler spectroscopy, fewer than one-third of stars have Jupiter-mass planets out to Jupiter-like distances (5 AU) from their stars.
- The giant exoplanets are thought to be rich in hydrogen and helium, though for only two planets (HD209458b and OGLE-TR-56b) is there direct observational evidence for this. They probably formed beyond the ice-line, in which case those inside the ice-line, the great majority, must have got there through migration. Migration through gravitational interaction with the circumstellar disc has a plausible theoretical foundation.
- Migration of giant planets makes it less likely that Earth-mass planets could form in the habitable zone of the star, though at least 10% of the known exoplanetary systems could have such planets.
- Perhaps in one-third of the known exoplanetary systems, Earth-mass planets, provided that they can form, could still be in the habitable zones.
- In the exosystems yet to be discovered there is likely to be a small but nevertheless significant proportion that more resemble the Solar System, with the giants well beyond the habitable zone and Earth-mass planets safe within the habitable zone.
- In the Drake equation, the probability p_p of planets forming around a suitable star is not vanishingly small, and could be as high as a few tens of percent. The average number n_E of suitable planets in a habitable zone per planetary system is not vanishingly small, and (speculatively) could be at least one-third, i.e. one per three systems, on average. (The rate R at which suitable stars are born, was estimated in Chapter 6 as about 30 per year in our Galaxy.)

CHAPTER 8
HOW TO FIND LIFE ON EXOPLANETS

In Chapter 7 we were essentially concerned with the first three factors in the Drake equation that was introduced in Section 6.1.1, namely, the rate R at which suitable stars are formed, the probability p_p of planets forming around a suitable star, and n_E the average number of suitable planets in a habitable zone. We now move on to consider how we could determine the next factor, p_l, the probability of life appearing on a suitable planet in a habitable zone.

In this chapter we will concentrate on the detection of life based on complex carbon compounds and liquid water, i.e. on carbon-liquid water life. We thus concentrate on life that resembles life on Earth. But we are not assuming that alien life is based on the *same* carbon compounds as terrestrial life. It might use a carbon compound other than DNA to carry genetic information, and it might use carbon compounds other than proteins to carry out the various functions performed by proteins in terrestrial life. But it is still carbon-liquid water life. Only in Chapter 9 will we be free of the carbon-liquid water restriction. In the search for extraterrestrial intelligence (SETI) we are searching for function, for evidence of technological civilization, regardless of the chemical basis of the life-forms.

It might seem biased to concentrate on carbon-liquid water life. One justification is that the only life we know to exist has this basis. Another justification stems from fundamental chemistry. No other element has anywhere near the same facility as carbon to form compounds of sufficient complexity, diversity, and versatility to support the many processes of life (Section 1.1.2). Few liquids approach water in their usefulness as solvents and reactants. Ammonia is a possible alternative to water at low temperatures, but it is pure speculation whether any life out there could use ammonia in this way. A third justification is that we know how to detect evidence of carbon-liquid water life. Apart from SETI, we have a far poorer idea how to detect life that has an entirely different chemical basis from ours, particularly as we are restricted to detection from afar.

8.1 Potential planetary habitats

From the summary of Chapter 7 it can be concluded that at least a few percent of the stars in our region of the Galaxy could have Earth-mass planets in their habitable zones, if they can form there. If we rather arbitrarily adopt a figure of 3%, then, with an average distance of about 6 light-years between stars in our neighbourhood, the average distance between stars that could have Earth-mass planets in their habitable zones is at most about 20 light-years. This value is more by way of illustration than a soundly based prediction, but it could be the case that Earth-mass planets are not far away.

The average volume per star is about (6 light-years)^3. The average volume per star with an Earth-mass planet is then about 33 times larger (1/3%). The average distance between such stars is then $(33 \times 6^3)^{1/3}$ light-years, i.e. 20 light-years or so.

If, as well as having a mass of order M_E, a planet also resembles the Earth in composition, then it could be a potential habitat. This is because it would then have a predominantly rocky composition (including iron as rocky). It is indeed likely that an Earth-mass planet in a habitable zone would be rocky – one Earth mass is too small for it to be like Jupiter, rich in hydrogen and helium.

- Why could such a planet not be icy in composition, i.e. completely covered in a huge depth of ice?
- The temperatures in the habitable zone are (by definition) too high for a planet there to be an icy body.

But even if such a planet were rocky it still might not be habitable. To be habitable one requirement is for volatiles that include water and carbon compounds that would form an atmosphere and liquid water at the surface. In the case of the Earth, water and carbon were present in the planetesimals and planetary embryos that made it, and there were comets and other volatile-rich bodies that arrived late and increased the endowment. It is likely that such volatiles were similarly available in other planetary systems, but perhaps not in all. There are yet further requirements for habitability. For example, if there are too many impacts, then life might never develop. In the Solar System, Jupiter's gravity shields us from excessive impacts, but there could be systems where there is no giant planet to fulfil this role. Other requirements will emerge below. Thus, a proportion of the Earth-mass planets in habitable zones will be uninhabitable, though it is difficult on present knowledge to put a figure on this.

On the other hand there are alternative possible habitats. These include satellites of giant planets that are in a habitable zone (Figure 8.1). We must also remember (Section 2.3) that the habitable zone is a conservative definition of the life-zone, confining attention to the surface of a planet. But even if the surface is too cold for life, the interior can be warm enough. Giants beyond a habitable zone could have habitable satellites if there were some tidal heating, as in the case of Europa. They would need to be sufficiently close to the giant, and in non-circular orbits. Even in the absence of tidal heating, the interior can be warm enough through internal heat sources such as the decay of radioactive isotopes, as is the case in the Earth's crust. The body must be sufficiently old for life to have emerged, perhaps older than about a few hundred Ma, but not so old that its near-surface regions have become cold. Even if stellar heating is negligible, the temperatures at a depth of a few kilometres in an Earth-mass planet could remain suitable for liquid water for the order of 10 000 Ma, i.e. twice the present age of the Solar System.

To see whether an exoplanet (or a satellite of a giant exoplanet) is not just a potential habitat but an actual habitat, we must learn more about it than the sort of properties discussed in Chapter 7 (see Question 8.1). We need to be able to analyse the electromagnetic radiation we get from it, preferably by having sufficient resolution to isolate the planet's radiation from that of the star. This will be possible in the decade 2011–2020 when the various new ground-based and space telescopes described in Section 6.3 become available. It will also be possible even further into the future when we send probes to the stars. But whether our instruments are on a probe approaching a planet, or orbiting the Sun or Earth, or even at the Earth's surface, we need to ask the question – how could remote observations like these find life out there?

QUESTION 8.1

What do we learn about an exoplanet from each of the following: radial velocity, astrometric, and transit methods of detection? Present your answer in the form of a table, and comment on what this information tells us about life on these planets.

Figure 8.1 An impression by the artist Julian Baum of a large satellite around a giant planet in a habitable zone. (Copyright © Take 27 Ltd)

8.2 How to find biospheres on exoplanets

A test of our capability to find life on other planets by remote observations, is whether such techniques can reveal life on Earth. This is the subject of Section 8.2.1. We then look at the fruitful techniques in more detail, in particular the use of spectroscopy.

8.2.1 Is there life on Earth?

In 1989 the Galileo spacecraft was launched by NASA (Figure 8.2). Its primary mission was to study Jupiter and its satellites (Section 4.1.5), and in Section 4.2 you saw how it has boosted the idea that there might be life on Europa. It reached Jupiter in December 1995, but before that, in December 1990 and again in December 1992, it came close to the Earth (Figure 8.3) in order to gain kinetic energy through gravitational interaction with the Earth, in the same general way that the Voyager spacecraft moved among the giant planets (Box 4.1). This gain enabled Galileo's modest rockets to raise the massive payload to faraway Jupiter. Advantage was taken of these close encounters to see whether life could be detected on the Earth and the Moon. In 1990 the Earth was the object of study, and in 1992 the Moon. Even though the instruments on board were not designed specifically to detect life, it was hoped that the outcome would help astronomers design ways of detecting life elsewhere.

The answer that Galileo gave to the question: 'Is there life on Earth?' is a resounding 'Yes!'. The conclusion itself came as no surprise, but it was encouraging that Galileo's instruments could give it. There were three instruments that provided the evidence.

Figure 8.2 An artist's impression of the Galileo spacecraft, leaving Earth-orbit (to which it was delivered by the Space Shuttle) on its circuitous journey to Jupiter. (NASA)

Figure 8.3 Galileo spacecraft views of the Earth and Moon, 16 December 1992. (NASA)

First, there was the near-infrared spectrometer NIMS (Section 4.1.5). This enabled astronomers to identify atmospheric substances through the imprint they placed on the spectrum of infrared radiation emitted by the Earth – an emission spectrum (to be discussed in Section 8.2.2). The substances detected included ozone (O_3) and methane (CH_4). One of the major effects of the Earth's biosphere is that it sustains molecular oxygen (O_2) as a major component of the atmosphere. O_2 has only a weak spectral signature in the infrared, but through the action of solar UV radiation, O_2 gives rise to an appreciable trace of O_3, and this has such a strong spectral signature in the infrared that it is readily detected. You will see in Section 8.2.2 that it is difficult to envisage any process other than photosynthesis that could generate sufficient O_2 to yield the amount of O_3 seen. But one can't be quite sure of this. The clincher is CH_4. Like O_2 this is also generated by large organisms in the biosphere and by certain bacteria. It is very readily oxidized by O_2 to give CO_2 and H_2O, and as a result it only accounts for about 1 molecule in every 600 000 in the Earth's troposphere. This however was sufficient to give NIMS a clear if small infrared signature. The crucial point is that without a huge rate of release of CH_4 into the oxygen-rich atmosphere there would be far less atmospheric CH_4 and it would have been undetectable by NIMS. A warm planet like Earth has no reservoirs of methane-ice to supply the CH_4, and volcanic emissions are insufficient. Without a prodigious production rate by the biosphere the quantity of CH_4 would be many orders of magnitude less, i.e. the presence of O_2 and CH_4 *together*, far from chemical equilibrium, puts the existence of a biosphere on Earth beyond reasonable doubt.

Second, Galileo measured the amount of solar radiation that the Earth *reflected* at various wavelengths – a **reflectance spectrum**. Around 0.8 μm, just into the infrared beyond the red end of the visible spectrum, a sharp rise in reflectance was detected, particularly over the continents. This 'red-edge' as it is called, is due to green vegetation, and is associated with chlorophyll and with structures that reject radiation not utilized by chlorophyll. However, though we know how to interpret this spectral feature on Earth, it is not at all certain that photosynthesis in an alien biosphere would look the same. Perhaps the best that can be hoped for is to see features that could not be readily accounted for by common minerals. Overall, this is a less certain indicator of life than pairs of atmospheric gases a long way out of chemical equilibrium, such as O_2 and CH_4.

Chlorophyll absorbs radiation from the Sun predominantly from the blue and red regions of the spectrum.

Third, the radio receiver on Galileo detected strong radiation confined to a set of very narrow wavelength ranges. Moreover, the radiation at each of these wavelengths was not constant but was modulated in an intricate way that could not be explained by natural processes.

■ What was the source of this radiation?

❑ These were the various terrestrial radio and television transmissions.

The modulation was the information that carried the programme content – a steady wave carries no soap operas. This shows that the Earth not only has a biosphere but that it has evolved in a very particular, possibly very rare manner, with the appearance of technological civilization. Unambiguous images of cities and other artefacts were not obtained by the relatively small Galileo cameras.

In December 1992, Galileo's instruments were turned to the Moon, with entirely negative results! There were no pairs of atmospheric gases a long way out of chemical equilibrium, indeed, hardly any gases at all, no characteristic red reflection, no radio or TV broadcasts. If there is life on the Moon it must be deep in the crust, and this is extremely unlikely given the scarcity of water.

We will now look at infrared spectra and visible spectra in more detail, deferring to Chapter 9 the detection of radio transmissions. Spectra are of huge importance (Figure 8.4) because (like radio transmissions) they could reveal a biosphere from a great distance, and could therefore be used to investigate Earth-mass exoplanets with the sort of instruments we will have in the near future. We start with the infrared spectrum.

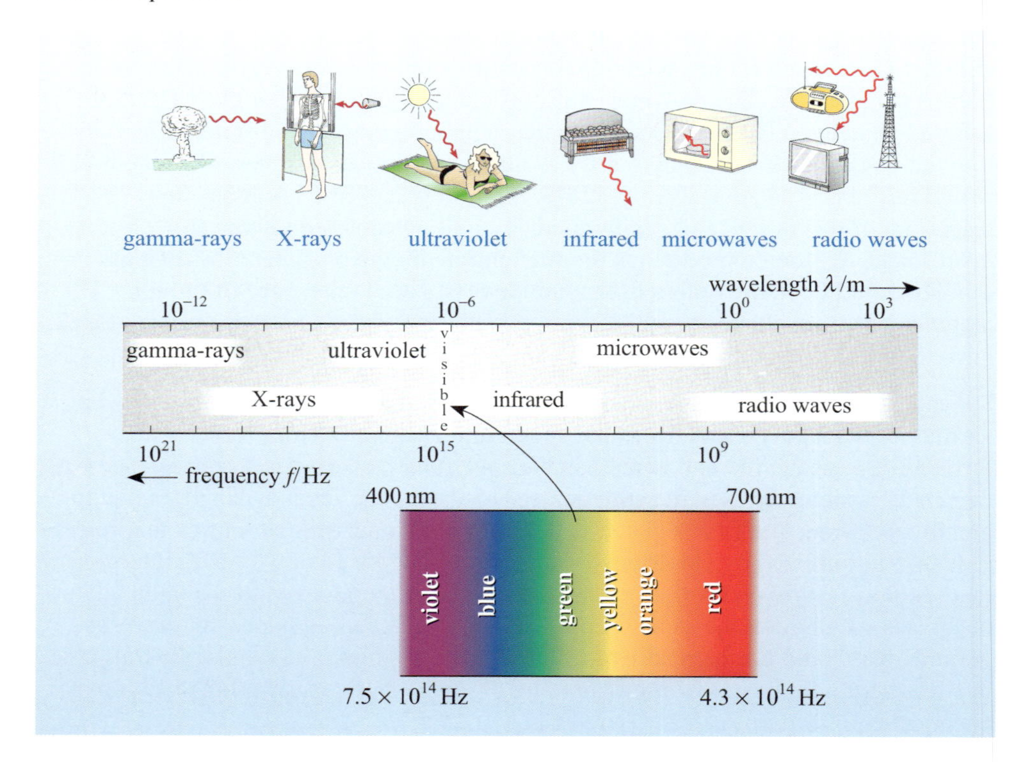

Figure 8.4 The electromagnetic spectrum.

8.2.2 The infrared spectrum of the Earth

Spectral lines and their interpretation are central to this section and to subsequent sections. The basic science is given in Box 8.1, and you should be familiar with its contents before you move on.

BOX 8.1 SPECTRAL LINES

Imagine a molecule (or atom) in its lowest energy state, called its ground state. It is then exposed to electromagnetic radiation of a certain wavelength. For many wavelengths there is no dramatic response by the molecule. At some wavelengths however, the molecule will absorb a parcel of energy from the radiation, called a photon. This will increase the energy of the molecule, putting it in what is called an excited state. After a short time the molecule will emit a photon with the same energy as the absorbed photon, returning the molecule to its ground state. The direction of emission is random, so it could emerge, for example, as in Figure 8.5. If you were observing this molecule along the direction of the emission of the

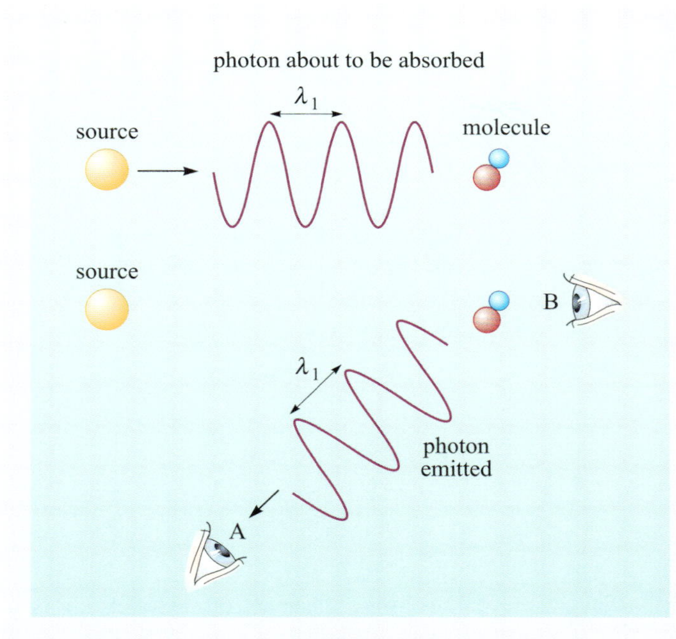

Figure 8.5 Photon absorption and emission by a molecule.

photon (position A in Figure 8.5) you would see a bright flash of radiation at the photon wavelength (λ_1 in this case), the same as in the incident radiation. If you were observing along the direction to the source of the incident radiation (position B in Figure 8.5) you would see a brief dip in the intensity.

Imagine now that there are lots of these molecules. Photons would emerge from the group as if from a scatter-gun, in all directions. From A there would be a randomly spaced sequence of photons, and unless the detector had a very rapid response it would record steady emission at the photon wavelength λ_1. From B it would record less radiation than if there were no molecules between the detector and the source.

Now imagine that the source contains a continuous spread of wavelengths. Imagine also that the detection system displays the radiation it receives at different wavelengths in different places, so that we can see how much radiation is present at each wavelength, as in Figure 8.4. From A there will be no change – we were only receiving radiation at λ_1 and we continue to do so, because there is no other radiation coming from this direction.

- What do we see from B?
- We see less radiation at λ_1 and no change at other wavelengths.

These two cases are shown in Figure 8.6. These displays are called spectra (singular, **spectrum**). The bright line seen from A is called an **emission spectral line**, and the dark line seen from B is called an **absorption spectral line**. So, what we see depends on our point of view, though the wavelength is the same in both cases.

The collection of molecules between an observer and the source will have a temperature that represents the random motion of the molecules. The collisions that arise from these motions will produce radiation across a wide and continuous range of wavelengths in accord with the temperature, and irrespective of the source. This is **thermal emission** from the molecules. In many cases the range of wavelengths in the thermal emission would be appreciable in the region of λ_1. In this case the emission adds appreciable background that makes the absorption line less dark.

So far, we have referred to a single spectral line. A molecule will, in fact, have very many absorption lines, spread across the spectrum from the UV to microwaves. Their wavelengths will be characteristic of the atomic constituents of the molecule and the way that the atoms are connected. It is this that makes spectroscopy such a powerful way to establish compositions.

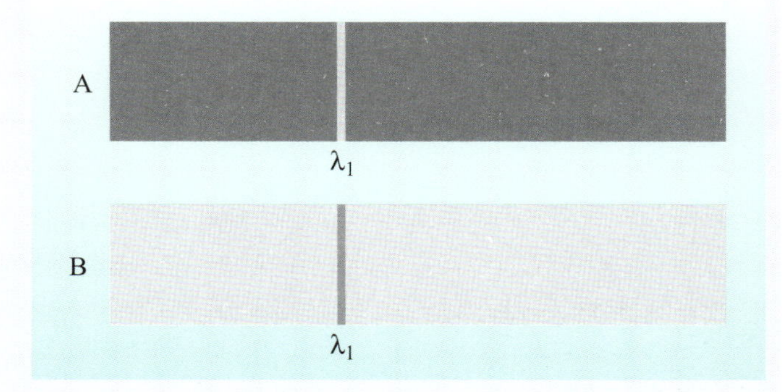

Figure 8.6 The spectrum seen from A and B in Figure 8.5.

Spectral lines come in different strengths. Thus, at some wavelengths a molecule will absorb photons at a low rate, and at other wavelengths at a high rate. It turns out that diatomic molecules consisting of two identical atoms absorb very weakly in the infrared. Consequently, for these so-called **homonuclear diatomic molecules**, the infrared is not a good place to look for their spectral lines.

- Why would you look for ozone O_3 in the infrared, but not the common molecular oxygen O_2 that we breathe?
- O_3 is not a homonuclear diatomic molecule, whereas O_2 is.

Roughly speaking, in homonuclear diatomic molecules the distribution of electric charge across the molecule is symmetrical around its centre. As a result, electromagnetic radiation cannot exert much net force on the molecule. The O_3 molecule has a different symmetry from O_2, and the influence of the electromagnetic force is correspondingly stronger for this reason. It is also stronger in **heteronuclear molecules**, diatomic or otherwise, such as HCl, CO_2, or H_2O, and so these also have strong infrared absorption.

Figure 8.7 shows the infrared spectrum of the Earth, as seen from space. This is a different way of displaying a spectrum from that in Figures 8.4 and 8.6. The relative quantity of radiation at each wavelength is now shown in the form of a graph. This is not the spectrum obtained by the Galileo spacecraft but a more detailed one obtained by the Nimbus-4 satellite in the 1970s. The particular spectrum in Figure 8.7 was acquired in day-time above the western Pacific Ocean, and has been chosen because it resembles the sort of spectrum that would be obtained from a cloud-free Earth from a great range, when the light from the whole planet would enter the spectrometer. Let's see how we can use Figure 8.7 to infer that the Earth is not only habitable, but is in fact inhabited.

In Figure 8.7, note that the intervals on the wavelength scale get more crowded at longer values. By contrast, in the frequency scale at the top of the frame the intervals are equally spaced. Frequency f and wavelength λ are related via $\lambda = c/f$, where c is the wave speed, the speed of light in this case.

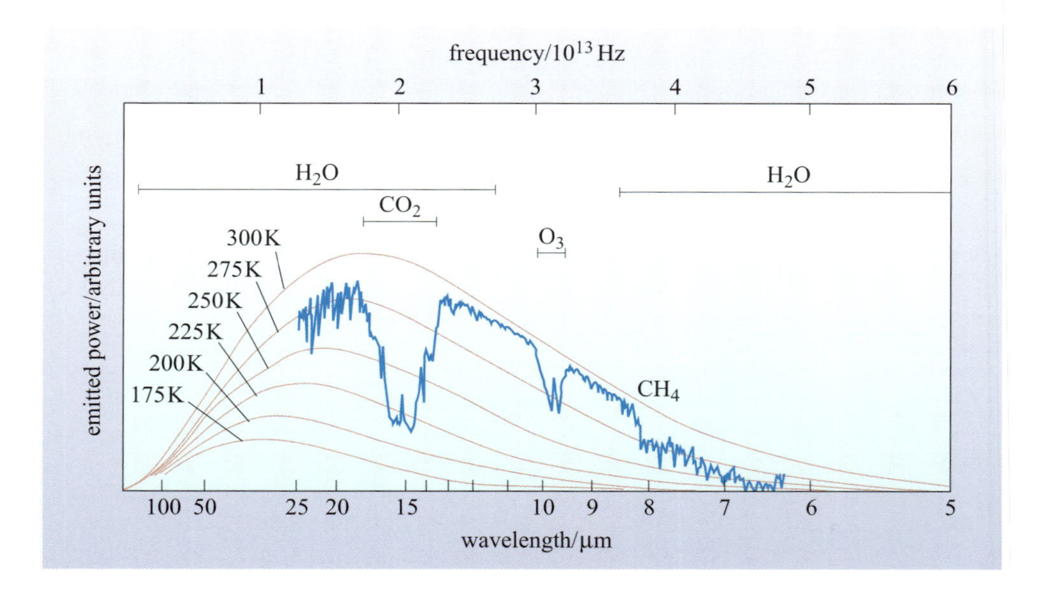

Figure 8.7 The Earth's infrared spectrum, as obtained in daytime by the Nimbus-4 satellite over a cloud-free part of the western Pacific Ocean in the 1970s.

■ Taking the value of c to be $3.00 \times 10^8\,\mathrm{m\,s^{-1}}$ (the speed of light), calculate λ for $f = 1.00 \times 10^{13}\,\mathrm{Hz}$, $2.00 \times 10^{13}\,\mathrm{Hz}$, $3.00 \times 10^{13}\,\mathrm{Hz}$.

❑ The values are respectively 30.0 μm, 15.0 μm, 10.0 μm.

Therefore, if the frequency intervals are equally spaced, the wavelength intervals cannot be. It is common in the infrared to use a scale in which the frequency intervals are equally spaced, but we shall refer to the wavelength scale alone. The vertical scale is proportional to the power emitted. There are several smooth curves, each labelled with a temperature. These correspond to thermal emission from matter at temperatures equal to those shown. There is also a jagged curve displaying much detail. This is the power emitted by the Earth. It is the detail in this curve that provides the evidence that the Earth is inhabited.

Surface temperature

Between 12 μm and 8 μm, except for the dip around 9.6 μm, a smooth curve labelled 300 K follows the spectrum closely. This curve corresponds to emission at a temperature of 300 K. The absence of strong spectral lines, such as would be generated by gases, indicates that the radiation received by Nimbus-4 in the 12 μm to 8 μm wavelength range has been emitted by the Earth's surface or by clouds rather than by a layer of gas in the atmosphere. It has in fact been emitted by the surface, the area being cloud-free. The surface temperature is a little less than 300 K.

■ What does a temperature around 300 K indicate about the possible phase of water?

❑ At this temperature, water can exist as a liquid.

Water can exist as a liquid from about 273 K to a higher temperature depending on the atmospheric pressure. At the surface of the Earth this pressure is about 10^5 Pa, and enables water to be liquid up to about 373 K. So we can conclude that the Earth has a surface temperature at which surface water could be liquid. For an exoplanet we would not know immediately whether clouds or the surface was responsible for the 8 to 12 μm spectrum. If the cloud cover were variable, or if there were spectral lines or other spectral features that could be linked to cloud particles, then probably we could tell.

The other important inference from the surface temperature is that it is well within the range for complex carbon compounds to exist. Most biological compounds such as proteins and DNA break up at temperatures much above 400 K (Section 2.5), so most of the Earth's surface is safely cool. On the other hand it is not so cold that biochemical reaction rates, which decrease rapidly as temperature falls, are too low to favour life.

Water

That water is *actually* present on Earth, at least as vapour in the atmosphere, is indicated by much of the fine structure in the spectrum in Figure 8.7. The H_2O molecule has a great many narrow absorption lines in the infrared spectrum, so many that they overlap and blend together to form **absorption bands**. It is these bands that are seen in abundance in Figure 8.7 rather than the individual lines. At the ends of the wavelength range in Figure 8.7, the water vapour absorption is particularly marked, and the smooth curve with a temperature of 275 K represents the spectrum fairly well at the ends.

- Could this part of the spectrum be coming from the surface?
- No – the surface temperature is nearer to 300 K.

The water vapour at these wavelengths must therefore be stopping the surface emission from reaching space. The water vapour has absorbed all the radiation emitted from the surface, and has re-emitted it at the lower temperature of its own location. We can thus infer that the atmospheric temperature at the general altitude of the water vapour emission to space is lower than at the surface. The actual altitude of 275 K cannot be obtained from Figure 8.7. To obtain this we need the variation of atmospheric temperature with altitude above the Earth, and this is shown in Figure 8.8, obtained from direct measurements and averaged over the Earth's surface. You can see that 275 K occurs at altitude of only a few kilometres. Therefore, there must be enough water vapour below this altitude to hide the ground from being seen from space at this wavelength. This is in accord with direct measurements that show water vapour to be heavily concentrated in the lowest few kilometres of the Earth's atmosphere.

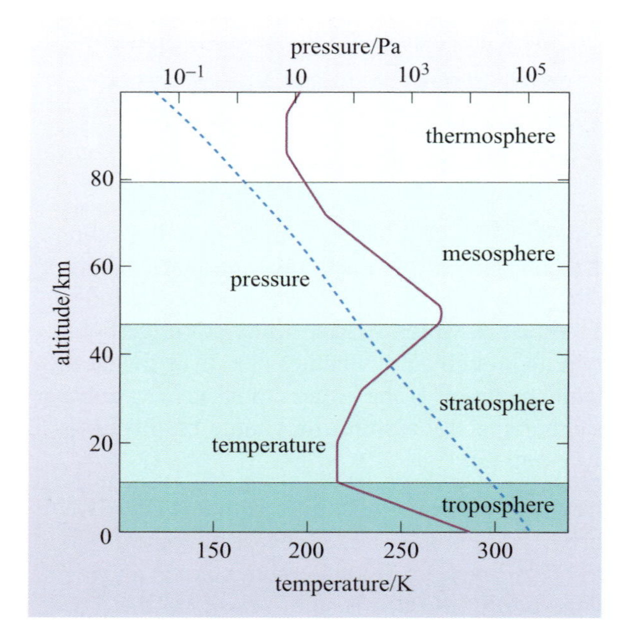

Figure 8.8 The variation of atmospheric temperature (solid line) and pressure (dashed line) with altitude above the Earth's surface (averaged over the Earth).

Carbon dioxide

Around 15 μm the Earth's spectrum in Figure 8.7 has a large dip corresponding to a deep and broad set of overlapping absorption bands due to carbon dioxide, CO_2. The most heavy absorption corresponds to a temperature of about 220 K, and Figure 8.8 shows that this value occurs in the upper troposphere at about 10 km. This is not because CO_2 is concentrated there – it is not – but because this is as deep as we can see at 15 μm into an atmosphere in which there is sufficient CO_2 at and above the upper troposphere to hide deeper levels. From Figure 8.7 alone we can conclude that CO_2 is clearly present and therefore that the planet has carbon, which is essential for biomolecules. The actual amount is about 0.035% (as a proportion of all molecules). That CO_2 has such a clear infrared signature is because the molecule has more than two atoms and is heteronuclear.

Oxygen and methane

The dip around 9.6 μm in Figure 8.7 is due to ozone O_3. This is derived from O_2 by photolysis (Section 4.2.1), specifically by the action of solar UV radiation on O_2. The depth of the O_3 feature shows that O_2 must be present in considerable quantity in the Earth's atmosphere. O_2 is not directly detectable in the infrared because it absorbs infrared radiation only very weakly (Box 8.1). The temperature that fits the central region of the O_3 absorption is about 270 K.

- Where does this occur in the Earth's atmosphere?

- Figure 8.8 places this either in the lower troposphere or in the upper stratosphere.

Given that around 9.6 μm we are only seeing radiation from O_3, these altitudes are candidate locations for the O_3 that is able to radiate directly to space. In fact, O_3 is concentrated in the stratosphere, so this radiation is coming from the upper stratosphere rather than the lower troposphere. The more elevated location for O_3 can be deduced alternatively on the basis that solar UV radiation would create O_3 well above the Earth's surface, and it would then screen the lower atmosphere from the UV that has produced it.

The other question that arises from the presence of O_3 is the source of the O_2 that gave birth to it – is this source a biosphere, or something else? A non-biogenic origin is the photolysis of water (Section 4.2.1), in which a UV photon splits a water molecule into OH and H. The hydrogen, being of low mass, is lost to space. Reactions of OH lead to the formation of O_2. Photolysis has always been generating some O_2 on Earth. However, the O_2 oxidizes surface rocks and volcanic gases, and consequently the quantity of O_2 in the atmosphere from photolysis alone would be far less than it is in the Earth's atmosphere. This indicates that the actual quantity is largely sustained by a biogenic origin through photosynthesis.

There are, however, conditions under which photolysis of water could give rise to oxygen in abundance. First, it could do so if the rate of generation through photolysis were far higher. This will be the case in 1000 Ma or so from now when the luminosity of the Sun will have increased to the point where the Earth's upper atmosphere will be warm enough to hold a lot more water vapour than it does now. More water means more photolysis which means more oxygen. However, this supply cannot last long. Photolysis leads to loss of water, and in a few Ma all the water will be lost, and the Earth will be dry. In a comparable time the O_2 will disappear through oxidation of surface rocks and volcanic gases. Thus, unless the Earth were caught in the act of losing its water, photolysis could not account for the high O_2 abundance. Venus seems to have already lost its water. This would have happened early in its history because it is closer to the Sun than the Earth is.

Second, photolysis could give rise to oxygen in abundance if oxygen were being removed geologically at a far lower rate than it is. It would then build up slowly over hundreds of Ma. In this regard, size matters. Large rocky planets like the Earth can sustain considerable geological activity (with plate tectonics at its heart in the Earth's case). Small planets are unlikely to sustain such activity. Therefore, if a small planet is sufficiently near its star so that its atmosphere is not dry, it could gradually build up a lot of O_2 from the photolysis of water. Mars, though it is a small planet and not very active geologically, has presumably been too far from the Sun to have had the continuously non-dry atmosphere that would have provided a lot of O_2.

Thus, for the Earth, a combination of size and the unlikelihood of catching our planet at the moment it loses nearly all its water, would make it seem likely to an observer from space that oxygenic photosynthesis was at work. As discussed in Section 8.2.1, the presence of CH_4 and O_2 *together* in the Earth's atmosphere put the existence of a biosphere beyond reasonable doubt. The spectral band due to methane is rather feeble in Figure 8.7. Though CH_4 is a strong absorber of infrared it is only present at about 1.6 p.p.m. by fraction of molecules in the atmosphere, with a smaller proportion above the troposphere. Nevertheless, it *is* detected, and with it comes the strong conclusion that the Earth is inhabited!

QUESTION 8.2

Suppose that an otherwise Earth-like planet was entirely covered in cloud, and that its effect was to block all radiation to space from altitudes less than 10 km. Discuss briefly whether its infrared spectrum would reveal whether it was inhabited. You will find the pressure curve in Figure 8.8 useful for part of your answer.

8.2.3 The infrared spectrum of Mars

Mariner 9 went into orbit around Mars in November 1971.

Figure 8.9 shows the infrared spectrum of Mars, at mid-latitudes in daytime under clear conditions, obtained by an orbiting spacecraft, Mariner 9. Some of the interpretation of this spectrum is left as an exercise (Question 8.3), but it is clear that there is no evidence of O_3, and so O_2 must be present as a trace at most. This is indeed the case. There is therefore no evidence for a biosphere in Figure 8.9. But this does not mean that there *is* no biosphere! There could be one that has left no atmospheric signature. Indeed, until oxygen was present in substantial quantities in the Earth's atmosphere from about 2000 Ma ago (Section 2.4) it would have been difficult to find convincing evidence of life on Earth from its infrared spectrum. However, there is evidence from landers that there is no life on the Martian surface today (Section 3.3).

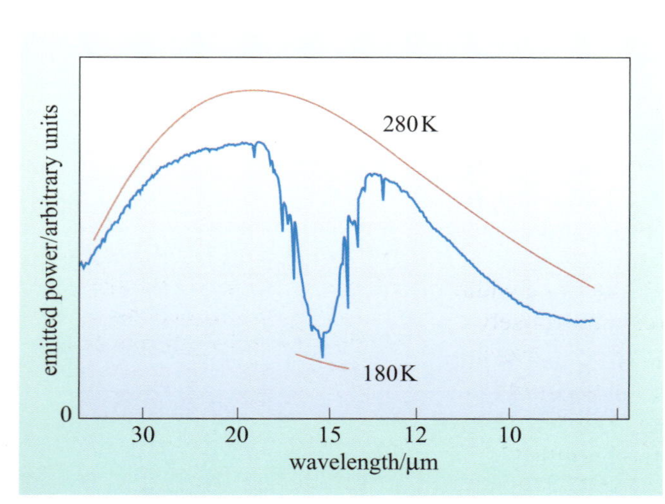

Figure 8.9 The infrared spectrum of Mars, at mid-latitudes in daytime under clear conditions, obtained by an orbiting spacecraft, Mariner 9. (Wallace, 1977)

Therefore, if there is life on Mars it must be under the surface. Indeed this is where one might still find liquid water continuously present, in contrast to the surface, where water might only appear for brief periods, if at all. This is because the temperatures in all planetary interiors increase with depth. In Mars the increase with depth is not well known because of our uncertain knowledge of the interior heat sources and thermal properties of the crust.

In the Martian tropics the maximum depth before the temperature reaches 273 K is 11 km, and it could be less than 1 km. The best current estimate is 2.3 km. At the poles the best estimate is 6.5 km. We know that water in some phase is abundant near much of the Martian surface, from Mars Global Surveyor and Mars Odyssey evidence (Section 3.4.1). It could well be liquid at depths of a few kilometres.

QUESTION 8.3

Explain how it can be deduced from Figure 8.9 that

(i) the surface temperature at mid-latitudes in daytime on Mars is about 270 K,

(ii) there is a considerable proportion of CO_2 in the Martian atmosphere,

(iii) there is very little water vapour in the Martian atmosphere?

8.2.4 The infrared spectra of exoplanets

We now look ahead to that happy time when Darwin or some comparable telescope, perhaps OWL (Section 6.3), has obtained an image of an Earth-mass planet in an exoplanetary system. Until we send our instruments to exoplanets (Section 8.2.6) this will be no more than a dot or smudge of light, with no discernible disc or surface features. But, dot or disc, the light can be passed into an infrared spectrometer, and can therefore be analysed for signs of life. Figure 8.10 shows the sort of spectrum that might be obtained.

OWL is the OverWhelmingly Large telescope.

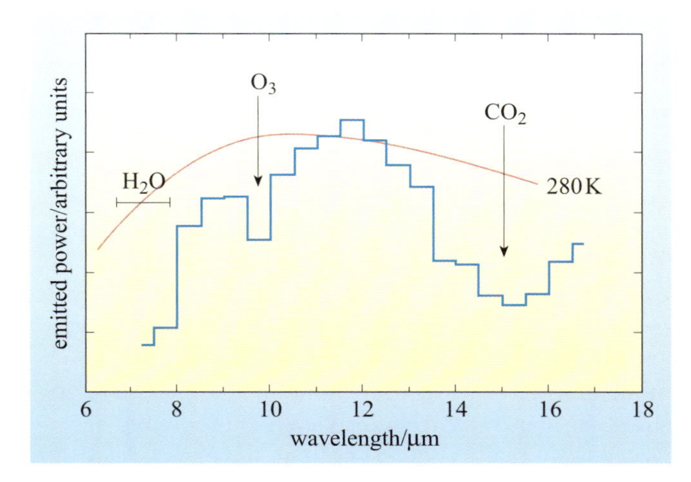

Figure 8.10 A plausible infrared spectrum that would be obtained by a Darwin-sized infrared space telescope for an Earth-like planet in the habitable zone of its star. Spectral resolution $\Delta\lambda$ is about 0.5 μm, exposure time Δt is about 40 days, and the star is about 30 light-years away. (ESA)

The first thing to notice is that the **spectral resolution** is very low, i.e. the spectrum is divided into rather wide buckets along the wavelength axis. The width of each bucket is $\Delta\lambda$, and we cannot discriminate spectral features more closely spaced than this. It is clearly desirable to make $\Delta\lambda$ as small as possible. However, the smaller the wavelength interval $\Delta\lambda$ that we split the light into, the lower the rate at which we will accumulate photons in each interval, so we have to accumulate photons for a longer time to build up the spectrum. There is therefore a trade-off between decreasing $\Delta\lambda$ and increasing the exposure time Δt for which a spectrum has to be built up. The hypothetical case in Figure 8.10 is for an Earth-like planet in the habitable zone of its star, about 30 light-years away, observed with a Darwin-sized infrared space telescope, with $\Delta\lambda \approx 0.5$ μm and $\Delta t \approx 40$ days.

The symbol Δ, delta, denotes an interval of the parameter that follows it. Thus, $\Delta\lambda$ denotes an interval of wavelength, and Δt an interval of time.

Note that the step heights in the spectrum are subject to statistical fluctuations – a second measurement will display differences, though with the main features still discernible. This is because of the finite number of photons accumulated into each bucket. In the example of Figure 8.10, rather few photons would be accumulated, and so the fluctuations from measurement to measurement would be clearly discernible.

From Figure 8.10 we can obtain the surface temperature from the window at 8–12 μm (excluding the O_3 feature).

- What condition has to be met for this to be the surface temperature?
- The atmosphere has to be fairly transparent at these wavelengths (except for the O_3 feature).

- What is the surface temperature, and what can you deduce from it?
- About 280 K, which is suitable for complex carbon compounds, and for water to be liquid if the atmospheric pressure is sufficiently high.

We can also discern the CO_2 and O_3 absorptions. Furthermore, the reduced temperature at 6–8 μm is suggestive of water vapour low in the atmosphere. CH_4 absorption is beyond detection. Nevertheless, it could be concluded that there probably is an active biosphere on this exoplanet.

- Under what circumstances would this conclusion be shaky?
- If the photolysis of water was going through a spurt, or if geological activity had been at a low level (Section 8.2.2).

The case of no oxygen

Consider now an exoplanet spectrum like Figure 8.10, except that the O_3 absorption is absent. We are now in the tantalizing position that the planet seems *habitable* – right temperature, presence of carbon, likely presence of liquid water – but there is no evidence that it is *inhabited*. At this point we must recall an important dictum: 'Absence of evidence is not evidence of absence'. There are several reasons why there could be a biosphere but no detectable oxygen. It could be that there is plenty of O_2, but too little UV flux from the star to form O_3, or the O_3 is efficiently removed in some way. But even if there really is no O_2 present, the planet could still be inhabited. There are at least three possible reasons.

- There *is* a surface biosphere that includes oxygenic photosynthesis, but it is in the state that the biosphere was on Earth before about 2000 Ma ago – it has not been able to build the oxygen content sufficiently for detection from afar.
- There *is* a surface biosphere, but it either performs photosynthesis in a manner that releases no oxygen, or it does not rely on photosynthesis at all. There are terrestrial organisms that act in both of these ways (Section 2.5.9).
- The biosphere is deep under the surface and does not influence the atmosphere, as might be the case on Mars today, or under an icy carapace such as on Europa.

But even if there is little atmospheric oxygen, there are other ways in which we might detect life.

First, we might detect in the infrared spectrum abundant gases other than O_2 that defied explanation by non-biological processes. But whereas one gas is suggestive, we must always be aware of plausible non-biological explanations, such as the photolysis of water for O_2. Two gases that should readily react to virtually eliminate the presence of one or both of them would be a much stronger indication of life.

- What is one such pair in the Earth's atmosphere?
- In the case of the Earth such a pair is O_2 and CH_4.

The less abundant CH_4 is driven by biochemical production to be far more abundant than it would be if it were in simple chemical equilibrium with O_2. In fact CH_4 and O_2 are an example of a **redox pair**, a name derived from the **reduction** of one atom or molecule and the **oxidation** of another in a chemical reaction between them. In reduction–oxidation, electrons are transferred from one atom or molecule to the other.

The species that loses the electron is the donor, and the one that gains the electron is the acceptor. The donor is said to be oxidized and the acceptor reduced.

In this case CH_4 is the donor and O_2 is the acceptor. Any other redox pair far from chemical equilibrium with each other could be evidence of an alien biochemistry – oxygen does not have to be the acceptor.

There is potentially a great variety of redox pairs, for example with sulfur rather than oxygen as the oxidizer (which some terrestrial bacteria actually use).

There could also be pairs not involving oxidation and reduction. If any out-of-equilibrium pair of gases were identified, then strenuous attempts would be made to find plausible non-biological explanations. If all of these failed, then there would be a case for an alien biochemistry.

QUESTION 8.4

By referring to Figure 8.10, describe the changes that would take place in the infrared spectrum of an Earth-like exoplanet with carbon-liquid water life, from just before the biosphere has had much effect on the composition of the atmosphere, to when the biosphere is at its most active, to when the biosphere has retreated to a few areas near the poles. The retreat is caused by the increase in its star's luminosity as it ages.

8.2.5 Exoplanet spectra at visible (and near-infrared) wavelengths

The infrared spectrum of an exoplanet over the wavelength range in Figure 8.10 is generated by emission from the planet's surface and atmosphere, and by the absorption of some of these emissions by atmospheric constituents. As well as using such a spectrum to search for life, there are other ways in which we might do so.

- Thinking back to the Galileo detection of life on Earth, name two other ways of detecting life?
- We might detect the red-edge of green vegetation at the planet's surface, or radio transmissions from a technological civilization.

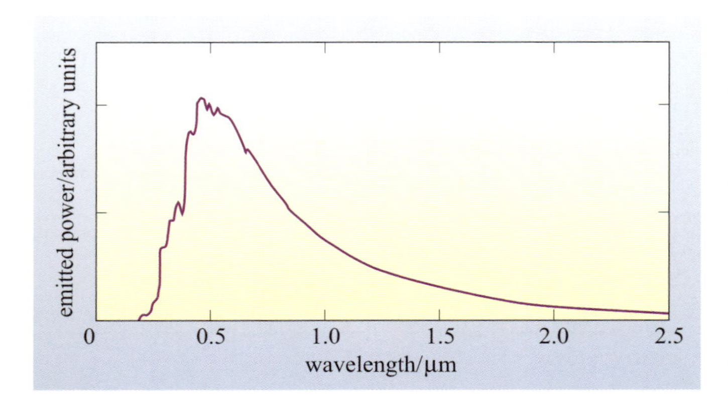

Figure 8.11 The spectrum of a Sun-like star.

The detection of radio transmissions will be discussed in Chapter 9. The red-edge of green vegetation would be detected at near-visible wavelengths. Near-visible is often called the **near-infrared (NIR)**, and extends to about 2 μm. Figure 8.11 shows that, for a Sun-like star, most of its emission is at visible (0.38–0.78 μm) and NIR wavelengths, and that these are considerably shorter than those in Figure 8.10. This section is concerned with spectra at visible and NIR wavelengths.

The red-edge of green vegetation is associated with chlorophyll. This is the molecule in green plants (and in some unicellular organisms) that is central to photosynthesis. It absorbs photons of solar radiation and thus enables the organism to capture solar energy and fix carbon to perform life's functions. Chlorophyll absorbs photons only at specific wavelengths, notably red and blue, which is why many plants look green. In green plants, chlorophyll is associated with structures that reflect away NIR radiation, presumably to prevent overheating. This gives the red-edge in the spectrum. The combined effect is shown in Figure 8.12 for the case of a deciduous leaf, typical of green plants. Take care not to confuse this reflectance spectrum with the emission spectra we have been considering before. In a reflectance spectrum the vertical axis shows the fraction of light from some separate radiation source (such as the Sun) that is being reflected by a substance. In an emission spectrum the radiation originates within the substance.

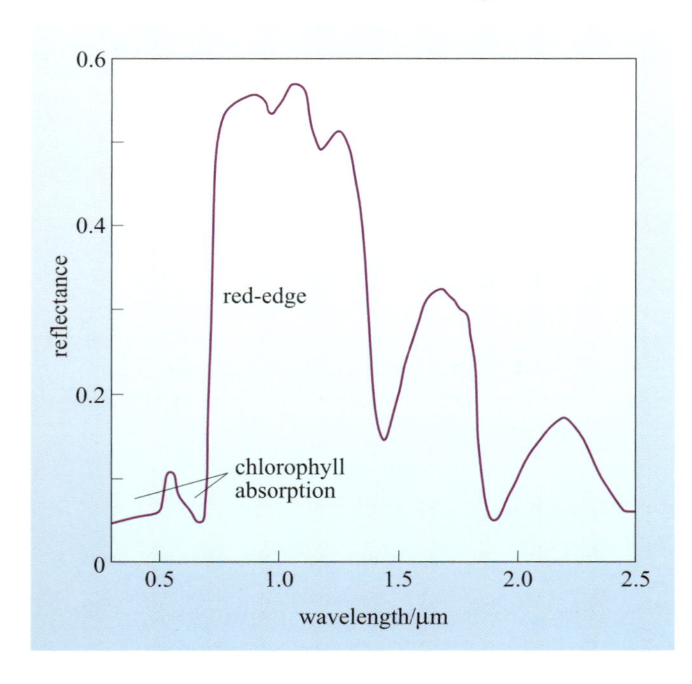

The red-edge is the most readily detectable feature associated with chlorophyll. Nevertheless, for an exoplanet, a space telescope about 10 metres in diameter will be equired, so this is for the decade 2011–2020. Moreover, unless the spectral features of the exoplanet were rather similar to the terrestrial features, we might well be unable to decide whether the features were of biological or non-biological origin.

Figure 8.12 The reflectance spectrum of a deciduous leaf.

Figure 8.13 Visible-light images of (a) Mercury, (b) Venus, (c) the Earth, (d) Mars, (e) the Moon. (NASA)

There is another way we could use the reflection from a planet, and this could be exploited by the not-so-far-off Darwin/OWL generation of telescopes. Examine the visible-light images of Mercury, Venus, the Earth, Mars, and the Moon in Figure 8.13. Imagine the outcome of the repeated measurement of the total sunlight (not the spectrum) reflected by each of these bodies at intervals of a few hours. Figure 8.14 shows a notional outcome.

- Why has the more variable light curve in Figure 8.14a been labelled 'Earth'?
- This is because Figure 8.13 shows that the Earth's surface and atmosphere is the most highly varied of these planetary bodies in its reflectivity from place to place.

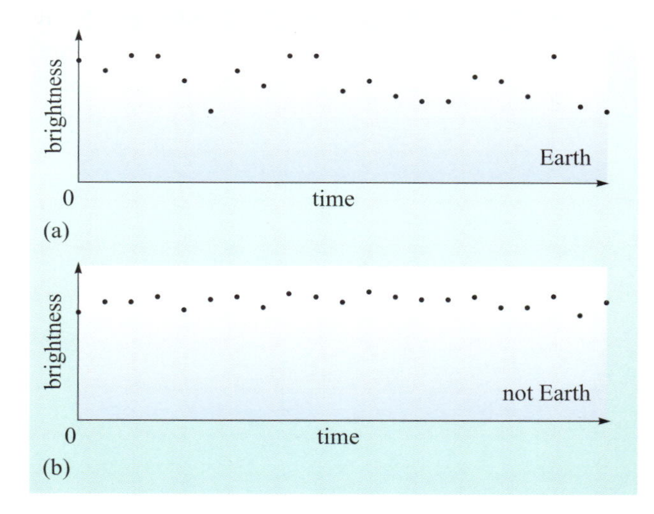

Figure 8.14 Light curves that typify the sunlit hemisphere of (a) the Earth and (b) a terrestrial body with a more uniform surface.

Oceans reflect less than 10% of incident light, whereas clouds, snow and ice reflect more than 60%, and deserts and vegetation somewhere between. As the Earth rotates, the different contributions would be presented to a distant observer in different proportions, to give a daily variation in the brightness of the illuminated hemisphere of up to a factor of two. The variation would be even greater if, instead of the total light reflected, light in certain wavebands were isolated. Also, there are longer-term terrestrial changes of up to 20%, due to variation in cloud cover, that are larger than for the other planets in the Solar

System. Such variations would not prove the existence of a biosphere, but would be indicative, and combined with other observations, such as the infrared spectrum, they could help build a convincing case.

As well as the reflectance spectrum of the planet, we also have the absorption spectrum of its atmosphere at visible wavelengths. This is the outcome of the light reflected by the planet's surface being absorbed at certain wavelengths by the atmospheric constituents. Like the infrared spectrum, the visible absorption spectrum can reveal the atmospheric constituents, notably H_2O, CO_2, and O_2. Note that O_2 now has a strong spectral signature, at the red extremity of visible wavelengths, unlike the weak spectral signature that it has in the infrared. The feasibility of this approach has been demonstrated by examining the Earth's visible spectrum reflected off the dark side of the Moon (Earthshine), though to apply it to 'exoEarths' we will need 10 metre space telescopes with optical designs that will suppress the light of the planet's star. This capability might be with us by 2015.

As an aside, you might wonder why O_2 has a strong spectral signature at visible wavelengths, given that it is the symmetry of the molecule that prevents it having a strong infrared signature. This is because in the infrared it is the entire O_2 molecule that is influenced by the radiation, and so the symmetry is sensed. At the much shorter visible wavelengths a single electron in one of the atoms is involved and the molecular symmetry is less apparent.

QUESTION 8.5

Discuss the possibility of detecting an aquatic biosphere on a Europa-like satellite of an exoplanet, through emission and reflection observations made from the Earth.

8.2.6 Interstellar probes

Figure 8.15 An artist's impression of an interstellar probe. (Copyright © 2001 by Don Dixon)

The obvious advantage with an interstellar probe (Figure 8.15) is that you can get instruments much nearer the exoplanet. Once there, it would be much easier and quicker to gather spectral data than from a remote Earth-based vantage point, and it also makes it possible to gather spectral information that would be too weak to get from Earth. If the probe got really close, entered the atmosphere and landed on the surface, then direct sampling and close-up imaging would be possible, and any but the most deeply buried biospheres would be detected. So, as we have sent probes to the outer Solar System, why don't we send them to exoplanets? Consider sending a probe to the nearest Sun-like star, Alpha Centauri A, 4.40 light-years away. The Sun is 1.6×10^{-5} light-years away and so Alpha Centauri A is about 275 000 times further away than the Sun, and that makes it difficult to reach.

- At an average speed of 10% of the speed of light, how long would it take a probe to reach Alpha Centauri A?
- It takes light 4.40 years to reach us, so at 10% of the speed of light it would take 44 years for a probe to reach Alpha Centauri A.

This is rather a long journey. Moreover the energy required to accelerate a probe to 10% of the speed of light is huge, and would need a propulsion method other than chemical.

What have we achieved with space probes so far? Since our first steps into space in the 1950s many probes have been launched, and though none of them have been aimed at the stars, there are four NASA spacecraft that will leave the Solar System, all of them having completed their missions to the outer planets. These are Pioneer 10, Pioneer 11, Voyager 1, and Voyager 2. Though Pioneer 10 was the first of these to be launched, in 1972, it is Voyager 1, launched in 1977, that is now furthest away. In 2003, it was about twice as far as Pluto, though this puts it only about one-thousandth of a light-year away! It has been slowed by the Sun's gravity to a few kilometres per second, and so it will be tens of thousands of years before it gets among the nearer stars, and much longer before it accidentally gets close to one. Nevertheless, the possibility that it will be found by aliens has not been ignored. Voyager 1 carries a long-playing gramophone record, bearing sounds and images of Earth. Whether any aliens could play and understand it is a matter of debate.

It is likely to be the end of this century before we find a feasible, affordable way to launch a probe that could achieve an average speed of 10% of the speed of light. Even if we could, there is still the 44 year travel time to Alpha Centauri A, plus the 4.4 years needed for information from the probe to reach us at the speed of light. There is also the problem of slowing the probe down so that it does not whizz past any planet in a few seconds. One possibility is to use radiation pressure from the solar photons acting on huge sails to accelerate the craft out of the Solar System, perhaps supplemented by an Earth-based laser beam, and then slow the craft down with the photons from its destination star. However, it is for a distant successor to this course to give interstellar probes more space. Fortunately, developments in spectroscopy from ground-based and space telescopes give us a real prospect of finding life beyond the Solar System in the next few decades.

QUESTION 8.6

Suppose that an exoplanet has an extensive biosphere, but that it is very different from Earth's in that

- it is not based on water,
- the photosynthesis that takes place releases a gas other than oxygen into the atmosphere.

(i) List the clues that could have otherwise been used to establish that an Earth-like biosphere existed, that are not relevant in this case.

(ii) Discuss how it could be established from Earth that a biosphere existed on this exoplanet.

8.3 Summary of Chapter 8

- To determine if an exoplanet has a biosphere, we could perform direct sampling and imaging at its surface – if we were able to land a probe on the planet. That possibility is a century or more off. In the foreseeable future we have to rely on observations from within the Solar System, and the first requirement is that we can analyse the electromagnetic radiation that we receive from the planet.
- The infrared spectrum of an exoplanet can reveal the temperature at its surface if there are spectral windows to the surface, and within its atmosphere otherwise. It can also reveal the atmospheric composition. If evidence of water vapour and CO_2 is found, and if the surface temperature is within the range for liquid water and complex carbon compounds, then the conditions would probably be suitable for carbon-liquid water surface life.
- That a biosphere were actually present would be indicated by a strong O_3 (ozone) absorption feature, because this would point to considerable quantities of O_2, such as could be sustained readily by oxygenic photosynthesis. Photolysis of water as a source of abundant O_2 would be possible if the planet had long been geologically inactive, or if it were caught in a brief moment of water loss.
- Much stronger atmospheric evidence of a biosphere would be provided by redox pairs far from chemical equilibrium with each other, such as O_2 (as indicated by O_3) and CH_4 in the Earth's atmosphere.
- Lack of abundant O_2 in the atmosphere would not necessarily imply absence of a biosphere.
- The visible and NIR spectrum of an exoplanet might also reveal habitability or a biosphere, by detecting atmospheric gases or the effects associated with chlorophyll and other biological substances, or the variability of the light curve.

CHAPTER 9
EXTRATERRESTRIAL INTELLIGENCE

9.1 Introduction

What do we mean by intelligence? This is not an easy question to answer in a philosophical sense. However, if we are willing to accept a subjective answer, then we can start by declaring that humans are intelligent. This still leaves a question mark over many other species on the Earth. If we are willing to be more specific still, then we can consider only intelligence that is capable of communicating signals across space. At present this is the only type of extraterrestrial intelligence we can hope to discover. So, for the purposes of this chapter, when we refer to intelligent life it should be understood that we mean life which can engage in interstellar communication. (In this pragmatic usage, humans have only been 'intelligent' for a few decades and Newton, Darwin and many other great scientists don't even qualify as being intelligent life.)

With our pragmatic interpretation of intelligence, **SETI**, the Search for Extraterrestrial Intelligence, is a well-defined problem: we must search for signals from life elsewhere in the Universe. Given *our* current technology only signals transmitted as electromagnetic radiation are detectable (though there is also the possibility that we may be visited by an alien spacecraft). The distances between stars and galaxies are vast, which means that electromagnetic signals, which travel at the speed of light, can take anywhere from a few years to millions of years to travel between an alien world and our planet. Also the sheer number of stars to search through, and the weakness of the signals, makes SETI a monitoring task that may need to be sustained over many generations. We can improve our chances by sending out signals ourselves, targeting certain regions of space to maximize our chances of success. The action of trying to initiate contact in this way is sometimes called **CETI**, Communication with Extraterrestrial Intelligence. It is interesting to consider what we should broadcast, as our decisions will help us anticipate the signals we might expect while searching. If we do detect signals, we might then attempt to begin a dialogue, which must begin on common ground by identifying knowledge that is recognizable to both humans and aliens. Once contact is made we need to exercise great care: what should we say, how should we present the human race, and who should represent our planet? Underlying the whole issue of CETI is one important question: should we even attempt to make contact? To do so involves openly attracting more advanced, possibly hostile aliens to meet our young, possibly infant civilization that has been 'intelligent' for just a few decades.

9.2 Searching – SETI

Why should we search? This is the common question voiced by everyone from astronomers, who are eager to devote resources to their own specialist interests, to politicians and tax-payers, who are more concerned with the costs in both time and money. Those opposed to SETI might argue that if there is life out there, specifically more advanced life, surely sooner or later it will come and find us, perhaps by landing in the heart of a major capital city. Whether to search or wait is a tough social, political and economic question that we will leave you to ponder. For now, let's assume that we do want to search, and consider how we might go about it.

At present, the search for extraterrestrial life is confined to hunting through the stream of electromagnetic radiation arriving at the Earth. Perhaps more advanced civilizations can use gravitational radiation, particle beams or something else that we cannot yet conceive of, but we cannot detect such signals at present. One day we may have the ability to travel the distances between the stars, but even the present prospect of sifting through electromagnetic radiation is in itself a challenge to our technology. Not only must we perform a search across the sky, but we must search across a wide range of wavelengths. Fortunately science gives us some clues as to where (in space, wavelength and time) to concentrate our efforts. The presumed universality of physical law gives us the added hope that alien scientists might have arrived at the same conclusions.

9.2.1 What frequency?

On Earth, all creatures with sight have evolved to see in the visible part of the electromagnetic spectrum. This is for two reasons: air is transparent and the Sun is bright in the visible part of the spectrum. The Earth's atmosphere is also transparent to radio waves, which has allowed the human race to develop radio communication and the science of radio astronomy. (No creature has evolved 'radio eyes' because there is no strong source of illuminating radio waves, and because giant eyes would be needed to achieve a useful spatial resolution – see Box 6.3, Equation 6.7.) At present, for purely economic reasons, it would seem that visible and radio communication are preferable for SETI because searches and broadcasts can be conducted from the Earth's surface. If one supposed that intelligent life elsewhere is similar to humans, then one could further argue that they lived on Earth-like planets, with Earth-like atmospheres that were also transparent to visible and radio frequencies. Irrespective of whether this is correct, the bottom line is that, at present, SETI is effectively restricted to visible and radio frequencies.

The spectrum of the Sun, and for that matter all stars, peaks in the infrared to visible part of the electromagnetic spectrum. If you consider two equally wide frequency ranges, one in the visible and one in the radio, the energy emitted by a star in the radio band will be orders of magnitude smaller than in the visible band. So, to avoid signals competing with the dominant, visible emission of stars, early SETI pioneers decided to search at radio frequencies. (By convention, as with our household radios, radio astronomers generally talk about frequencies rather than wavelengths.) To make signals even more noticeable, we can concentrate them over a narrow range of frequencies and in a narrow beam, i.e. a signal with a narrow bandwidth and narrow beamwidth. Such a transmission *from* a radio telescope on a planet orbiting a distant star could cause us to see the star apparently brighten by a factor of a million or more in the radio part of the spectrum. Such a signal would be unmistakable. However, to see it we would need to be looking at a band of frequencies that corresponded to the alien's transmission. At what frequency should we search for such transmissions?

The answer to the question of frequency relies on the ubiquity of hydrogen in the Universe. Astronomers have been able to map out the shape of our Galaxy by detecting emission from hydrogen atoms at a wavelength of 21 cm, corresponding to a frequency of 1420 MHz (1 MHz = 1 million cycles per second). Any intelligent aliens interested in mapping out the Galaxy will probably have discovered the utility of this spectral line and so they too might conclude that 1420 MHz is a good frequency for monitoring and broadcasting. Of course, being a source of much natural emission, signals broadcast at this line will have to be given some obvious 'artificial' quality. A further complication is that Doppler shifts arise because hydrogen clouds rotate with our Galaxy (with rotation speeds of up to about $150\,\mathrm{km\,s^{-1}}$). This makes the line somewhat blurred, extending from 1419 MHz to 1421 MHz. To avoid signals being lost in the blur of natural emissions, Carl Sagan (Figure 9.1) suggested that we might want to broadcast and monitor at (1420π) MHz or ($1420/\pi$) MHz. Since π is a universal number (see Section 9.3.1), alien scientists are likely to have this same idea. The lines would still be subjected to a Doppler shift due to the radial velocity of the emitter with respect to the receiver, but there is little natural emission at either frequency, so there would be no blur. Another notable frequency is 1720 MHz, the highest emitted by the OH molecule (oxygen and hydrogen), which together with another hydrogen atom makes up water, H_2O. If we believe that water (or at least oxygen and the most common element hydrogen) is essential to life it makes sense to monitor the 1720 MHz line.

Figure 9.1 Carl Sagan (1934–1996) was a key figure in the development and promotion of SETI. He was also an active planetary scientist, being instrumental in discovering the hot, dense nature of Venus's atmosphere and establishing that Titan's oceans could be stocked with molecules necessary for life. (© SyracusePhotographer.com)

- Why would we multiply and divide a significant frequency by π rather than add and subtract π?
- If we were to add and subtract π, then we would have to rely on aliens knowing our units of megahertz.

Assuming that we have decided on a frequency to use, what bandwidth is best? There are two bandwidths involved in the problem: the bandwidth of the emitted signal, and the bandwidth of the receiver used to detect the signal. For two signals of the same total emitted energy, a narrow bandwidth will produce a relatively larger increase in flux around one frequency in the spectrum, effectively concentrating its energy in a small part of the spectrum. However, such a narrowband signal is most easily detected if the monitoring receiver is operating at a similarly narrow bandwidth centred on the correct frequency. An emitted signal with a wider bandwidth makes it less crucial to be monitoring at the correct frequency with a narrow bandwidth, but the signal will produce a more modest increase in the spectrum for the same emitted energy. Figure 9.2 illustrates the bandwidth problem for a given signal observed by two receivers with different bandwidths.

The terms 'narrowband' and 'narrow bandwidth' are used interchangeably by radio astronomers. The same applies to 'wideband' and 'wide bandwidth'.

- An alien transmits at 1420 MHz with a bandwidth of 10 Hz. A radio telescope can be set to monitor a channel at this frequency with bandwidths of 1 Hz, 5 Hz or 100 Hz. Which bandwidth would you select to receive the largest signal relative to the background? (Assume that there are no Doppler shifts present.)
- You should use 5 Hz, since 1 Hz would pick up only one-tenth of the signal, and the 100 Hz setting would include 90 Hz of background noise.

The problem of choosing which frequency to monitor has eased with improving technology. An early radio telescope would simply be set to a particular frequency and it would receive radio waves within its bandwidth around that frequency. These telescopes could be 'tuned' to other frequencies, much like you would tune a radio to different frequencies. To get a spectrum like that shown in Figure 9.2, one would have to **scan** slowly through the frequencies of interest. In contrast, modern radio telescopes can monitor hundreds of millions of narrowband (less than 1 Hz) **channels** simultaneously, greatly speeding up the search process.

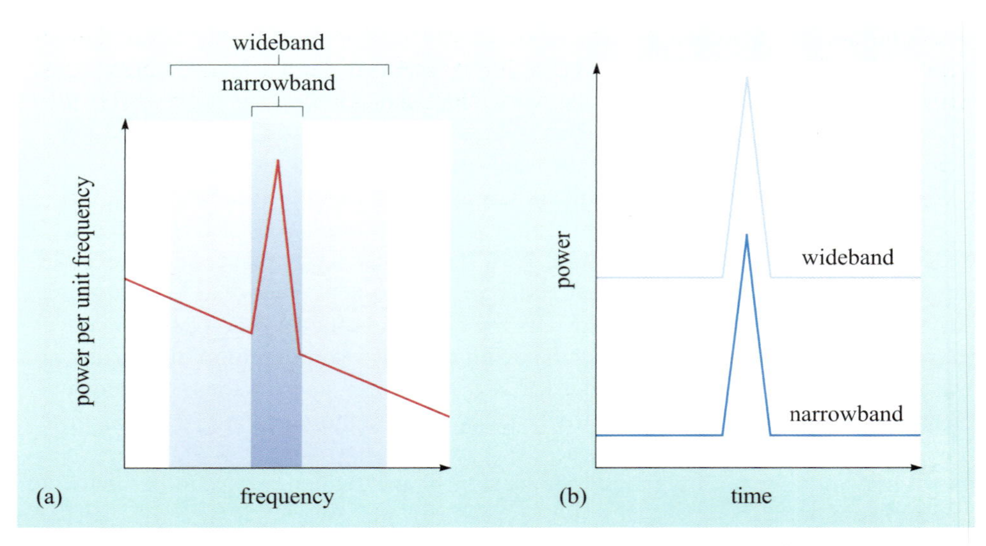

Figure 9.2 (a) The spectrum of a narrow bandwidth radio pulse, in the form of power emitted per unit frequency. The power in the emitted signal bandwidth is equal to the area under the spike. In the same way, the power detected by a receiver is equal to the area under the spectrum within the relevant bandwidth. (b) The time profiles that would be observed with a narrow bandwidth and wide bandwidth receiver centred around the correct wavelength. The narrow bandwidth only covers part of the frequency range of the emitted signal, but includes none of the natural (or background) spectrum on either side where there is no signal. As such the jump in detected power is relatively large. The wide bandwidth receiver dilutes the signal power because of inclusion of the background spectrum either side of the emitted bandwidth. Although the overall power is greater in the time profile for the wide receiver bandwidth, the jump in received power relative to the background is smaller.

Of course, if observations of this kind are performed regularly – which they are – then we have the further problem: analysing and searching through an enormous amount of data. This problem challenges our modern computer technology. By clever use of the internet, one project, SETI@Home, has tapped into the unused processing power of idle computers across the world, of which more later.

9.2.2 Where to look?

The practicalities of observing mean that we must concentrate on one patch of sky at any given time. Although we are faced with a laborious search, this does have an advantage in that, just as we saw with wavelength, looking at too large a patch of sky will tend to wash out any signal with background. The question of where to look depends on what we are looking for. Are we eavesdropping on aliens' local communications, or are we searching for purposefully transmitted advertisements of their existence?

Unintentional transmissions

Let us consider unintentional transmissions first. Since these are not directed at us, the signal that we could detect would be much weaker than a directed beam as it spreads out through the three dimensions of space. The worst case is an isotropic signal, because the radiation spreads out equally in all directions. We could only hope to detect unintentional transmissions from a handful of systems in our neighbourhood of the Galaxy. Clearly, if it is eavesdropping we want to do, then our choice is made for us, and we have a limited number of targets – the nearest ones. However, we would have to monitor many frequencies because, based on the example of radio emissions leaking out from the Earth, we might expect that such transmissions will be sprinkled haphazardly across the radio spectrum.

Intentional transmissions

Now consider that an alien civilization is intentionally trying to make contact with us. It's a startling fact that we currently possess the technology to conduct a dialogue with any other civilization in the Galaxy by using narrow beaming and bandwidths, provided we accept long delays between messages leaving and arriving. If distant aliens wished to attract our attention with an electromagnetic signal, then they would concentrate their emitted energy in a narrowband around some special frequency (as discussed in Section 9.2.1) and transmit a beam concentrated at the Solar System. To discover the alien's signal we have to hunt across the sky with our narrow beamwidth receiver, targeting billions of stars. Clearly this is a very time-intensive task if we have no idea where the alien is broadcasting from. Also, we have to ask why an alien would choose to target the Earth orbiting an ordinary star such as the Sun. We can give no definite answer to this. We could hope that such aliens were far more advanced than us and so could use their science and technology to identify the Sun as having a planet that could sustain life.

If we are searching for signals that we are intended to detect, then we have a huge number of potential targets, and must be selective. Of course, nearby stars would remain at the top of our list. Firstly, these targets offer a reasonable message exchange time (a few years to decades). Secondly, in coming decades we may be able to detect the presence of Earth-like planets (and perhaps even signs of life) around the nearest stars. Many star systems are binary systems and a significant number are multiple-star systems. Although it is easy to envisage problems for life evolving in such systems (stability of planetary orbits and limited habitable zones), we really have no compelling, scientific reason for dismissing them from a search. That said, if resources are limited and we are forced to be very selective, it is natural to follow our one, known example – the Earth – and prioritize single, Sun-like star systems. With the rapid development in planet detection techniques, we could go further and restrict attention to systems with Jupiter-like planets, in Jupiter-like orbits.

- Why might we expect stars with Jupiter-like planets in Jupiter-like orbits to contain other planets that harbour life?
- The reason is that such planets serve to protect inner Earth-like planets from some of the bombardment by comets and other planetary system debris.

One strategy would be to make best use of the radio telescope's beam size – the patch of the sky that it is sensitive to. A nearby (meaning about 100 light-years or so) star cluster might seem like a good target. For example, the Pleiades (Figure 9.3) contains some 150 stars spread over about 0.5° of sky (the same angular size as the Sun or full

Figure 9.3 The Pleiades. A stellar nursery 400 light-years distant of some 150 stars that are only about 100 Ma old. (8444611 Photographed by David Malin. Copyright © UKATC/AAO, Royal Observatory, Edinburgh)

Moon). Unfortunately, these stars are too young, hot and luminous for life as we know it to evolve, being typically 100 Ma old. In comparison, intelligent life, in the sense we've defined it here, took about 4.6 Ga to appear in our Solar System. Being the birthplace of stars, such **open star clusters**, as they are called, do not make obviously good targets.

There is another type of cluster: the **globular star cluster** (Figure 9.4). In contrast to open star clusters, these contain thousands, possibly millions of old stars, most aged over 10 Ga old. However, there are several good reasons why we should not expect there to be life in these clusters. Firstly, being so old, these stars are poor in the heavier elements which were created in the core of large stars and then blasted out into space for successive generations of stars to form from. Secondly, the short distances between stars in a globular cluster are such that interactions would seriously limit the number of stable planetary orbits. Still, it is important to remember, that if we are keeping an open mind about what life-forms may exist, then these reasons, or the ones given against open clusters may not preclude the presence of life.

Figure 9.4 M13, a globular star cluster of ancient stars. (US Naval Observatory, Washington DC)

9.2.3 What to look for?

Imagine sitting outside on a summer's day. You'll hear a myriad of sounds. If you pick any particular sound, even if you've never heard it before and do not know what it is, you'll probably be able to tell whether it is natural (from a bird or due to the wind) or artificial (from a car or a lawnmower). Most artificial sounds have a simple repetitive structure due to the regular movement, most usually rotation, of some working part. Nature on the other hand produces sounds with a more intricate structure across the spectrum of frequencies. The sound of wind through trees contains almost all audible frequencies in equal proportion (known as *white noise*), whereas the almost musical birdsong produces a highly structured but irregular frequency spectrum.

If we compare the spectral structure of natural and artificial radio waves we can reach similar conclusions. Our radio and television transmissions have distinctive, regular features in both frequency and time, unlike most natural phenomena. There are exceptions of course, where natural frequencies arise (e.g. the 1420 MHz hydrogen line) but these can be easily predicted from atomic physics. Naturally, there are occasional confusions. The most striking example in astronomy was when a very regular pulse was detected by a radio telescope in Cambridge, UK in 1967. At first, the phenomenon, later known as a *pulsar*, was nicknamed LGM – Little Green Men – because no natural emission mechanism could be imagined to pulse so regularly. It turned out that there was a theoretical explanation for it – a rapidly rotating star, called a neutron star, that had contracted from the radius of a normal star down to a few tens of kilometres. The story of pulsars serves as a warning, but such confusion between natural and artificial signals is the exception rather than the rule (and the confusion over pulsars was actually resolved very quickly). A more common problem in concluding that a radio signal with regular structure is a sign of alien intelligence is that the vast majority of such signals originate here on Earth.

- ■ Why would it be unfortunate if an alien, by coincidence, were to broadcast a signal with a period of a second, minute, hour, day or year?
- ❏ Any signal that showed a period close to one of our units of time would probably be regarded as being of terrestrial origin. A period of a day or year could well arise from an error of failing to correct fully for the rotation or orbit of the Earth.

Intentional first contact

Assuming that false alarms from terrestrial radio, television, communications and even secret military installations can be ruled out, what structure might we expect in a radio signal of alien origin intended for first contact? Firstly, unless they are inept or trying to tease us, we might expect the signal to contain some obvious regular structure, either in time or in the frequency spectrum, that looks unlike any natural phenomenon. One simple example noted above was the suggestion of transmission at a frequency that is a factor of π greater than or less than the 1420 MHz hydrogen line. Another example is depicted in the time profile in Figure 9.5, where π is coded into the arrival time of pulses. Such a pattern would be easy to detect and hard to explain in

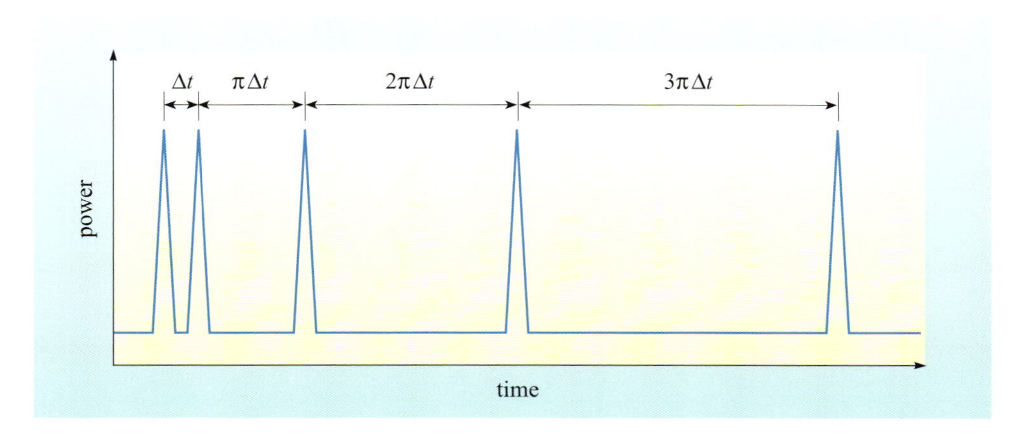

Figure 9.5 A time profile in which pulses arrive in a structured way. The time between the first and second pulses is Δt, the time between the second and third is $\pi\,\Delta t$, the time between the third and fourth is $2\pi\,\Delta t$, that between the fourth and fifth is $3\pi\,\Delta t$ and so on. There is a variety of fast, easily automated methods for scanning radio signals and detecting such structure.

terms of any natural phenomena. In Section 9.3 we will consider in more detail how aliens might attempt to contact us.

Unintentional first contact

In order to consider whether we can detect other civilizations by their own, unintentional transmissions into space, let us consider how the Earth might be picked up by an alien's SETI. The myriad of weak radio transmissions from inside the Earth's atmosphere is, in principle, detectable from space. The first problem is that an alien would not know which frequency to monitor. Also these radio transmissions would be broadband in SETI terms. Furthermore, this radiation is not beamed in any particular direction (in fact it is intended only to reach small patches on the Earth's surface). As such, the strength of the signal (its flux) decreases as it spreads out into space and will ultimately become lost in the natural background of radio emissions. Depending on an alien's technology it may be possible for them to detect it if they are close-by. What 'close-by' means in quantitative terms depends on the frequency, strength of signal compared to natural background and the signal bandwidth, but the range would typically be no more than a few tens of light-years.

Let's suppose that aliens do detect our inadvertent transmissions into space and have noticed the artificial structure of the spectrum. Their suspicions will be heightened if they can identify regular variations with a 24 hour period. This is because the Earth's surface is not uniformly covered with transmitters; there are particularly noticeable concentrations on the East and West coasts of the USA and Canada, and across Western Europe. When these sites are positioned so their transmissions are directed towards the aliens' home system, the aliens will register a noticeable increase in the emitted signal, as shown in Figure 9.6.

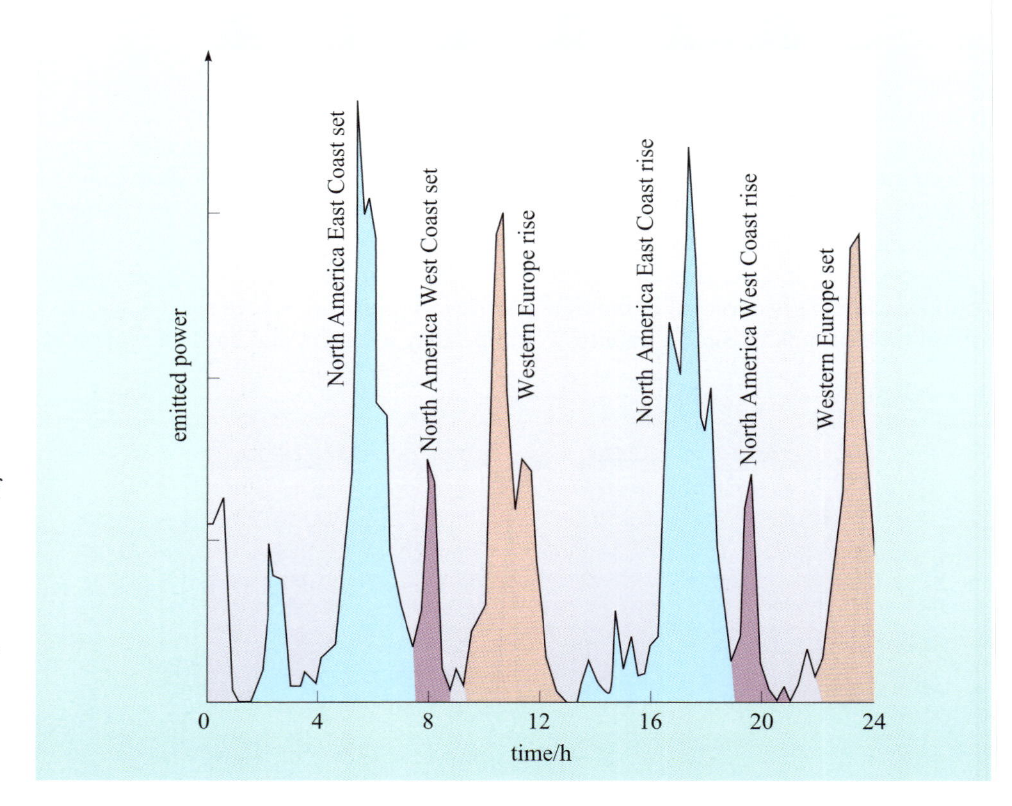

Figure 9.6 The time profile of signal power that might be detected by an alien SETI search. If the aliens could observe the Earth directly with some super-telescope (one that is currently far beyond our technology), then they would see certain heavily populated areas of the Earth rising and setting at times corresponding to the peaks. (Copyright © Woodruff T. Sullivan III, University of Washington)

- In Figure 9.6, why are the peaks closely associated with the *rising and setting* of heavily populated areas?
- Firstly, heavily populated areas contain the greatest number of radio and television stations. Secondly, these stations are intended for terrestrial reception, so they transmit radio waves in directions parallel to the Earth's surface. So, the strongest signal occurs when such areas are edge on to the aliens, i.e. when they are either rising or setting.

- What other encouraging sign might aliens discover if they monitored patiently over several years?
- They would be able to detect the orbital period of the Earth around the Sun, confirming that they have almost certainly detected an orbiting planet.

A more sinister method of detection, one that might make us think twice about initiating contact, is the radiation signature of nuclear explosions. A flash of X-rays, γ-rays and other high-energy particles coming from a star could at first be mistaken for a small stellar flare. Even our own Sun's small flares exceed a nuclear weapon's energy output by several orders of magnitude. Nevertheless, the signature of the energy released from a nuclear weapon would presumably be quite different (unless stellar flare physics inspires a weapons research program). So, if nearby aliens exploded very powerful nuclear weapons (designed to destroy entire planets!) then we might just be able to detect their presence.

Our current rate of population growth on the Earth will, if unchecked, cause serious problems at some point in the not too distant future. With advancing technology we can hope to offset this problem by placing colonies on the Moon, Mars or even asteroids. Ultimately these bodies have limited resources and we might be forced to look outside the Solar System. For a species whose instinct is to continually extend, we will need to find a solution to an on-going living space problem. One ingenious idea is that sufficiently advanced civilizations might construct enormous rings or even complete spheres around stars – called **Dyson spheres** – with radii comparable to planetary orbits. Civilizations living on these spheres could enjoy all the benefits of the star's life-supporting radiative energy, while having an immense living space. The enormous number of stars and their billion-year lifetimes offer a very long-term solution to the problem of accommodating an expanding civilization. An even more radical idea is that a Dyson sphere be constructed around a black hole, into which all the civilization's waste products would be dropped. Energy will be released by nuclear reactions as this waste matter encounters the hot, dense material swirling into the black hole. The energy would ultimately be emitted as radiation and would be absorbed by the inner surface of the Dyson sphere, providing a manageable energy source for the inhabitants existing on the outer surface of the sphere. This is the ultimate in energy recycling programmes and, assuming the technology can be developed, is perhaps the most efficient way imaginable of sustaining a large population.

- Assuming that aliens, like us, prefer their environment at temperatures of a few hundred kelvin, what would be the tell-tales signature of a Dyson sphere?
- At temperatures of a few hundred kelvin, like the Earth, and other planets in the Solar System, a Dyson sphere would emit in the infrared. Being much larger

9.3.2 Encoding

Coding, or encryption, usually conjures up images of confidential messages whose contents must be kept secret from some third party. Encoding in the present context is concerned with the opposite problem: how to encode so anyone, specifically extraterrestrial intelligent life, can decode it. If you were to ask almost any scientist, mathematician, engineer or computer specialist what the simplest, most universal code system was, they would probably all reply 'binary'. The **binary** system only involves two numbers: 0 and 1 (off and on, or no and yes, or nothing and something). This simplest of all systems can be used to encode almost every type of communication. Some examples of binary are discussed in Box 9.1. For our purposes we are interested in 1 representing a pulse in a radio signal, and 0 representing no pulse.

BOX 9.1 BINARY

All information can be encoded as a series of 0s and 1s, known as binary digits or bits. In everyday life we receive numbers, sounds and images in digital form. (In this context, 'digital' simply means encoded in binary.) We'll be concerned with binary numbers and images for SETI purposes.

Binary numbers

Our usual number system (decimal) has ten digits that can be placed in an order to represent adding multiples of powers of 10 (1, 10, 100, 1000...). Likewise it is possible to create a binary system by ordering binary digits to represent powers of 2 (in decimal these are 1, 2, 4, 8, 16, 32, 64, 128, 256...). Consider the following examples:

First of all, the number 13:

Decimal: $13 = (1 \times 10) + (3 \times 1)$

Binary: $1101 = (1 \times 8) + (1 \times 4) + (0 \times 2) + (1 \times 1)$

Secondly, the number 307:

Decimal: $307 = (3 \times 100) + (0 \times 10) + (7 \times 1)$

Binary: $100110011 = (1 \times 256) + (0 \times 128) + (0 \times 64) + (1 \times 32) + (1 \times 16) + (0 \times 8) + (0 \times 4) + (1 \times 2) + (1 \times 1)$

Humans have naturally adopted a decimal number system based on the number 10 because we have ten fingers, which is an aid to counting while we are learning basic arithmetic.

Binary images

Images can be represented in binary form. For example, consider Figure 9.7 which shows two representations of an image. The presence of a 1 indicates that a pixel (picture element) should be white, whereas the presence of 0 indicates that it should be black. This image can be transmitted as one long series of binary digits as long as the recipient has some way of knowing the image dimensions.

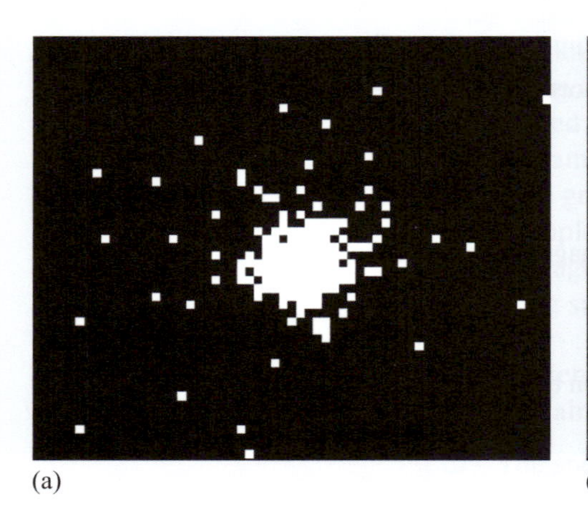
(a)

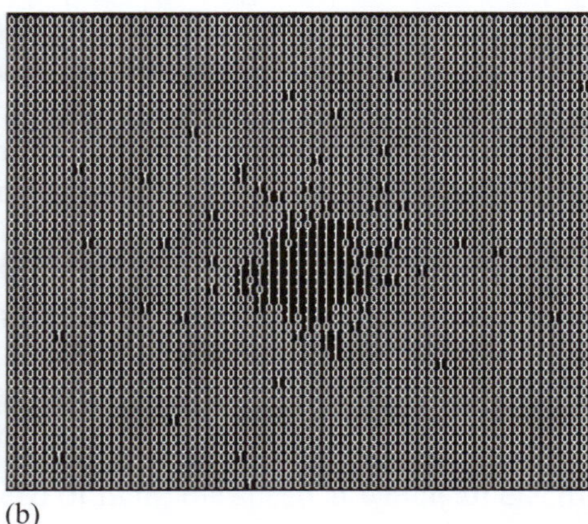
(b)

Figure 9.7 The Globular cluster M13 (based on the image in Figure 9.4): (a) a black and white image. (b) binary form. (David Parker, 1997/Science Photo Library)

- Translate the following binary numbers into decimal: 101, 1010, 100011011.
- $4 + 0 + 1 = 5$

  $8 + 0 + 2 + 0 = 10$

  $256 + 0 + 0 + 0 + 16 + 8 + 0 + 2 + 1 = 283$

QUESTION 9.2

You notice that groups of four pulses are being sent repeatedly separated by gaps of three time units, i.e.

11110001111000111100011110001111000...

Occasionally you notice some missing pulses in the groups of four and you find that each time this happens the pattern after 000 is

...00110000000100001000000000100001010001001000...

What is the message being communicated? Is it likely to be natural? How many fingers might these aliens have?

If we accept that binary is a truly universal system, is it a straightforward matter to use it in sending our messages? Well, almost. The only trouble is that although the concept of binary must be universal (there isn't a simpler system!), the precise system of encoding it is still a convention that could be unique to us. For example, another civilization might read from the last to the first digit (some written languages on Earth read from right to left), in which case 1011 would be 13 and not 11. This is where natural sequences of mathematical numbers come in. If we transmit the following

1, 10, 11, 101, 111, 1011, 1101, 10001, 10011...

an alien intelligence might try decoding these in a number of ways. If the alien's system of binary happened to be right to left, this would read

1, 1, 3, 5, 7, 13, 11, 17, 25...

which wouldn't seem significant. The alien might notice that all the numbers are prime except for $25 = 5 \times 5$. This fact might encourage them to try another system of binary encoding. The correct translation (left to right) would tell the alien that this transmission was no natural phenomenon, as it gives the sequence of prime numbers:

1, 2, 3, 5, 7, 11, 13, 17, 19...

Once the aliens have seen this sequence they will know our binary encoding system and will be able to send a simple message in reply. Of course, we might well find ourselves in the position of the alien if and when our SETI is successful. By considering the problems involved in encoding for CETI, even if we do not wish to actually broadcast a signal, we can learn valuable lessons for our SETI effort. For example, the binary encoding example here tells us that we should search for general signs of structure in a message rather than, for example, particular binary numbers.

9.3.3 The Arecibo message

Attempts to broadcast our existence have been made. Surprisingly, given the potential dangers involved (see Section 9.5), the decision to broadcast was taken without the explicit consent of most people on this planet. The most famous message sent was from the Arecibo radio telescope (Figure 9.8) by Frank Drake (Figure 6.1) in 1974. The then Astronomer Royal, Sir Martin Ryle, wrote to Drake questioning the wisdom of this act. Drake's response was that we had been leaking radio transmissions into space for many years without worrying about it. Irrespective of this, it is safe to say that there is no immediate threat, as we shouldn't expect a response for 50 000 years. In addition, its destination is the globular cluster M13 (Figure 9.4), which as discussed previously, isn't necessarily an ideal home for life, at least as we know it. Also, the message was only sent twice, for two minutes each time, so the aliens will need a much better SETI program than ours to identify it. The 1974 Arecibo message is therefore of more symbolic than real importance, though it provides a classic example of how to encode information for an unknown recipient.

Figure 9.9a presents the raw data stream of bits that was transmitted at two radio frequencies near 2380 MHz (chosen for a practical technical reason, rather than a scientific one), each with a bandwidth of 10 Hz. In total there are 1679 bits of information. This sentence, stored using a standard system of encoding would take up 800 bits of computer memory. Nonetheless, it is surprising how much information can be crammed into this number of bits. But why 1679? Why not 1000 or 2000 or a power of two like 1024 or 2048? The number 1679 is a product of two prime numbers, 23 and 73, which means that those two numbers are the only two whole numbers that it can be divided by to give another whole number.

Figure 9.8 The Arecibo radio telescope in Puerto Rico. (NAIC, Arecibo Observatory)

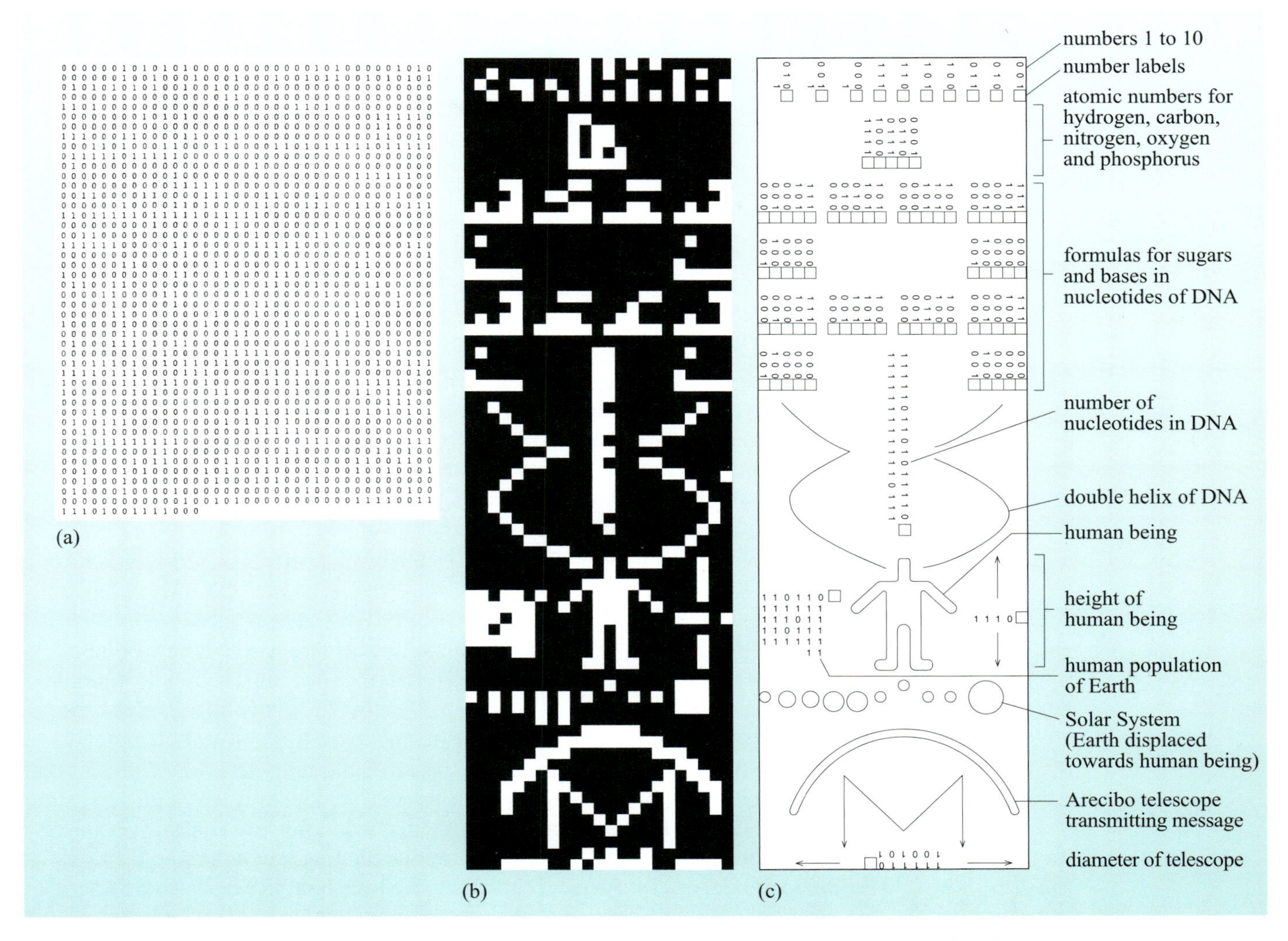

Figure 9.9 (a) The binary message, (b) the two-dimensional pixel image, (c) the interpretation of the image. (Sagan and Drake, 1975)

- What clue does this give you, bearing in mind what was said in Box 9.1?

- The message is designed to be viewed as a two-dimensional image. The two prime factors tell you that the image dimensions must be 23 by 73.

Figure 9.9b shows the two-dimensional black and white, chunky pixel image that can be constructed from the binary stream in Figure 9.9a. When Drake presented the binary bits to colleagues, only some of them succeeded in partially decoding the message. Not one of them, however, succeeded in decoding all the information as presented in Figure 9.9c.

- Two length measurements are given in the message: the height of a human and the diameter of the Arecibo dish. What length unit do you think Drake has assumed?

- The frequency of 2380 MHz corresponds to a wavelength of 12.6 cm. Both the height of the human (given as 14 units) and the diameter of the telescope are given as multiples of 12.6 cm.

9.4 The search to date

It is instructive to consider the history of SETI and CETI. You now know of the famous Arecibo message, but that was not in fact the first recorded method for communicating with extraterrestrial intelligence. Surprisingly, an idea for CETI was suggested as long ago as 1826 and, ironically, it is attributed to Karl Gauss who would go on to help lay the foundations of electromagnetism, the branch of physics that underlies radio astronomy. Gauss suggested that a giant right-angled triangle be drawn by cutting down trees in a Siberian forest. This, he speculated, would allow beings living on the Moon to conclude that there is intelligent life down here on Earth. This story is perhaps sobering; 150 years after Gauss, are our attempts at radio communication any less futile?

The modern age of SETI began in 1959 with a paper published in Nature, penned by physicists Giuseppe Cocconi and Philip Morrison at Cornell University in the USA. They concluded that radio waves were most suitable for the job and that 1420 MHz would be a good frequency to monitor. Apart from making such firm, and scientifically grounded suggestions of how to proceed, their work gave the first real method of detecting extraterrestrial intelligence without betraying our own existence. The following year, Frank Drake began real radio SETI with Project Ozma. Unaware of the Cocconi and Morrison paper, but confirming the wisdom of their important suggestion, Drake's receiver operated across a 100 Hz band centred on the 1420 MHz line. Drake may not have found anything in this initial search, but he did raise public interest, and importantly, opportunities for further funding, not least from the fledgling National Aeronautics and Space Administration (NASA).

The ambitious plan that NASA ultimately developed, led by Bernard Oliver (who was Hewlett–Packard's head of research), was to build up a network of about 1000 radio telescopes, each of diameter 100 m. Project Cyclops, as it was called, was too costly for the liking of US politicians and it remained unfunded. Hopes were raised in the early 1990s when NASA agreed to serious funding (amounting to 0.1% of NASA's total budget) for a project called the High Resolution Microwave Survey, but a year later politics intervened once again and the money was withdrawn.

From the ashes of previous attempts at organized, large-scale radio SETI came the SETI Institute and, in time, its Project Phoenix. Supported by private money, this project has access to several of the world's major radio telescopes: Arecibo, Jodrell Bank in the UK, Parkes in Australia (which relayed telemetry from the first moon-landing) and Greenbank, West Virginia (where Drake had begun SETI). Plans are afoot for something called the Allen array (formerly known as the one-hectare telescope) composed of some 500 standard satellite dishes. This array will be primarily used for SETI, but also for radio astronomy, the reverse of the situation with almost all other radio telescopes.

SERENDIP is an acronym for Search for Extraterrestrial Radio Emissions from Nearby Developed Intelligent Populations.

The University of California at Berkeley has an on-going project called SERENDIP. This project makes serendipitous use of the constant observations made by radio astronomers at the giant Arecibo dish. It provides them with data every second or so from 168 channels (each of bandwidth 0.6 Hz) around the 1420 MHz line. There are two catches. Firstly, they must simply accept random pointing to different parts of the sky; this is partly eased because after several months they will have a comprehensive survey of the sky as afforded by Arecibo's fixed dish. Secondly, the rate of incoming data to be searched exceeds their available computational capacity. The solution to this problem is called SETI@Home. The idea is that computers across the world are idle for much of the time, often wasting CPU cycles on bouncing balls, scrolling banners, surreal patterns and irritating dancing paper clips. If you prefer your CPU to be put to use in SETI then you can download a screen-saver from the SETI@Home project, which will collect data over the internet and analyse it for structured signals. Your computer can accomplish in seconds what might have taken Frank Drake months or even years during Project Ozma in 1960.

Table 9.2 lists a selection of radio searches that have taken place. Notice that the more recent attempts have vastly improved frequency ranges and bandwidths, though the antenna sizes used have not increased that much over the last 40 years. The improvement in radio technology also has a negative influence: artificial, terrestrial signals now swamp the radio spectrum, causing serious problems for radio SETI and astronomy alike.

Optical SETI, sometimes called **OSETI**, is SETI performed in the visible range of the electromagnetic spectrum and was inspired by the invention of the laser in 1961. Earlier in this chapter we explained why radio has historically been favoured for SETI work. However, it is being increasingly acknowledged that OSETI has been underestimated. With improvements in the technology producing more powerful lasers, we now know that it is feasible to compete with the Sun's tremendous radiation output inside a narrow laser beam. This competition can only be sustained with a laser during a brief pulse (lasting typically 1 nanosecond), and in a specific direction, but it is enough to be detectable, and distinct from remote planetary systems. Even though the laser's radiation is concentrated in a narrow wavelength band, the total power (energy per unit time) can still be greater than the total power *over all wavelengths* of the solar radiation. This means that there is no need for a narrowband detector to be centred on the correct wavelength as there is in radio SETI. If aliens were to direct such a laser at us, the apparent jump in brightness of their star would be obvious to any of our broadband optical detectors.

Table 9.2 A chronological list of searches for extraterrestrial radio signals from alien civilizations.

Year	Investigator(s)	Antenna diameter/metres	Observation frequency/MHz	Frequency resolution/kHz	Total frequency band/MHz
1960	Drake	26	1420	0.1	0.4
1968–82	Troitskii	14	100, 1800, 2500	0.013	2.2
1972–76	Zuckerman & Palmer	91	1413–1425	4	12
1972	Verschuur	43, 91	1420	7	20
1972–76	Bridle & Feldman	46	22, 235	30	–
1973–86	Dixon et al.	53	1420	30	0.4
1975–76	Drake & Sagan	305	1420, 1653, 2380	1.0	3
1976–85	Bowyer et al.	26	variable	2.5	20
1988	Bania & Rood	43	8665	0.3	–
1992–93	NASA	305	1300–2400	1, 7, 28	1100
1992–93	NASA	26, 34	1700, 8300–8700	0.019	400
1992–	Bowyer et al.	305	424–436	0.0006	10
1995–	Horowitz	26	1400–1720	0.0005	320
1995	SETI Institute	64, 22	1200–1750	0.001	550
1996–	Werthimer	305	1370–1470	0.0006	100

So, in principle at least, it is not unreasonable to think that we can perform an OSETI search with our own eyes (perhaps aided by a broadband optical filter, i.e. a form of sunglasses) by simply looking for brief flashes from stars in the night sky!

9.5 Final thoughts

What use is a null result? So far, in the forty-two years of SETI, we have failed to detect any sign of extraterrestrial civilizations. This, in science and statistics, is known as a **null result**. As with the micro-lensing search for planets (Chapter 6), a null result doesn't tell us nothing – it can be used to establish upper limits on the number of extraterrestrial civilizations. Actually, it only places limits on the number of civilizations that are *broadcasting*. For this reason, unlike the case of existence of Jupiter-like planets (which cannot decide not to be detected!) from micro-lensing surveys, no firm limits have been placed on the existence of extraterrestrial intelligence.

Our current technology precludes any of us from travelling beyond the Solar System in a human lifetime. However, with our present technology, or at least the technology that might come in the next century or so, we can envisage an amazing prospect. Consider a ship that leaves the Earth and takes 90 years to traverse the 10 light-years to a nearby star. The ship departs carrying 1000 people, with equal numbers of male and female volunteers from a variety of key professions (engineers, scientists, crew, teachers, doctors etc.). Upon arrival at a suitable stellar system, let's call it Stellar One, it has the resources to create a copy of itself, exploiting as much material as it can from planets, asteroids and comets in the host system. Ten years after arrival, one hundred years after its departure from Earth, two ships can depart Stellar One, leaving behind a self-sustaining colony if a suitable location is available. In another 100 years, another two ships will be constructed at each of two new locations about 10 light-years distant from Stellar One.

- Assume that humans embarking on the voyage are 25 years old on departure and live to an age of 75. Further suppose that couples on the ship have an average of four children each, and that their children (and grandchildren etc.) have offspring after 25 years, how many generations are alive upon arrival and what are their populations?

- The first generation of 500 couples produce $4 \times 500 = 2000$ children. After 25 years these children will produce $4 \times 1000 = 4000$ grandchildren. Fifty years into the voyage the first generation dies out, but $4 \times 2000 = 8000$ great-grandchildren are born. In 75 years the number of additions now doubles to 16 000 great-great-grandchildren, but the second generation have now died out. This leaves a population of 24 000, composed of two generations: 16 000 humans in their mid-twenties and 8000 humans around the age of fifty.

Covering 10 light-years in 100 years, the human race can start colonizing the opposite side of the Galaxy (which is about 100 000 light-years away) after 1 Ma. By this time, given sufficient resources, the initial population of 1000 on the first ship to Stellar One will have been doubled 10^6 years/25 years = 40 000 times – that's an unimaginably large number, namely 2^{40000}. The number of ships will double every 100 years, and after about 30 000 years there will be as many ships as there are stars in the Galaxy. In practice, this tells us that it is the availability of resources rather than technology that will limit galactic colonization.

The point of this fantasy is that we could colonize the Galaxy with sub-lightspeed technology on timescales much shorter than that of the billions of years that characterize the evolution of Sun-like stars, or the 4.6 Ga it has taken for our civilization to appear on Earth. This remains true even if the above colonization rate is reduced by a couple of orders of magnitude, because say, resources are short, or it really takes 1000 years to cover 10 light-years.

Now consider what is sometimes called the **Fermi paradox** (credited to physicist Enrico Fermi). If intelligent life is not rare, and if it is able to colonize the Galaxy within a few millions years of becoming an intelligent civilization, why isn't the Galaxy obviously teeming with such life? Of course, this is only a paradox if you already believe the two 'ifs' in the question are true. The obvious way out is to suppose that either intelligent life is rare, or that the resources to sustain it are in short supply. Clearly these suppositions are not independent, as a shortage of resources will place limits on the progress of intelligent life. Even given plentiful resources, civilizations might be quite capable of annihilating themselves by other means, for example by nuclear war. Natural disasters could also limit the lifetime of civilizations, from impacts of asteroids and comets to radiation, such as γ-rays, from stellar flares or nearby supernova explosions.

Less obvious, but perfectly plausible explanations of the Fermi paradox exist even if you believe that intelligent life has proliferated throughout the Galaxy. Given the fact that something like our Solar System could have easily formed elsewhere several billions of years before our Sun did (earlier than that might be difficult because of a lack of heavy elements, which are essential for life as we know it), then other civilizations could be millions to billions of years ahead of us in terms of technology, and evolution. Looking at our own history, we can see the tremendous difference even one hundred years makes in our technological progress. So even tiny differences by stellar standards in the ages of intelligent civilizations, say several thousand years, could mean tremendous differences in technology and culture. If so, then much of the intelligent life in the Galaxy may have no more interest in communicating with us than we have in talking to bacteria.

It has been argued that we are the most advanced civilization, and that is why we are apparently alone. That would appear to violate the **Copernican principle**, which states that humans occupy no special place or time in the Universe. This principle has been repeatedly vindicated, despite many belief-systems that wished humans a special position in the overall scheme of things. Copernicus's placement of the Sun, not the Earth at the centre of the Solar System, and Galileo's subsequent battle with the Catholic church is perhaps the most famous example. This said, the Copernican principle is not a law of nature and so cannot be used in itself to arrive at a firm scientific conclusion. Debate on these philosophical points will not continue once sufficient evidence has emerged to answer the question of whether there is life elsewhere. This evidence could be a thorough exploration of the Galaxy that yields no evidence for life elsewhere, or it might be communication with an alien race.

An important question is how we should handle a first contact situation. If aliens arrived on Earth today, like Caesar arriving in ancient Britain, they would find no central authority to negotiate with, but instead many coexisting peoples, some at peace and some at war. To overcome this problem, and to ensure that the Earth is not simply represented by the most powerful or wealthy nation at the time of the alien's visit, a protocol must be specified for first contact. A protocol is an agreed code of conduct that will vary depending on the parties that are meeting.

For example, a greeting such as 'Hello, how are you?' followed by a handshake is a commonly employed social protocol. The exchange of business cards is a common business protocol. More specific protocols exist for the use of technology. For example, computers have protocols for security, in which usernames and passwords must be transferred.

QUESTION 9.3

What do you think a first contact protocol should include?

There are plenty of examples of advanced (in terms of technology and law) civilizations meeting less advanced civilizations in our history. Perhaps there is something to be learned from this in anticipating our first encounter with extraterrestrial civilizations who, if they come to us in the near future, will almost certainly be far more advanced in their technology. Our examples of first contact amongst terrestrial civilizations are not promising. Indigenous peoples, such as the native Americans and the Australian aborigines have, in extreme cases, been simply exterminated. Many natives also died from the introduction of new diseases, to which they had no immunity. More fortunate peoples had their distinct population interbred out of existence, or at best forced to live in territories that were not wanted by the more advanced settlers. Perhaps the most hopeful story is that of the New Zealand Maoris who managed to find some sort of acceptable coexistence with the European settlers. This, at least in part, was due to the strong Maori warrior tradition and the fact that they quickly gained possession of the newcomers' weapons (perhaps demonstrating that development of more advanced technology does not necessarily imply greater intelligence).

Figure 9.10 Captain Robert Fitzroy (1805–1865) was commander of HMS *Beagle* which carried Charles Darwin on his famous world voyage of discovery. In fact, part of the original reason for Darwin being on the ship was to provide intellectual company for Fitzroy. As well as his association with Darwin, Fitzroy is also remembered as a pioneer in meteorology and its application to seafaring.

A telling encounter is recounted by Charles Darwin from his famous voyage aboard HMS *Beagle*, which was commanded by Captain Fitzroy (Figure 9.10). It landed on Tierra del Fuego (Land of Fires), an inhospitable land on the southern tip of South America. Darwin's candid opinion of the natives reflects the vast gulf between their primitive, and apparently savage culture and his Victorian background:

> 'I believe if the world was searched, no lower grade of man could be found.'
>
> Charles Darwin, *Beagle Diary*, R. D. Keynes ed., 1988

In addition to his mission of mapping the region, Captain Fitzroy, being very religious, wished to bring civilization to the 'savages' and to convert them to Christianity. To this end he left a member of his party, Matthews, to live for a while with the natives while he, Darwin and other members of the crew spent a few days exploring. Upon their return, they found the would-be missionary had been robbed of all his possessions:

> 'From the time of our leaving, a regular system of plunder commenced, fresh parties of the natives kept arriving... Matthews (lost) almost everything which had not been concealed underground. Every article seemed to have been torn up and divided by the natives.'
>
> Charles Darwin, *Voyage of the Beagle*

It seems likely that the natives did not see it this way, as they had surprisingly equitable rules for governing possessions in their society. However, Darwin saw this as a possible reason for their backwardness:

> 'The perfect equality among the individuals composing the Fuegian tribes must for a long time retard their civilization... At present, even a piece of cloth given to one is torn into shreds and distributed; and no one individual becomes richer than the other.'
>
> Charles Darwin, *Voyage of the Beagle*

The difference between societies was so great that even firing guns into the air did not worry the natives – they simply didn't understand the implied threat:

> 'Nor is it easy to teach them our superiority except by striking a fatal blow... Captain Fitzroy on one occasion being very anxious, from good reasons, to frighten away a small party, first flourished a cutlass near them, at which they only laughed; he then twice fired his pistol close to a native. The man both times looked astounded, and carefully but quickly rubbed his head, he then stared a while, and gabbled to his companions, but he never seemed to think of running away.'
>
> Charles Darwin, *Voyage of the Beagle*

In the end Captain Fitzroy was left with two choices. Either stay and shoot a native to demonstrate the potency of their advanced weapons, or leave these people and avoid bloodshed:

> 'Captain Fitzroy, to avoid the chance of an encounter, which would have been fatal to so many of the Fuegians, thought it advisable for us to sleep at a cove a few miles distant.'
>
> Charles Darwin, *Voyage of the Beagle*

The Beagle was on a mission of discovery and science. Other ships captained by men with other motivations have of course chosen the stay-and-shoot option, which ultimately led to the complete annihilation of the native people of Tierra del Fuego, partly through premeditated murder, but also through the introduction of disease.

The story of HMS *Beagle* gives us hope, but also urges caution. Advertising our position may be quite dangerous, because if the Fermi paradox is explained by the rarity of life and the resources that support it, we could find ourselves visited by aliens wishing to ensure their survival, or even just their wealth, at the expense of our very existence. Hopefully our first visitors will be like HMS *Beagle*, and will be led by someone with the ethics of Captain Fitzroy, who is attended by a great scientist like Darwin.

QUESTION 9.4

The correspondence between Ryle and Drake was mentioned in Section 9.3.3. What would be your response to Drake's reply?

QUESTION 9.5

A short (few seconds) radio transmission at a frequency of 90.0 MHz is broadcast unintentionally into space. It is isotropic (equal signal strength in all directions) and is emitted with 10^{20} times more power in its bandwidth than the natural galactic background at that frequency when it is 1 km from the antenna.

(a) What direction will receive the greatest blue-shift due to the Earth's orbital motion?

(b) What is this greatest shift in wavelength?

(c) Extraterrestrials are on a planet orbiting a distant star that happens to be travelling directly towards the Sun at 100 km s^{-1} (this is due to galactic rotation and so will not change on a timescale of a few years). What is the shift due to this and how could this be distinguished from the Earth's orbital shift?

(d) Is the signal powerful enough to be detected by extraterrestrials living on a planet around another star, assuming the alien is monitoring using a similar bandwidth to the transmission?

9.6 Summary of Chapter 9

- SETI – the Search for Extraterrestrial Intelligence – currently involves searching for signals transmitted as radio and visible radiation, from life elsewhere in the Galaxy.
- The radio transmissions are searched for at certain naturally occurring frequencies such as the 1420 MHz of hydrogen, spanned by as many as hundreds of thousands of narrowband channels (less than 1 Hz).
- The top priority is to target nearby, Sun-like single stars. Although life may well exist in multiple-star systems and star clusters (which can be covered by a single radio beamwidth), there are reasons why life like that on Earth might find it difficult to evolve in such environments.
- We can either search for intentional transmissions, or eavesdrop on alien civilizations' internal communications. We can in principle detect the former from anywhere in the Galaxy, though the latter is restricted to stars within a few tens of light-years of the Earth.
- Artificial transmissions can be identified by certain structures in both their spectrum and time-profile. Filtering out terrestrial transmissions with such artificial structure is a major problem for radio SETI.
- Communication with other civilizations requires that we first establish some common ground: universal constants such as the value of π, or the ratio of physical constants with the same units, such the proton and electron mass, are possibilities.
- It is likely that transmissions will be encoded in binary form.
- Humans have, on more than one occasion, broadcast messages intended to declare our existence to an attentive extraterrestrial. The methods and wisdom of doing this were questioned by Astronomer Royal Sir Martin Ryle.
- The first SETI attempt was Project Ozma in 1959. At present the SETI institute's Project Phoenix and the University of California, Berkeley's Project SERENDIP, incorporating SETI@HOME, lead the way in SETI research.
- Given that we can conceive of colonizing the Galaxy in about 1 Ma, it is surprising to some that we have not been visited by some pre-existing alien civilization – this is known as the Fermi paradox.
- Encounters between terrestrial civilizations of different technological advancement provide an interesting, and perhaps worrying precedent for the arrival of a more advanced alien civilization at the Earth. Protocols are necessary to deal with such an event.

ANSWERS AND COMMENTS

QUESTION 1.1

Noble gas elements are insignificant constituents of living organisms as they are unreactive and difficult to combine into molecules by biological activity. It is for this reason that they are absent from the 'humans' column in Table 1.1.

QUESTION 1.2

(a) **Table 1.7** Accretion rates on Earth today (completed).

Sources	Mass range /kg	Mass accretion rate (estimated) /10^6 kg yr^{-1}	Carbon %	Carbon accretion rate /10^6 kg yr^{-1}
meteoritic matter				
meteors (from comets)	10^{-17} to 10^{-1}	16.0	10.0	1.6
meteorites	10^{-2} to 10^5	0.058	1.3	7.5×10^{-4}
crater-forming bodies	10^5 to 10^{15}	62.0	4.2	2.6
unmelted material contributing organic matter				
meteors (from comets)	10^{-15} to 10^{-9}	3.2	10.0	0.32
meteorites, non-carbonaceous	10^{-2} to 10^5	2.9×10^{-3}	0.1	2.9×10^{-6}
meteorites, carbonaceous	10^{-2} to 10^5	1.9×10^{-4}	2.5	4.7×10^{-6}

Note: carbon accretion rate = mass accretion rate $\times \dfrac{\text{carbon \%}}{100}$

(b) The greatest source of meteoritic carbon is crater-forming bodies. These objects are unlikely to arrive steadily over time.

(c) For organic matter the greatest source of carbon is meteors from comets. These objects will be arriving on the Earth relatively constantly.

(d) The accretion rate for total meteor carbon is much larger than that for meteor organic carbon. The carbon in meteors must be mainly inorganic.

(e) 42×10^6 kg of meteoritic carbon arrives in 10 years; 420×10^6 kg in 100 years and $420\,000 \times 10^6$ kg arrives in 100 000 years.

(f) 3.2×10^6 kg of meteoritic organic carbon arrives in 10 years; 32×10^6 kg in 100 years and $32\,000 \times 10^6$ kg in 100 000 years.

QUESTION 1.3

The carbon in the current biosphere is 6.0×10^{14} kg. At present-day rates, meteoritic materials would supply:

(a) a similar amount of carbon in

$$\frac{6.0 \times 10^{14}\ \text{kg (biosphere carbon)}}{4.2 \times 10^6\ \text{kg (meteoritic carbon yr}^{-1})} = 1.4 \times 10^8\ \text{yr.}$$

(b) a similar amount of organic carbon in

$$\frac{6.0 \times 10^{14}\ \text{kg (biosphere carbon)}}{0.32 \times 10^{6}\ \text{kg (meteoritic organic carbon yr}^{-1}\text{)}} = 1.9 \times 10^{9}\ \text{yr.}$$

Rates of meteoritic infall would have been much higher earlier in the Earth's history.

QUESTION 2.1

The star is 10 000 times as luminous as the Sun. The effective temperature will therefore increase by the fourth root of 10 000 or 10. This implies an effective temperature of 2550 K. This is about 2300 °C.

QUESTION 2.2

To maintain the same temperature on Earth, we would have to move our planet out to a distance square root of 10 000 times larger, that is, to a distance of 100 AU from the Sun. This distance is about 2.5 times further out than the orbit of planet Pluto.

QUESTION 2.3

The reason that Mars is presently cold is that controlling the level of CO_2 in the atmosphere through the weathering of silicates (Box 2.2), and thereby regulating climate, requires both a substantial inventory of carbonate rocks and some mechanism for recycling them to CO_2. Even if carbonate rocks are abundant on the surface of Mars (in early 2003 they have not yet been identified spectroscopically by spacecraft or observed by landers), there is apparently no mechanism present to recycle them to CO_2. As you saw in Box 2.2, on Earth this process, termed decarbonation, occurs when oceanic carbonates are subducted, resulting in the decomposition of carbonate and the eventual release of CO_2 back to the atmosphere. Mars, being a small planet, has a cooler interior than Earth and shows no signs of global tectonic activity or recent volcanism.

QUESTION 2.4

The total mass of BIFs older than 2.5 Ga is 3.3×10^{16} kg. Of this mass, 30% is Fe_2O_3, thus $3.3 \times 10^{16} \times 0.3$ kg is Fe_2O_3, which is 9.9×10^{15} kg.

The relative atomic masses of oxygen and iron are 16 and 56, respectively, so the relative molecular mass of Fe_2O_3 is $(2 \times 56) + (3 \times 16) = 160$. Therefore the amount of oxygen incorporated in the BIF deposits is $(48/160) \times 9.9 \times 10^{15}$ kg $=$ 2.97×10^{15} kg.

QUESTION 2.5

If we look at both the geological information you examined in this chapter and biochemical information you examined in Chapter 1, then several key points have emerged that suggest a possible scenario for the emergence of life on Earth:

The oldest rocks so far examined were apparently deposited under water, suggesting the presence of oceans. They contain sedimentary rocks, indicating that the processes of weathering and erosion must have been active. Thus the earliest geological record supports the idea that familiar geological and geochemical processes were operating extremely early in Earth history.

Models of planetary accretion, differentiation and mantle convection suggest that plate tectonics was operating on the early Earth and that up to five times more internal heat was being produced. The Earth's early atmosphere appears to have been composed primarily of CO_2, N_2 and water vapour.

Hydrothermal systems are a key environment that can provide energy to thermophilic and hyperthermophilic organisms that populate the deepest and shortest branches of the phylogenetic tree.

Thus, it seems increasingly difficult to find support for the idea you first met in Sections 1.5 and 1.8.2 that it was the input of external sources of energy into a reduced atmosphere that created a 'prebiotic soup' from which the first organism appeared as a heterotroph. Instead, we have a scenario for the emergence of life that includes internal forms of geochemical energy that result in the formation of environments in which autotrophic reductive metabolisms are nurtured.

QUESTION 2.6

It appears that there are significant ways in which conditions on the early Earth affected the potential for the emergence of life in hydrothermal systems: the higher magnitude of heat flow, a hydrosphere, and upper mantle and crustal rocks that could host hydrothermal systems. Taken together, these apparent differences between the present and the early Earth suggest that conditions in hydrothermal systems on the early Earth were more favourable to the emergence of life than they are at present.

QUESTION 2.7

Some extrapolations would seem to have a basis from the overall trends in evolution observed on Earth. If we assume that life elsewhere has a cellular basis, then increases in the size of organisms, their complexity and diversity from some initial starting point would seem likely to occur. However, we should not forget that for the first 3 Ga of life on Earth there were very few large species. As you'll see in Chapter 9, this has a significant effect on the probability of intelligent life elsewhere in the Universe.

QUESTION 2.8

Methanopyrus, and *Sulfolobus* occur close to the root of the archaea part of the tree, *Thermoplasma* occurs slightly further up the archaea lineage. The phylogenetic tree suggests that the last common ancestor may have been similar to heat-loving chemosynthetic organisms that populate hydrothermal vents today.

QUESTION 3.1

$g_E = GM_E/R_E^2$ and $g = GM/R^2$ where the subscript 'E' refers to values for Earth.

Therefore, $$\frac{g}{g_E} = \frac{M}{M_E}\left(\frac{R_E}{R}\right)^2 \frac{G}{G}$$

So G, the gravitational constant, cancels out and you are left with:

$$\begin{aligned}\frac{g}{g_E} &= \frac{M}{M_E}\left(\frac{R_E}{R}\right)^2 \\ &= (0.1)\times(2)^2 \\ &= 0.4\end{aligned}$$

i.e. Mars's surface gravity is around 40% of Earth's.

QUESTION 3.2

Average conditions on the surface of Mars correspond to a pressure of 6.3 mbar and a temperature of around –60 °C. This would correspond to a point to the left and below the triple point, marked O in Figure 3.6, falling in the region marked 'ice'. Thus liquid water cannot exist under normal or average Mars conditions. However, a typical daily temperature range might be –100 °C to +15 °C. This would correspond to a region that straddles the line OA in Figure 3.6 and passes into the region marked 'water vapour'. This explains why water-ice in the polar ice-caps passes into the atmosphere during the warmer summer period. It's interesting to consider whether conditions on the surface of Mars can enter the region marked 'liquid' in Figure 3.6. To achieve this, we need to look for the possibility of excursions towards higher temperature and pressure. You saw in Section 3.2 that the average pressure is 6.3 mbar with a variation of 2.4 mbar due to seasonal factors. Furthermore, due to altitude variations over the surface, low-lying regions will experience higher pressures. Coupled with the fact that occasionally the temperature rises to maybe +20 °C, this means that sporadically conditions may fall in the region marked 'liquid' – however, this will be only be a temporary circumstance.

QUESTION 3.3

(a) The Viking biology experiments were clearly only deployed at the two Viking lander sites. There are many types of environment on Mars, some probably more favourable for extant life or relics of extinct life, that were not tested. In addition, soil samples were only collected from a short distance below the surface. In view of the oxidizing nature of the surface, we might have expected these samples to be sterile. Samples from greater depths may have produced different results. These are perhaps two of the most compelling arguments which could be used against the notion that the Viking biology experiments ruled out all possibilities of life on Mars.

(b) The results from the GEX, LR and PR experiments from the Martian samples were respectively that oxygen was emitted, labelled gas was emitted and that carbon was detected. These are superficially the same results that would be expected from terrestrial life samples (see Table 3.3). So without the control samples, which would have shown similar results, at least for the GEX and PR experiments, it might have been that the Viking biology experiment results would have been interpreted as indicative of the existence of life.

QUESTION 3.4

(a) In either direction, across or down the image, the resolution is given by the image size or scale divided by the number of pixels.

So, top to bottom, resolution = (4.5/512) km = 8.8 m.

And, side to side, resolution = (12.7/1024) km = 12.4 m.

(Note that it is possible to have different figures for resolution in different directions.)

(b) (i) The resolution in both directions is much smaller (and therefore better) than 500 m. So impact craters of this size could be distinguished (resolved).

(ii) Conversely, in this case, 1 m is below the figure for resolution in both directions. Therefore, 1 metre-scale boulders would not be resolved.

(c) In this case, the area covered by the image would be larger and thus the resolution would be greater (i.e. worse).

QUESTION 3.5

(a) The average speed depends on the molecular mass, m. More specifically, the speed varies as $(1/\text{molecular mass})^{1/2}$. So the more massive a molecule, the lower the average speed (as one would intuitively expect). So lighter molecules are more likely to have speeds above the escape velocity of a planetary body and therefore it will be harder to retain light gases (e.g. hydrogen) in an atmosphere.

(b) From Table 3.2, we see that the two most common constituents in the Martian atmosphere are CO_2 and N_2. From Appendix C, Table C1, we have the following values (Table 3.9) for the appropriate relative atomic masses:

Table 3.9 For Question 3.5(b).

Element	Relative atomic mass
C	12
O	16
N	14

So the relative atomic mass of CO_2 is 44 and of N_2 is 28. Therefore the ratio of the average speeds of these two species is:

Average speed $(CO_2/N_2) = (28/44)^{1/2} = 0.8$, i.e. the average speed of the CO_2 molecules is 80% of that of the N_2 molecules.

QUESTION 3.6

(a) Using appropriate values for masses and radii of Mars, Venus and the Moon, and substituting into Equation 3.8, we obtain the following values (Table 3.10) for the escape velocities:

Table 3.10 For Question 3.6(a).

	Escape velocity/km s^{-1}
Mars	5.0
Venus	10.4
Moon	2.4

So, in ascending order of the escape velocity, we have the Moon, Mars and Venus.

(b) However, this is not the only factor that dictates the likelihood of material from these bodies reaching the Earth. Other factors include:

1 The existence or not of an atmosphere on the relevant body. In the case of Venus, for example, the combination of the high escape velocity and the thick atmosphere means that some of the material ejected as a result of a surface impact will be vaporized during passage through the atmosphere of Venus.

2 Distance can also affect the likelihood of ejecta reaching the Earth.

3 Position in the Solar System. For example, proximity to the Sun (e.g. Mercury) or to a large planet such as Jupiter (e.g. as in the case of Io) can also adversely influence the chances of material reaching the Earth due to gravitational effects.

QUESTION 3.7

The answer is given in Table 3.11.

Table 3.11 Answer to Question 3.7.

	Category	Reasons
(i)	IV	Since comets are of interest for understanding the origins of life and contamination could jeopardize future experiments, Category IV is suggested. However, since there are many comets, it might be argued that a lower category (and thus lower level protection) is warranted.
(ii)	I	Mercury is not of direct interest for understanding the process of chemical evolution so Category I is appropriate.
(iii)	II	Since only a fly-by of Mars is planned, the concern here is primarily over unintentional impact which places it into Category II.
(iv)	IV	This category covers lander missions to targets of interest for understanding the origins of life and for which contamination could jeopardize future experiments. Mars is in this category.
(v)	V	All Earth-return missions are in Category V.

QUESTION 4.1

(a) There are various ways to work this out – here is ours. The value we are looking for is x, so we need to rearrange Equation 4.1 to isolate all the terms involving x on the same side. First, expand the bracket, to get:

$$\rho_{\text{av}} = x\rho_{\text{dense}} + \rho_{\text{light}} - x\rho_{\text{light}}$$

Next, subtract ρ_{light} from each side:

$$\rho_{\text{av}} - \rho_{\text{light}} = x\rho_{\text{dense}} - x\rho_{\text{light}}$$

Rearranging this equation:

$$\rho_{\text{av}} - \rho_{\text{light}} = x(\rho_{\text{dense}} - \rho_{\text{light}})$$

We can now divide both sides by $(\rho_{\text{dense}} - \rho_{\text{light}})$ to get:

$$\frac{(\rho_{\text{av}} - \rho_{\text{light}})}{(\rho_{\text{dense}} - \rho_{\text{light}})} = x$$

Now we can simply insert the density values we were given. Callisto's average density is ρ_{av}, ice density is ρ_{light} and rock density is ρ_{dense}, so:

$$x = \frac{(1.83 \times 10^3\ \text{kg m}^{-3}) - (0.95 \times 10^3\ \text{kg m}^{-3})}{(3.10 \times 10^3\ \text{kg m}^{-3}) - (0.95 \times 10^3\ \text{kg m}^{-3})}$$

$$x = \frac{0.88 \times 10^3\ \text{kg m}^{-3}}{2.15 \times 10^3\ \text{kg m}^{-3}} = 0.41$$

The fraction of Callisto's volume occupied by rock is about 0.41.

(b) One reason the value may be unreliable is that the densities used are for rock and ice at low pressure. In the interior of a large icy satellite the pressure might be high enough for self-compression to lead to significantly higher densities. Another reason is that the method assumes rock and ice only, and ignores the possibility that there could be an even denser component such as an iron-rich inner core.

QUESTION 4.2

Box 4.2 states that tidal force is inversely proportional to the cube of the orbital radius. Thus (tidal force on Europa)/(tidal force on Io) = (Io orbital radius)^3/(Europa orbital radius)^3 = $421.6^3/670.9^3$ = 0.249. Thus the tidal force on Europa is a quarter that on Io. (Note: the amount of tidal heating as a result of this force depends on other factors such as the amount of forced eccentricity and the body's internal properties.)

QUESTION 4.3

This is an exercise in reading values of a logarithmic scale. The concentration of Cl^- in terrestrial seawater is shown as 0.6 moles per litre. The concentration of Cl^- in Europa's ocean is shown as 0.02 moles per litre. The ratio between the two is 0.6/0.02 = 30. Thus the concentration of Cl^- in terrestrial seawater is thirty times that in Europa's ocean.

QUESTION 4.4

Answering this question was part of your learning process. Do not worry if you found yourself at a loss. However, we hope that after reading the answer you will be able to tackle a similar task better in the future.

(a) Most of the surface area appears fairly featureless and mid-grey. This is cut by a large number of linear features (bands), up to several tens of kilometres in width. Most of the bands are dark. Some consist of joined segments of straight lines and some are curved. There is one prominent curved bright band near the lower left. The surface pattern is different in the upper right (northeast), where the pattern of bands disappears and the surface takes on a mottled appearance. Topography becomes apparent only near the right hand (eastern) edge of the view, where the Sun was low in the sky. It is difficult to trace the dark bands into this region, but instead a series of curved ridges shows up.

(b) The dark bands must be younger than the pale (mid-grey) surfaces that they cut. The mottled terrain in the upper right is probably younger than most of the bands, because these disappear when they reach the mottled terrain. Some of the curved ridges in the lower right-hand corner appear to run over the bands, and so these curved ridges must also be younger.

QUESTION 4.5

Pwyll is circular in outline, which is to be expected, but its topography appears to be extremely subdued, even on this image that was recorded when the Sun was very low in the sky (to judge form the shadows in the surrounding area). The rim is very poorly expressed, and there is a cluster of central peaks rather than a single central peak such as you might expect in a crater of this size.

QUESTION 4.6

Although Figure 4.22 includes more variation in size of ridges and grooves than in the comparable sized area shown in Figure 4.19b, this and most of the area of Figure 4.21 has the basic 'ball of string' texture.

QUESTION 4.7

Feature A cuts across feature B, and so feature A must be the younger of the two. Moreover, the parts of feature B on either side of feature A are no longer aligned. They have been displaced to the right by nearly 1 km. The simplest explanation of this is that feature A is a fault with about 1 km of sideways movement across it (a geologist would describe it as a dextral (or right-lateral) strike-slip fault). You can get the same impression of displacement to the right where feature A offsets the edge of the relatively smooth surface in the lower third of the image. (Note that although A is younger than B, A is certainly not the youngest ridge or groove in this area: for example, an even younger groove cuts through A at right angles near the top of the image.)

QUESTION 4.8

(a) It is obvious that the 'ball of string' texture once covered the whole of the area shown in Figure 4.23. However, there are many patches about 10 km across where this texture can be seen in various degrees of disruption. For example:

(i) in square D4 the 'ball of string' surface has been warped upwards into a gentle dome, with a zig-zag fracture where its roof has been stretched apart;

(ii) in squares D/E–5/6 the 'ball of string' surface has been destroyed in a roughly rectangular area, except for a 4 km × 2 km fragment that survives near the southwest edge of the disrupted area. The surface of this whole disrupted area is domed upwards, but its edge must be lower than the surrounding terrain because it is surrounded by an inward-facing cliff;

(iii) in squares D/E–1/2 there is a dome that looks like a mushy extrusion across the original surface within which no identifiable traces of 'ball of string' texture remain.

Sites (i)–(iii) can be regarded as progressively more disrupted examples of 'ball of string texture'. A dome intermediate in character between those at sites (ii) and (iii) occurs at B1–B2. You may also be able to make out several more subtle domes within which the surface has not been fractured at all (in Figure 4.23 and nearby parts of Figure 4.21).

(Note: there are no patch-like depressions in this image; they are all domes. If you cannot perceive them as such, despite being told that the illumination is coming from the right, try rotating the page 90° anticlockwise, so that the illumination now comes from above. This is a more 'natural' illumination direction, and your brain may now be able to make better sense of the topography in the image.)

(b) Throughout Figure 4.24, the 'ball of string' surface has been fractured into slabs, which are bounded by cliffs and so stand higher than the intervening surface, which is occupied by hummocks a few hundred metres across. The slabs still retain their 'ball of string' texture, and by matching prominent ridges and grooves on adjacent slabs it is possible to see that the slabs in the northwest (top left) of the image have been jostled apart by distances of about 1 km. However, the further southeast you look in this image, the harder it is to identify matching slabs and the greater the proportion of new, low-lying, hummocky surface.

QUESTION 4.9

The groove cuts through rafts and matrix alike. Its appearance on the rafts is unremarkable, and it would be taken for just another element of each raft's 'ball of string' texture if we did not see it also cutting the matrix. Generally speaking, the groove's course is not deflected where it crosses from one surface type to another. This groove must have formed at a time when the matrix had become virtually as rigid as the rafts. It is seen cutting the matrix near the right-hand edge of Figure 4.26. There are at least two other grooves cutting the matrix in Figure 4.24. One runs parallel to the first groove, about 5 km to its southwest. The other is at right angles to the first groove, which cuts it about 5 km from the northwest corner of the image.

QUESTION 4.10

(a) Because it is w that we are trying to find, we need to get all the terms involving w into the same side of the equation.

Equation 4.3 can be expanded as:

$$\rho_1 h + \rho_1 w = \rho_2 w$$

Subtracting $\rho_1 w$ from both sides of this equation, we get:

$$\rho_1 h = \rho_2 w - \rho_1 w = w(\rho_2 - \rho_1)$$

And to find w we need to divide both sides by $(\rho_2 - \rho_1)$:

$$w = \frac{\rho_1 h}{(\rho_2 - \rho_1)}$$

(b) It might not be immediately obvious whether the maximum raft density will give the maximum or the minimum raft thickness, but it has to be one or the other. Inserting the value of 1126 kg m^{-3} as ρ_1 in this equation and using 100 m as h and ρ_2 as 1180 kg m^{-3}, we get:

$$w = \frac{(1126\,\text{kg m}^{-3} \times 100\,\text{m})}{(1180\,\text{kg m}^{-3} - 1126\,\text{kg m}^{-3})} = \frac{(1126\,\text{kg m}^{-3} \times 100\,\text{m})}{54\,\text{kg m}^{-3}} = 2085\,\text{m}$$

The raft thickness is $(h + w)$, and so we need to add 100 m to this value, giving a raft thickness of 2185 m.

Inserting 927 kg m^{-3} as ρ_1 in the same expression we get:

$$w = \frac{(927\,\text{kg m}^{-3} \times 100\,\text{m})}{(1180\,\text{kg m}^{-3} - 927\,\text{kg m}^{-3})} = \frac{927 \times 100\,\text{m}}{253} = 366\,\text{m}$$

and hence a raft thickness of 466 m.

The cliff height is certainly not known to three significant figures, so we should not quote these results to more than two significant figures. Thus, according to this method, the raft thickness is not less than about 470 m and not more than about 2200 m.

In fact, the less the density contrast between raft and fluid, the lower the height of the cliffs. If a raft has the same density as the fluid it barely floats at all. If a raft is very much less dense than the fluid, only a relatively small proportion of the raft's volume needs to be immersed in the fluid in order to displace an equivalent mass of fluid.

QUESTION 4.11

(a) Inserting the relevant values into Equation 4.2 (and remembering to convert from km to m), we get:

$P = 1030\,\mathrm{kg\,m^{-3}} \times 9.8\,\mathrm{m\,s^{-2}} \times 3.0 \times 10^3\,\mathrm{m} = 3.0 \times 10^7\,\mathrm{kg\,m\,s^{-2}\,m^{-2}} = 3.0 \times 10^7\,\mathrm{Pa} =$ $30\,\mathrm{MPa}$.

($\mathrm{kg\,m\,s^{-2}\,m^{-2}}$ is force per unit area, which is pressure. The SI unit of pressure is the pascal, abbreviated Pa. Note that $\mathrm{kg\,m\,s^{-2}\,m^{-2}}$ could be written as $\mathrm{kg\,m^{-1}\,s^{-2}}$ but this would obscure the significance of $\mathrm{kg\,m\,s^{-2}}$ being the SI unit of force.)

(b) Similarly, inserting the relevant values into Equation 4.2 we get:

$P = 1180\,\mathrm{kg\,m^{-3}} \times 1.3\,\mathrm{m\,s^{-2}} \times 10^5\,\mathrm{m} = 1.5 \times 10^8\ \mathrm{kg\,m\,s^{-2}\,m^{-2}} = 1.5 \times 10^8\,\mathrm{Pa} = 150\,\mathrm{MPa}$

(or 200 MPa if we treat the 100 km depth to be valid to only one significant figure).

QUESTION 4.12

Europa's annual biomass production is estimated to be at least eight orders of magnitude less than that of present-day Earth (a maximum of $10^6\,\mathrm{yr^{-1}}$ on Europa versus a total of about $10^{14}\,\mathrm{yr^{-1}}$ on Earth). Even if we compare only chemosynthetic biomass production, Europa is estimated to be at least ten thousand times (four orders of magnitude) less productive.

QUESTION 4.13

Perhaps the most obvious technique to use to find out more about Europa is extensive imaging of the surface, at high enough spatial resolution to identify chaos regions and with high enough spectral resolution to identify salts and other contaminants in the ice. You may also have thought of the use of a radar or laser altimeter to map the topography, and thereby contribute to Objective 3. Potentially, a radar instrument could also help significantly with Objectives 1 and 2, as discussed shortly in the text. Precise tracking of the orbiter's trajectory could give information about the details of Europa's gravity field, and hence its internal structure, which would also help with Objectives 1 and 2.

QUESTION 4.14

(a) The matrix between the rafts here is smooth and shows no sign of being cut by the groove. This means that the groove must be older than the matrix. (In case you found it hard to see the necessary detail on Figure 4.24, it is enlarged in Figure 4.34.) This is unlike what happens to the northwest and to the southeast, where the groove is clearly seen to cut the matrix (see Question 4.9).

Thus the matrix between the rafts here appears to be younger than the matrix elsewhere in Figure 4.24.

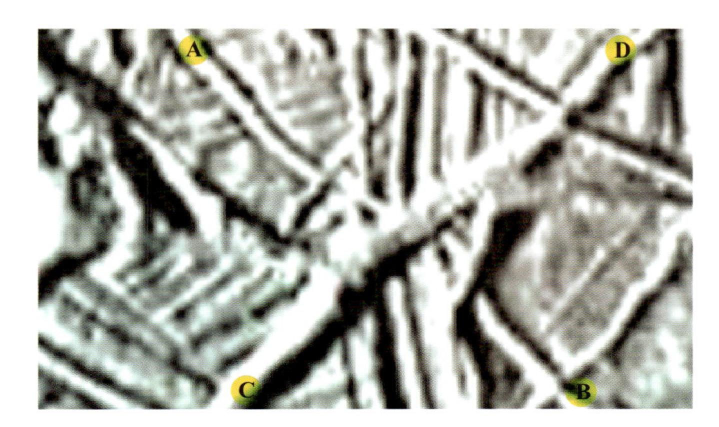

Figure 4.34 Enlargement of the area of interest for Question 4.14. The groove in question passes through A and B. The matrix-filled gap between rafts runs through C and D.

(b) If the matrix between the rafts is younger than the matrix elsewhere, the time sequence of events in the region as a whole must be: chaos formation, freezing and thickening of matrix until it becomes rigid enough for grooves to form on it, formation of this groove, local remobilization of matrix between the rafts erasing the groove at this location. We can conclude that this part of the matrix has been active over a protracted time period.

(c) The matrix between the rafts here is white, so Pwyll ejecta lies on top of it (see caption for Figure 4.21). Remobilization of the matrix would probably disrupt or destroy such a thin ejecta blanket, so probably the Pwyll impact post-dates this local remobilization event.

QUESTION 4.15

Inserting the new value of 1140 kg m^{-3} for ρ_2 into the method used to answer Question 4.10b, we get:

$$w = \frac{(1126\,\text{kg m}^{-3} \times 100\,\text{m})}{(1140\,\text{kg m}^{-3} - 1126\,\text{kg m}^{-3})} = \frac{1126\,\text{kg m}^{-3} \times 100\,\text{m}}{14\,\text{kg m}^{-3}} = 8043\,\text{m}$$

The raft thickness is $(h + w)$, and so we need to add 100 m to this, giving a raft thickness of 8143 m, which we ought to quote to no more than two significant figures, i.e. 8100 m.

QUESTION 4.16

Alternative implications could be:

1 The site has been contaminated by viable organisms accidentally brought from Earth by an earlier probe.

2 Life is indigenous to Europa and arose there independently.

3 Life is indigenous to Europa, and both Earth and Europa were seeded from the same external (for example, cometary) source.

4 Life is indigenous to Europa, but arrived there as contamination via a meteorite from Earth (or Mars).

Implication 1 would be unlikely if the pre-launch cleaning and sterilization of all the previous probes was believed with confidence to be sufficiently stringent. However, it could only be ruled out by detailed genetic study of the 'Europan' micro-organisms to prove that they were not closely related to terrestrial species (hard to do using a robotic probe) or if a sufficiently complex ecology (especially with multicellular organisms and heterotrophs preying on the autotrophs) were discovered that could not have had time to develop since the first possible contamination episode.

Implications 2–4 would all be taken as proof of extraterrestrial life, but detailed genetic studies would be necessary to try to establish which was correct. If Europan amino acids were discovered to have right-handed chirality in contrast to the left-handed chirality ubiquitous on Earth (Section 1.7), this would point towards implication 2. However, there is at least at 50:50 chance of left-handed chirality arising independently, so discovery of left-handed chirality on Europa would not help us to decide between any of these implications.

QUESTION 5.1

(a) Figure 5.19 shows the temperature profile for Titan with the labels 'troposphere' and 'thermosphere' added. The troposphere is the region near the surface where the temperature falls with altitude; the thermosphere is the high altitude region where the temperature increases with altitude. The curve has some similarity in shape to those for Mars and Earth.

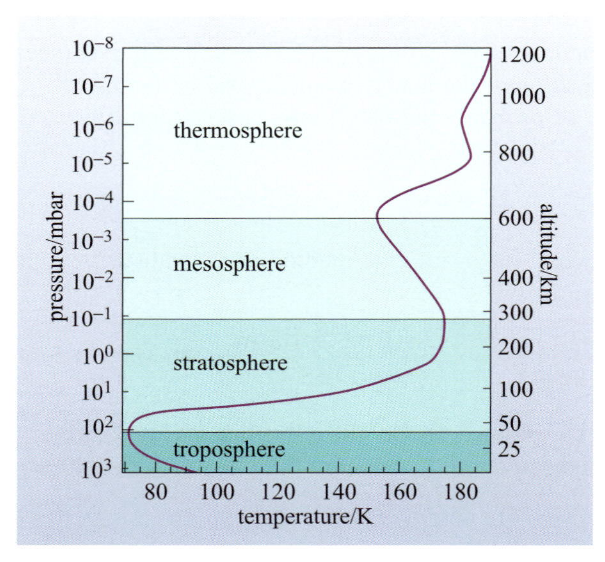

Figure 5.19 Temperature profile for Titan (as in Figure 5.7) with the troposphere and thermosphere labelled.

(b) By approximating the lowest part of the curve in Figure 5.7 with a straight line, and extending it across the axes in the figure, one can estimate the lapse rate. This is found to be approximately $1\,\mathrm{K\,km^{-1}}$. The value quoted in Section 5.4 for the lowest 10 km is $(1.38 \pm 0.1)\,\mathrm{K\,km^{-1}}$. Bearing in mind the coarseness of the plot, the agreement is as good as one might expect.

QUESTION 5.2

From Table 5.2, we have the following data for Titan:

$$\text{mass} = 1.346 \times 10^{23}\,\text{kg, radius} = 2.575 \times 10^{6}\,\text{m}.$$

For a sphere, volume = $(4/3)\pi R^3$ and density ρ = mass/volume.

Therefore, Titan's density is given by:

$$\rho_{\text{Titan}} = \frac{1.346 \times 10^{23}\ \text{kg}}{\frac{4}{3}\pi(2.575 \times 10^{6}\ \text{m})^{3}}$$
$$= 1.9 \times 10^{3}\ \text{kg m}^{-3}$$

By examining Appendix A Tables A1 and A2, we see that the densities of most of the planetary satellites lie in the range (1 to 2) × 10^3 kg m^{-3}, while the terrestrial planets fall within the approximate range of (4 to 5.5) × 10^3 kg m^{-3}. Therefore Titan, with an average density of 1.9 × 10^3 kg m^{-3}, is consistent with the majority of the (icy) planetary satellites.

QUESTION 5.3

Substituting values from Table 5.2, we obtain:

$$g = \frac{GM}{R^2}$$
$$= \frac{(6.67 \times 10^{-11}\ \text{N m}^2\ \text{kg}^{-2}) \times (1.346 \times 10^{23}\ \text{kg})}{(2.575 \times 10^{6}\ \text{m})^2}$$
$$= 1.35\ \text{m s}^{-2}$$

This confirms the value for gravity quoted in Table 5.2.

QUESTION 5.4

The volume of the ocean is given by $(4\pi R^2) \times 0.5 \times D$ where D is the depth, and R is Titan's radius.

Mass of ocean is $(4\pi R^2) \times 0.5 \times D \times \rho$ where ρ is density.

Mass of methane in the ocean is $(4\pi R^2) \times 0.5 \times D \times \rho \times 0.7$.

Mass of methane lost per second from the atmosphere is 4×10^{-12} kg m^{-2} s^{-1} × $(4\pi R^2)$.

Therefore the time T for which the ocean can resupply the atmosphere with methane is:

$$T = \frac{4\pi R^2 \times 0.5 \times D \times \rho \times 0.7}{(4 \times 10^{-12}\ \text{kg m}^{-2}\ \text{s}^{-1}) \times 4\pi R^2}$$
$$= \frac{0.5 \times D \times \rho \times 0.7}{4 \times 10^{-12}\ \text{kg m}^{-2}\ \text{s}^{-1}}$$
$$= 2.9 \times 10^{16}\ \text{s}$$
$$\approx 1000\ \text{Ma}$$
$$\approx 1\ \text{Ga}$$

This figure is of the same order of magnitude as the age of the Solar System so suggests that this may be a plausible mechanism for maintaining methane in Titan's atmosphere.

This assumes that methane lost from the atmosphere to space is immediately replaced by methane from the ocean.

QUESTION 5.5

The closest and the furthest possible distances from Earth to Saturn will occur when Saturn and the Earth lie on the same line from the Sun, in the first case when they are both on the same side of the Sun, and in the other when they are on opposite sides of the Sun (these situations are technically known as conjunctions). From Appendix A, Table A1, we see that Earth's orbit is almost circular ($e = 0.017$) while for Saturn $e = 0.055$. At least for an approximate answer, we can assume that the orbits are circular and use the mean distance of each planet from the Sun, namely 1.0 AU and 9.5 AU.

(a) The closest distance will therefore be (9.5 – 1.0) AU = 8.5 AU and (b) the furthest distance will be (9.5 + 1.0) AU = 10.5 AU. These values correspond to 1.275×10^9 km and 1.575×10^9 km respectively.

A slightly more accurate result could be obtained by taking account of the fact that Saturn's orbital eccentricity means that its closest distance to the Sun (perihelion) is 9.0 AU and furthest distance 10.0 AU. Therefore the smallest possible separation between Earth and Saturn is (9.0 – 1.0) AU = 8.0 AU and largest separation is (10.0 + 1.0) AU = 11.0 AU. These values correspond to 1.2×10^9 km and 1.65×10^9 km respectively.

The reason that the Cassini spacecraft travels more than twice these distances is because it does not travel on a direct (ballistic) trajectory. Instead it has to use the gravity assist technique to deliver its payload to the Saturnian system. This involves a fairly complex trajectory (see Figure 5.12) with fly-bys of Venus, Earth and Jupiter and explains the distance travelled of 3.2×10^9 km.

QUESTION 5.6

(a) We saw in Section 5.3.2 that the abundance of argon is a good indicator of the origin of nitrogen on Titan. If the abundance is similar to what it had been in the nebula from which Titan's atmosphere formed, namely an Ar/N_2 ratio of about 0.06 (i.e. 6%), then it would be expected that Titan's nitrogen had been directly captured (into a clathrate) and gradually outgassed to form the present nitrogen in the atmosphere. On the other hand, if argon's atmosphere were very much less than this value, then Titan's nitrogen was expected to come from the photodissociation of ammonia. In this question, we are told that argon's abundance was found to be about 0.05% so it is the second explanation, namely an ammonia source for nitrogen, that is favoured.

(b) The fact that waves are measured on an open liquid suggests that they are wind driven. We see from Table 5.5 that for a surface wind speed of 5 m s^{-1}, waves of significant height of 4.5 m are expected on Titan. If we measure a value of about 3 m, then we might expect the surface wind (at the time of the measurement) to be slightly less than 5 m s^{-1}, namely about 3 to 4 m s^{-1}.

(c) The Earth's average distance from the Sun is 1.0 AU while for Saturn (and Titan) this distance is 9.5 AU. Therefore, if all other factors were equal, the strength of sunlight at Titan compared to its strength at Earth would be reduced by a factor of $(1.0/9.5)^2 \approx 0.01 = 1/100$. However, the measured factor is 1/1000 instead of 1/100, i.e. it is 10 times weaker than we would expect. This is almost certainly due to absorption by Titan's cloud and haze layers.

QUESTION 6.1

Maximum brightness occurs when Jupiter's lit side faces the Earth. This occurs at the time when both planets are in a line on the same side of the Sun. At this time the Earth–Jupiter distance will be $d_{EJ} = 4.2\,AU$. From Equation 6.5 we have that the brightness of Jupiter from the Earth will be:

$$b_{JE} = \frac{L_J}{2\pi d_{EJ}^{\;2}}$$

where Jupiter's luminosity (of reflected radiation), L_J, can be calculated given the brightness of the Sun at Jupiter and Jupiter's albedo A_J and surface area S_J (Equation 6.4):

$$L_J = A_J b_{\odot J} S_J$$

The text above Equation 6.4 tells us that the brightness of Sun at Jupiter is $b_{\odot J} = 50.4\ W\ m^{-2}$, so we can work out the brightness of Jupiter from the Earth as being:

$$b_{JE} = \frac{L_J}{2\pi d_{EJ}^{\;2}} = \frac{A_J b_{\odot J} S_J}{2\pi d_{EJ}^{\;2}} = \frac{0.7 \times 50.4\ W\ m^{-2} \times 1.54 \times 10^{16}\ m^2}{2\pi(4.2 \times 1.50 \times 10^{11}\ m)^2} = 2.18 \times 10^{-7}\ W\ m^{-2}$$

This is roughly ten times brighter than a zero magnitude star (see Box 6.2) such as alpha Centauri so you have verified that Jupiter will seem like a very bright star in the night sky.

QUESTION 6.2

The brightness of Jupiter when its entire lit side faces alpha Centauri is:

$$b_{J\alpha} = \frac{L_J}{2\pi d_{\alpha J}^{\;2}} = \frac{A_J b_{\odot J} S_J}{2\pi d_{\alpha J}^{\;2}} = \frac{0.7 \times 50.4\ W\ m^{-2} \times 1.54 \times 10^{16}\ m^2}{2\pi(4.07 \times 10^{16}\ m)^2} = 5.22 \times 10^{-17}\ W\ m^{-2}$$

The magnitude of Jupiter when its entire lit side faces alpha Centauri is:

$$m_{J\alpha} = -2.5 \log_{10} \frac{5.22 \times 10^{-17}\ W\ m^{-2}}{2.29 \times 10^{-8}\ W\ m^{-2}} = -2.5 \log_{10} 2.28 \times 10^{-9} = 21.6$$

QUESTION 6.3

$R_s = 2950\ m$. Putting $d_{\odot\alpha} = 4.07 \times 10^{16}\ m$, the distance to alpha Centauri, yields an angular Einstein radius of $\beta_e = 3.81 \times 10^{-7} = 0.0784''$, larger than the $0.013''$ diffraction limit calculated above. In principle, therefore, the best telescopes could directly image an Einstein ring around our nearest star.

QUESTION 6.4

The angular size of the Einstein ring will be:

$$\beta_e = \sqrt{\frac{2R_S}{d_{*E}}} = \sqrt{\frac{2 \times 2950\ m}{100 \times 9.47 \times 10^{15}\ m}} = 7.89 \times 10^{-8}$$

The physical size of the Einstein ring at the distance of the lens will be $R_e = \beta_E d_{*E}$ $= 7.47 \times 10^{10}$ m (about 0.5 AU). The time t taken for the lens to traverse this distance at a speed $v = 20\ \mathrm{km\ s^{-1}}$ will be:

$$t = \frac{R_e}{v} = \frac{7.47 \times 10^{10}\ \mathrm{m}}{2.0 \times 10^4\ \mathrm{m\ s^{-1}}} = 3.74 \times 10^6\ \mathrm{s} = 43.2\ \mathrm{days}$$

QUESTION 6.5

(a) $v_r = v \cos\theta = 100\ \mathrm{m\ s^{-1}} \cos 0° = 100\ \mathrm{m\ s^{-1}} \times 1 = 100\ \mathrm{m\ s^{-1}}$

(b) $v_r = 100\ \mathrm{m\ s^{-1}} \cos 30° = 100\ \mathrm{m\ s^{-1}} \times 0.87 = 87\ \mathrm{m\ s^{-1}}$

(c) $v_r = 100\ \mathrm{m\ s^{-1}} \cos 60° = 100\ \mathrm{m\ s^{-1}} \times 0.5 = 50\ \mathrm{m\ s^{-1}}$

(d) $v_r = 100\ \mathrm{m\ s^{-1}} \cos 90° = 100\ \mathrm{m\ s^{-1}} \times 0 = 0\ \mathrm{m\ s^{-1}}$

(e) $v_r = 100\ \mathrm{m\ s^{-1}} \cos 120° = 100\ \mathrm{m\ s^{-1}} \times -0.5 = -50\ \mathrm{m\ s^{-1}}$

(f) $v_r = 100\ \mathrm{m\ s^{-1}} \cos 150° = 100\ \mathrm{m\ s^{-1}} \times -0.87 = -87\ \mathrm{m\ s^{-1}}$

(g) $v_r = 100\ \mathrm{m\ s^{-1}} \cos 180° = 100\ \mathrm{m\ s^{-1}} \times -1 = -100\ \mathrm{m\ s^{-1}}$

Your sketch should resemble the plot in Figure 6.16. Negative velocities indicate motion towards the observer.

QUESTION 6.6

(a) Your definitions should correspond to those given in Box 6.1, together with the notes in Part (b) on t_p, t_*, β_p and β_*.

(b) The equations relate to the following methods and have the following derivations:

$$a_p = \beta_p d_{*E} \tag{6.25}$$

Equation 6.25 relates to the direct observation of a planet by radiation either reflected from the star, or from the planet's own infrared emission. The observation of a planet at maximum angular distance β_p from its star is related to the actual radius of the orbit by Equation 6.6.

$$a_p = \frac{M_*}{M_p} \beta_* d_{*E} \tag{6.26}$$

Equation 6.26 relates to the astrometric method whereby the angular radius of the star's orbit, β_*, is observed directly. The equation comes from Equation 6.6 and the centre of mass Equation, either 6.13 or 6.21.

$$a_p = \left(\frac{GM_*P^2}{4\pi^2}\right)^{\frac{1}{3}} \tag{6.27}$$

Equation 6.27 relates to any method which tells us the period of the orbit, which could be: direct observation of the planet by emitted or reflected radiation; the astrometric method; the occultation method; or the Doppler spectroscopy method. This equation is in fact Kepler's third law, stated in Equation 6.20.

$$R_p = R_* \sqrt{f_p} \tag{6.28}$$

Equation 6.28 relates specifically to the occultation method where the planet blocks out part of the star's disc. This equation comes from rearranging Equation 6.9.

$$M_{\mathrm{p}} = M_*\left(\frac{t_{\mathrm{p}}}{t_*}\right)^2 \qquad (6.29)$$

Equation 6.29 relates specifically to the gravitational lensing method, where t_{p} is the time over which the planet is lensed and t_* is the time over which the star is lensed. It comes from rearranging Equation 6.11.

$$M_{\mathrm{p}} \sin i_0 = \left(\frac{M_*^2 P}{2\pi G}\right)^{\frac{1}{3}} (v_{\mathrm{r}})_{\mathrm{MAX}} \qquad (6.30)$$

Equation 6.30 relates specifically to the Doppler spectroscopy method, where the maximum radial velocity $(v_{\mathrm{r}})_{\mathrm{MAX}} = v_* \sin i_0$ is measured from wavelength shifts. It can be obtained from multiplying the centre of mass Equation 6.13 or 6.21 by $\sin i_0$ to give Equation 6.22, and then using Equation 6.17 to substitute for $a_* \sin i_0$ and Equation 6.20 to substitute for a_{p}.

QUESTION 6.7

(a) Remember, according to Kepler's third law (Equation 6.20), a planet that is close to the star will also have a short orbital period.

Direct: Close-in planets will be harder to detect in the glare of the star – this is the odd one out, as all other methods work best with close-in planets. The issue is not clear-cut however, as planets further out will also be dimmer, but as we've seen in Section 6.2, modern telescopes should still be able to pick out a Jupiter around a nearby star.

Occultation: the planet is more likely to pass across the disc of the star, and will do so more often than a planet with larger orbital radius.

Gravitational lensing: again the planet's proximity to the star makes it more likely for the planet to be lensed as well as the star.

Astrometric: a close-in planet of given mass will result in a greater and therefore more detectable change in the star's position than a planet of the same mass at a greater orbital radius (Equation 6.26).

Doppler spectroscopy: close-in planets have greater speeds and so can exhibit greater wavelength shifts, and so are easier to detect (Equation 6.24).

(b) Although the densities of planets do vary (by a factor of six in the Solar System), it is true to say in general that larger planets will also be more massive.

Direct: Larger planets (i.e. those with greater radii) are favoured as they will have greater surface areas to reflect or emit radiation.

Occultation: again larger planets will be favoured as they will block out more of their star's radiation (Equation 6.8 or 6.28).

Gravitational lensing: a more massive planet will have a larger Einstein ring (Equation 6.10) and so will be more likely to lens a background object and will also have a longer lensing timescale than a less massive planet (Equation 6.11 or 6.29).

Astrometric: A more massive planet will cause a greater movement (Equation 6.13, 6.21 or 6.26 rearranged so that a_* or β_* is on the left-hand side) of the star.

Doppler spectroscopy: A more massive planet will cause a greater radial velocity (Equation 6.24) and so result in greater, more easily detectable wavelength shifts.

(c) The occultation, gravitational lensing and Doppler spectroscopy methods (Equations 6.28–6.30) are independent of the distance to the planetary system.

QUESTION 7.1

(i) The amplitude is one-half the difference between the maxima and minima of the best-fit curve.

(a) 51 Pegasi: radial velocity amplitude = $60\ \text{m s}^{-1}$.

(b) 70 Virginis: radial velocity amplitude = $315\ \text{m s}^{-1}$.

(c) 16 Cygni B: radial velocity amplitude = $55\ \text{m s}^{-1}$.

(ii) To express the result as a multiple of the speed limit for cars (30 miles per hour), you need to divide the amplitude by $13.4\ \text{m s}^{-1}$.

(a) $60\ \text{m s}^{-1}/13.4\ \text{m s}^{-1} = 4.5$ times the speed limit.

(b) $315\ \text{m s}^{-1}/13.4\ \text{m s}^{-1} = 23.5$ times the speed limit.

(c) $55\ \text{m s}^{-1}/13.4\ \text{m s}^{-1} = 4.1$ times the speed limit.

Note: This illustrates how small the velocity changes are that are being measured. In proportion to the size of the object moving – a star – the changes are truly tiny.

QUESTION 7.2

The redrawn Figure 7.4 is shown as Figure 7.15 below. The main sequence lifetime of an M0 star is 200 000 Ma, so the luminosity (and surface temperature) and hence the habitable zone will hardly change in the first 5000 Ma of this lifetime. For a G2 star the main sequence lifetime is about 11 000 Ma (if you used 10 000 Ma that's OK). Therefore, a significant increase in the luminosity (and some change in surface temperature) will occur by 5000 Ma. Consequently the habitable-zone boundaries will have migrated outwards.

Note: Box 7.1 provides the information needed for this question. Figure 7.15 shows the accurate positions of the habitable zones at 5000 Ma. However, if you showed a small change or no change for the M0 star and a much larger outward shift for the G2 star, then that's fine.

QUESTION 7.3

The least massive exoplanet is quoted as having $M_\text{p} \sin i_0 = 38M_\text{E}$. The actual mass M_p is unlikely to be more than twice this value, so a likely upper limit is about $80M_\text{E}$. (See Section 7.2.2.)

An Earth-mass planet is much less massive than even the least massive exoplanet so far discovered, so would induce too small a radial velocity in its star for it to have been detected with current technology.

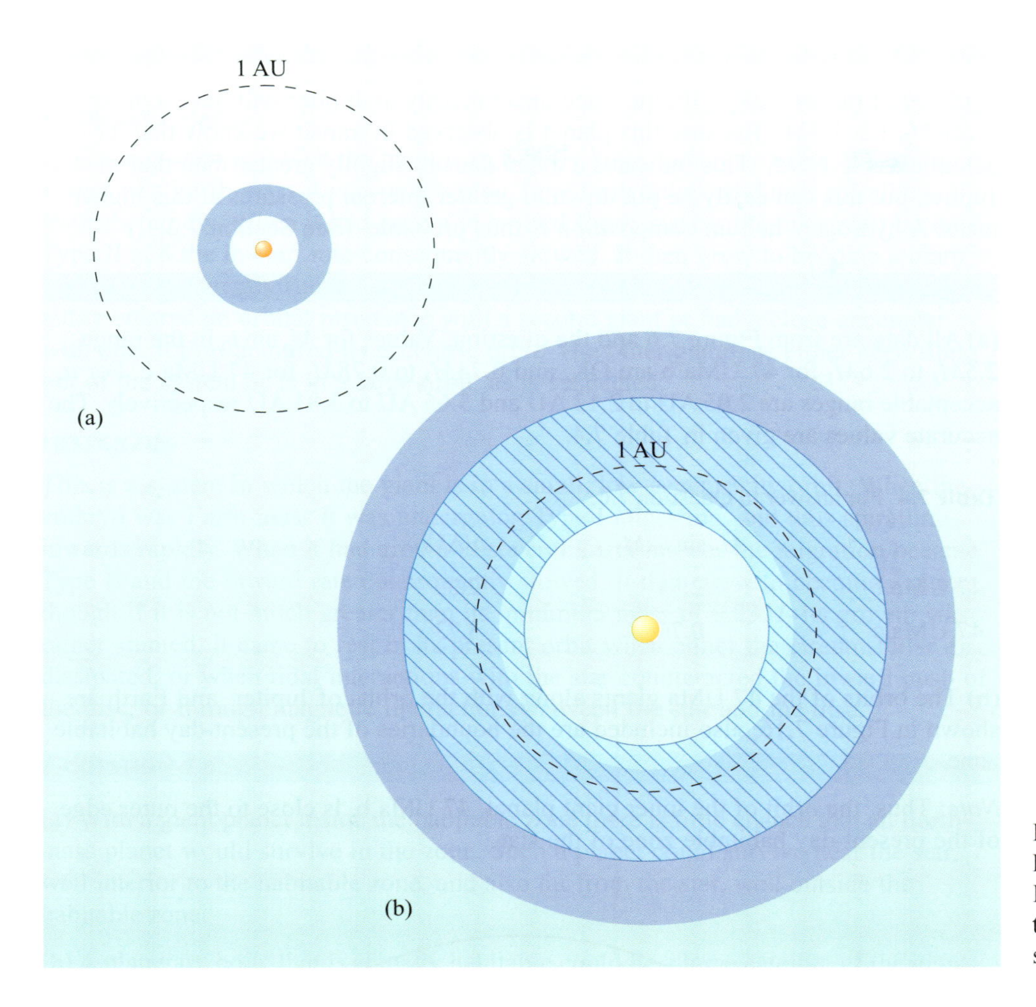

Figure 7.15 The redrawn Figure 7.4 for Question 7.2. In (b) hatching is used to indicate the habitable zone early in the star's life.

QUESTION 7.4

The planet obscures the starlight in proportion to the areas of their discs (see Figure 7.3 and Section 6.4). Equation 6.9 is

$$f_{\rm p} = (R_{\rm p}/R_*)^2$$

where $f_{\rm p}$ is the fraction of starlight blocked and is thus the fractional drop in apparent luminosity. $R_{\rm p}$ is the radius of the planet and R_* is the star's radius. The radius of the planet is $0.090R_{\rm J}$, and that of HD209458 is $9.7R_{\rm J}$. Therefore,

$$f_{\rm p} = (0.090R_{\rm J}/9.7R_{\rm J})^2 = 8.6 \times 10^{-5}$$

As a percentage this is 0.0086%. This is smaller than the 0.1% limit in the question, so it would not be possible to detect an Earth-radius planet.

(M dwarfs have radii about one-tenth that of the Sun, so transits in front of such stars would be detectable.)

(c) From Figure 7.13, which is for a Jupiter-mass planet, βd, at 1 AU for a solar-mass star, is 0.0031 arcsec light-year. For an Earth-mass planet βd = 0.0031 arcsec light-year/318 = 9.7×10^{-6} arcsec light-year. At d = 1 light-year this would be 9.7×10^{-6} arcsec, well under the 0.001 arcsec limit. To reach 0.001 arcsec the distance would have to be reduced by a factor $0.001/(9.7 \times 10^{-6}) = 103$. This would reduce the distance from 1 light-year to 1/103 light-years. There are no stars closer to us than this (except the Sun).

Note: The closest star (after the Sun), Proxima Centauri, is 4.2 light-years away, which illustrates how difficult it will be to detect Earth-mass planets astrometrically.

QUESTION 7.10

(i) These values can be read from the P and a scales in either Figure 7.12 or 7.13.

System A: P = 200 days for a $1M_{\odot}$ star corresponds to a semimajor axis of a = 0.67 AU.

System B: P = 2000 days for a $0.5M_{\odot}$ star corresponds to a semimajor axis of a = 2.5 AU.

(ii) *The radial velocity amplitude*

System A: With P = 200 days, Figure 7.12 shows that $v_{\text{rA}} = 35\ \text{m s}^{-1}$ for a $1M_{\odot}$ star with $M_{\text{p}} \sin i_0 = M_{\text{J}}$. With $i_0 = 89.9°$ and $M_{\text{p}} = 0.0030M_{\text{J}}$ we have $M_{\text{p}} \sin i_0 = 0.0030M_{\text{J}}$. Given that v_{rA} is proportional to $M_{\text{p}} \sin i_0$,

$$v_{\text{rA}} = 35\ \text{m s}^{-1} \times 0.0030 = 0.11\ \text{m s}^{-1}$$

System B: With P = 2000 days, Figure 7.12 shows that $v_{\text{rA}} = 26\ \text{m s}^{-1}$ for a $0.5M_{\odot}$ star with $M_{\text{p}} \sin i_0 = M_{\text{J}}$. With $i_0 = 30°$ and $M_{\text{p}} = 2.0M_{\text{J}}$ we have $M_{\text{p}} \sin i_0 = 1.0M_{\text{J}}$. Thus, v_{rA} remains $26\ \text{m s}^{-1}$.

(iii) *Astrometry*

System A: With P = 200 days, Figure 7.13 shows that $\beta d = 2.0 \times 10^{-3}$ arcsec light-years for a $1M_{\odot}$ star with $M_{\text{p}} = M_{\text{J}}$. Given that βd is proportional to M_{p}

$$\beta d = 2.0 \times 10^{-3}\ \text{arcsec light-years} \times 0.0030 = 6.0 \times 10^{-6}\ \text{arcsec light-years}$$

Now d = 25 light-years, so

$$\beta = 6.0 \times 10^{-6}\ \text{arcsec light-years/25 light-years} = 2.4 \times 10^{-7}\ \text{arcsec}$$

System B: With P = 2000 days, Figure 7.13 shows that $\beta d = 1.6 \times 10^{-2}$ arcsec light-years for a $0.5M_{\odot}$ star with $M_{\text{p}} = M_{\text{J}}$. Given that βd is proportional to M_{p},

$$\beta d = 1.6 \times 10^{-2}\ \text{arcsec light-years} \times 2.0 = 3.2 \times 10^{-2}\ \text{arcsec light-years}$$

Now d = 250 light-years, so

$$\beta = 3.2 \times 10^{-2}\ \text{arcsec light-years/250 light-years} = 1.3 \times 10^{-4}\ \text{arcsec}$$

(iv) System A has an Earth-mass planet in a small orbit. The values of v_{rA} and β are below the thresholds given in the question so it is not detectable by these means. It is however in an orbit presented almost edge-on and so it will transit its star. However, as it is an Earth-mass planet in front of a $1M_{\odot}$ main sequence star it will cause a dip in apparent luminosity of about 0.01% (Question 7.4 answer). Therefore, it will be below the given threshold of detection.

System B has a Jupiter-mass planet in a fairly large orbit around a low mass star. The value of v_{rA} is above the threshold given in the question so it is detectable by Doppler spectroscopy. The value of β is below the threshold so it is not detectable astrometrically. Its inclination shows that it will not transit its star.

(v) In System A the habitable zone is exterior to the planet (Question 7.2 answer) so the planet is not inhabitable. The planet in System B would be outside the habitable zone of a $1M_{\odot}$ star, so is even further outside the habitable zone of a lower mass star (Question 7.2 answer). The planet is thus not habitable. Moreover it is a giant, so it is an unlikely place for life even were it inside the habitable zone.

Note: Remember that the habitable zone is that range of distances from a star in which the stellar radiation on an Earth-like planet would sustain water as a liquid on at least parts of the surface. A planet or satellite can be habitable outside this zone if some other source of heat, such as tidal heating, is available.

QUESTION 8.1

The answer is given in Table 8.1.

Table 8.1 For answer to Question 8.1. M_p represents the mass of the planet and i_0 is the orbital inclination.

Method	Data revealed
radial velocity	$M_p \sin(i_0)$, orbital period, orbital eccentricity
astrometry	M_p, orbital period, orbital eccentricity
transit	Radius of the planet, orbital period

We can get the mean density (from M_p and radius), to see if it is a rocky body or hydrogen–helium rich body. We can also get the semimajor axis of its orbit (from its period and the mass of the star) to see if it is in a habitable zone. There is not enough information here to detect life.

QUESTION 8.2

Most of the ozone would be in the stratosphere, above the cloud, and so it *would* be detectable in the infrared spectrum. Turning to methane, the atmospheric pressure at 10 km would only be about one-tenth that at the ground (Figure 8.8), and so the amount of detectable methane would be much less than on Earth. Therefore, the detection of methane would be much less likely and so the infrared spectrum would reveal life as a possibility, but not a near certainty.

QUESTION 8.3

(i) Except for the absorption band, the general shape of the spectrum would fit quite well a 270 K curve that could be interposed in Figure 8.9. Figure 8.9 is for a cloud-free part of Mars, and so the atmosphere is likely to be transparent outside the absorption band, in which case 270 K would be the temperature of the surface.

(ii) The absorption band centred around 15 μm closely resembles the CO_2 absorption band in the Earth's atmosphere (Figure 8.7), so must be due to CO_2. Comparison also indicates that the quantity cannot be much less than in the Earth's atmosphere (if less at all). The Martian atmosphere is so much thinner than the Earth's atmosphere that CO_2 thus constitutes a considerable proportion of it.

We could then offer to share our knowledge. This might be difficult if they are far more advanced than us, but we can at least tell about our science, about life on Earth and the history of our species, planet and Solar System.

QUESTION 9.4

Leaking terrestrial radio signals could only alert a handful of nearby inhabited stellar systems, whereas the globular cluster M13 could, in principle, harbour a great number of listening civilizations. Also, the Arecibo message would give the recipient the impression that all humans sought contact, whereas aliens eavesdropping on our unintentional transmissions might be inclined to wait for an explicit invitation to make contact.

QUESTION 9.5

(a) The Doppler blue-shift (Chapter 6) will be greatest for a receiver in the direction of the Earth's motion at the time of transmission.

(b) The Earth's orbit is 1.50×10^{11} m in radius, which means it travels a distance of

$$2 \times \pi \times (1.50 \times 10^{11}\ \text{m}) = 9.42 \times 10^{11}\ \text{m}$$

in 1 year. Now 1 year = $365 \times 24 \times 3600\ \text{s} = 3.15 \times 10^{7}\ \text{s}$.

This means Earth is travelling at a speed of

$$v = 9.42 \times 10^{11}\ \text{m}/3.15 \times 10^{7}\ \text{s} = 2.99 \times 10^{4}\ \text{m s}^{-1}$$

A frequency of 90.0 MHz corresponds to a wavelength of

$$\lambda = c/v = 3.00 \times 10^{8}\ \text{m s}^{-1}/\ 9.00 \times 10^{7}\ \text{Hz} = 3.33\ \text{m}.$$

The Doppler shift formula (Equation 6.14) gives us

$$\frac{\Delta\lambda}{\lambda} = \frac{v_r}{c} = \frac{-2.99 \times 10^{4}\ \text{m s}^{-1}}{3.00 \times 10^{8}\ \text{m s}^{-1}} \times 3.33\ \text{m} = -3.32 \times 10^{-4}\ \text{m}$$

The negative sign indicates a blue-shift, i.e. the transmitter on the Earth is travelling towards the receiver.

(c) Now $v_r = +1.00 \times 10^{5}\ \text{m s}^{-1}$, so

$$\frac{\Delta\lambda}{\lambda} = \frac{v_r}{c} = \frac{1.00 \times 10^{5}\ \text{m s}^{-1}}{3.00 \times 10^{8}\ \text{m s}^{-1}} \times 3.33\ \text{m} = 1.11 \times 10^{-3}\ \text{m}$$

The star's motion introduces a constant red-shift (constant on the timescale of observation) whereas the shift from the Earth's orbit will oscillate between being a red-shift and a blue-shift with a period of 1 year.

(d) The signal power of the radio transmission falls as an inverse square law with distance. If the signal strength is S_1 at distance d_1 = 1 km, then the signal strength S_2 at a distance d_2 will be related by

$$\frac{S_1}{S_2} = \frac{d_2^2}{d_1^2}$$

If the signal has fallen to background strength then $S_1/S_2 = 10^{20}$, so we can calculate d_2 as

$$\begin{aligned} d_2 &= \sqrt{\frac{S_1}{S_2}} d_1 \\ &= 10^{10} \times 1 \text{ km} \\ &= 10^{13} \text{ m} \\ &= 1.06 \times 10^{-3} \text{ light-years} \end{aligned}$$

Since this distance is about one-thousandth of a light-year, the signal will be undetectable long before reaching another star.

Alternatively it is possible to answer this question by calculating the factor by which the signal strength is reduced when it reaches the nearest star (4.3 light-years). This factor is 1.65×10^{27} which is clearly much larger that 10^{20} and so the signal will have dropped well below the natural background before it reaches even the nearest star. So, either method tells us that no alien living on a planet around another star can possibly detect this signal above the background level.

Table A4 The largest known minor bodies in the Solar System (excluding Pluto and Charon).

Object	Semimajor axis/AU	Orbital period/yr	Orbital inclination	Orbital eccentricity	Mean radius[a]/km
Largest bodies in the asteroid belt:					
(1) Ceres	2.77	4.61	10.6°	0.079	457
(2) Pallas	2.77	4.61	34.8°	0.230	261
(4) Vesta	2.36	3.63	7.1°	0.090	250
(10) Hygiea	3.14	5.59	3.8°	0.121	215
(511) Davida	3.17	5.65	15.9°	0.180	163
Largest *known* (as of early 2004) bodies in the Kuiper Belt (excluding Pluto):					
2004 DW	44.2	293	2.6°	0.038	800?
2002 LM_{60} ('Quaoar')	43.2	284	8.0°	0.036	650
2002 AW_{197}	47.5	327	24.3°	0.128	400–650?
(28978) Ixion	39.3	246	19.7°	0.245	400–650?
2002 TX_{300}	43.3	284	25.9°	0.121	350–600?
(20000) Varuna	43.3	285	17.1°	0.054	450

[a] The mean radius is defined as the *volumetric* radius (i.e. the radius the body would have if it were a sphere of the same volume).

Table A5 Some selected comets.

Comet[a]	Perihelion distance/AU	Semimajor axis/AU	Orbital period/yr	Eccentricity	Inclination	Velocity at perihelion/km s^{-1}
2P/Enke	0.338	2.22	3.30	0.847	11.8°	69.6
46P/Wirtanen	1.059	3.09	5.44	0.658	11.7°	37.3
81P/Wild 2	1.590	3.44	6.40	0.539	3.2°	29.3
26P/Grigg–Skjellerup	1.118	3.04	5.31	0.663	22.3°	36.0
55P/Tempel–Tuttle	0.977	10.3	33.2	0.906	162.5°	41.6
1P/Halley	0.587	17.9	76.0	0.967	162.2°	54.5
109P/Swift–Tuttle	0.958	26.3	135	0.964	113.4°	42.6
153P/Ikeya–Zhang	0.507	51.0	367	0.990	28.1°	59.0
Hale–Bopp	0.925	184	≈2500	0.995	89.4°	43.8
Hyakutake	0.230	1490	≈58000	0.9998	124.9°	87.8

[a] Well observed periodic comets (i.e. short-period comets) are numbered, somewhat like asteroids, and this is indicated by the designation *number* P/, for example 2P/Enke.

Table A6 Major annual meteor showers.

Date of maximum rate	Name of shower	Hourly meteor rate	Parent comet
3 Jan	Quadrantids	130	unknown
12 Aug	Perseids	80	Swift–Tuttle
21 Oct	Orionids	25	Halley
17 Nov	Leonids	25[a]	Tempel–Tuttle
13 Dec	Geminids	≈90	(3200) Phaethon[b]

[a] This rate is usually what is observed, but every 33 years or so, this shower can display much higher rates.

[b] When discovered, Phaethon was assumed to be an asteroid as no cometary coma was observed. However it is likely that some activity has been present in the past.

Table A7 Some notable Solar System exploration missions.

Mission	Launch	Description
Sputnik 1 (USSR)	4 Oct 1957	First Earth-orbiting satellite. Remained in orbit for 92 days.
Pioneer 4 (USA)	3 Mar 1959	4 Mar 1959: lunar fly-by (within 60 000 km of Moon's surface).
Luna 2 (USSR)	12 Sep 1959	14 Sep 1959: first spacecraft to land (impact) on the Moon.
Venera 1 (USSR)	12 Feb 1961	19 May 1961: first Venus fly-by. (Contact lost before fly-by.)
Mars 1 (USSR)	1 Nov 1962	19 Jun 1963: first Mars fly-by. (Contact lost before fly-by.)
Venera 3 (USSR)	16 Nov 1965	1 Mar 1966: first spacecraft to land on Venus. (Contact lost before landing.)
Luna 9 (USSR)	31 Jan 1966	3 Feb 1966: First soft landing on the Moon. TV pictures returned to Earth.
Zond 5 (USSR)	14 Sep 1968	First spacecraft to both orbit the Moon (18 Sep 1968) and return a payload safely to Earth (21 Sep 1968). Payload included turtles, flies, worms and plants.
Apollo 8 (USA)	21 Dec 1968	First manned mission to orbit the Moon (24 Dec 1968). Returned 27 Dec 1968.
Apollo 11 (USA)	16 July 1969	First manned landing on the Moon (20 July 1969). Crew: Neil Armstrong, Edwin 'Buzz' Aldrin, Michael Collins (orbiter). Returned 24 July 1969.
Apollo 12 (USA)	14 Nov 1969	Second manned landing on the Moon (19 Nov 1969). Crew: Charles Conrad, Alan Bean, Richard Gordon (orbiter). Returned 24 Nov 1969.
Apollo 13 (USA)	11 Apr 1970	Moon mission aborted after onboard explosion on 14 Apr 1970. Crew: James Lovell, Fred Haise, John Swigert (orbiter). Returned 17 Apr 1970.
Luna 16 (USSR)	12 Sep 1970	First robotic sample-return from the Moon. Returned approximately 100 g of lunar material.
Apollo 14 (USA)	31 Jan 1971	Third manned landing on the Moon (5 Feb 1971). Crew: Alan Shepard, Edgar Mitchell, Stuart Roosa (orbiter). Returned 9 Feb 1971.
Mars 3 (USSR)	28 May 1971	2 Dec 1971: first spacecraft to survive landing on Mars. Soft landing. Images returned.
Apollo 15 (USA)	26 Jul 1971	Fourth manned landing on the Moon (30 Jul 1971). Crew: David Scott, James Irwin, Alfred Worden (orbiter). Returned 7 Aug 1971. First lunar rover used.
Pioneer 10 (USA)	3 Mar 1972	First outer Solar System mission. 3 Dec 1973: fly-by of Jupiter. Currently ≈80 AU from the Sun. Will reach the star Aldebaran in 2 million years!
Apollo 16 (USA)	16 Apr 1972	Fifth manned landing on the Moon (21 Apr 1972). Crew: John Young, Charles Duke, Thomas Mattingly (orbiter). Returned 27 Apr 1972.
Apollo 17 (USA)	7 Dec 1972	Sixth (and final) manned landing on the Moon (11 Dec 1972). Crew: Eugene Cernan, Harrison Schmitt, Ronald Evans (orbiter). Returned 19 Dec 1972.
Pioneer 11 (USA)	6 Apr 1973	4 Dec 1974: Jupiter fly-by. 1 Sep 1979: Saturn fly-by.
Skylab (USA)	14 May 1973	Manned orbiting 'space station'. Manned until 8 Feb 1974. Final usage of the Apollo Saturn V rocket.
Mariner 10 (USA)	3 Nov 1973	First (and only) spacecraft to go to Mercury. 5 Feb 1974: Venus fly-by. Mercury fly-bys on 29 Mar 1974, 21 Sep 1974 and 16 Mar 1975.

Table A7 continued.

Mission	Launch	Description
Viking 1 (USA)	20 Aug 1975	Mars orbiter and lander. 19 June 1976: reached Mars. 20 Jul 1976: lander touched down.
Viking 2 (USA)	4 Sept 1975	Mars orbiter and lander. 7 Aug 1976: reached Mars. 3 Sep 1976: lander touched down.
Voyager 2 (USA)	20 Aug 1977	First (only) spacecraft to undertake a tour of all the giant planets. 9 Jul 1979: Jupiter fly-by. 26 Aug 1981: Saturn fly-by. 24 Jan 1986: Uranus fly-by. 25 Aug 1989: Neptune fly-by.
Voyager 1 (USA)	5 Sep 1977	5 Mar 1979: Jupiter fly-by. 12 Nov 1980: Saturn fly-by.
ISEE-3/ICE (USA)	12 Aug 1978	11 Sep 1985: first spacecraft to 'distant fly-by' a comet (Giacobini–Zinner).
Venera 13 (USSR)	30 Oct 1981	1 Mar 1982: Venus landing. Returned colour images from the surface.
Giotto (ESA)	2 Jul 1985	13 Mar 1986: first close (600 km) fly-by of a cometary nucleus (comet Halley).
Magellan (USA)	4 May 1989	Venus orbit insertion 10 Aug 1990. Mapped Venus surface with radar (1990–1994).
Galileo (USA)	18 Oct 1989	First spacecraft to orbit one of the giant planets. 29 Oct 1991: fly-by of asteroid (951) Gaspra. 28 Aug 1993: fly-by of asteroid (243) Ida. 7 Dec 1995: Galileo reaches Jupiter and deployed probe enters the atmosphere of Jupiter. 21 Sept 2003: Galileo impacts Jupiter.
Ulysses (ESA)	6 Oct 1990	First spacecraft to leave the ecliptic plane and orbit around the Sun, passing over the north and south poles. 8 Feb 1992: Jupiter fly-by.
Near Earth Asteroid Rendezvous (NEAR) Mission (USA)	17 Feb 1996	First spacecraft to orbit and land on an asteroid. 27 Jun 1997: fly-by of asteroid (253) Mathilde. 14 Feb 2000: started orbiting near Earth asteroid, (433) Eros. 12 Feb 2001: spacecraft landed on Eros.
Mars Global Surveyor (USA)	7 Nov 1996	Highly successful Mars remote sensing mission. 12 Sep 1997: reached Mars. Mar 1999: began mapping planet.
Mars Pathfinder (USA)	4 Dec 1996	4 Jul 1997: landed on Mars. 6 Jul 1997: deployed the Sojourner rover.
Cassini–Huygens (USA + Europe)	15 Oct 1997	Mission to Saturn and Titan. 30 Dec 2000: Jupiter fly-by. 1 Jul 2004: Saturn orbit insertion. 14 Jan 2005: Huygens probe lands on Titan.
Deep Space 1 (USA)	24 Oct 1998	22 Sep 2001: close fly-by of comet Borrelly's nucleus. Images returned. 29 Jul 1999: fly-by of (9969) Braille.
Stardust (USA)	7 Feb 1999	Fly-by and cometary dust sample return mission to comet Wild 2. 2 Nov 2002: fly-by of asteroid (5535) Annefrank. 2 Jan 2004: fly-by of comet Wild 2. 15 Jan 2006: capsule carrying cometary dust lands on Earth for analysis.
2001 Mars Odyssey (USA)	7 Apr 2001	11 Jan 2002: entered Mars orbit. Acts as relay for 2003 rover missions.
Genesis (USA)	8 Aug 2001	Solar wind particle sample return mission. 3 Dec 2001: capture experiment deployed. Sep 2004: samples returned to Earth.
Rosetta (ESA)	02 March 2004	Comet orbiter and lander. Nominal mission plan: Nov 2005: Earth fly-by. Feb 2007: Mars fly-by. Nov 2007: 2nd Earth fly-by. Nov 2009: 3rd Earth fly-by. 2014: Comet Churyumov-Gerasiminko orbit entry. Nov 2014: lander deployed (Note: exact mission plan may change.)
Mars Express (ESA) + Beagle 2	2 Jun 2003	Mars orbiter and lander. Jan 2004: Mars Express entered Mars orbit, contact lost with Beagle 2 lander.

APPENDIX B SELECTED PHYSICAL CONSTANTS AND UNIT CONVERSIONS

Table B1 SI fundamental and derived units.

Quantity	Unit	Abbreviation	Equivalent units
mass	kilogram	kg	
length	metre	m	
time	second	s	
temperature	kelvin	K	
angle	radian	rad	
area	square metre	m^2	
volume	cubic metre	m^3	
speed, velocity	metre per second	$m\,s^{-1}$	
acceleration	metre per second squared	$m\,s^{-2}$	
density	kilogram per cubic metre	$kg\,m^{-3}$	
frequency	hertz	Hz	(cycles) s^{-1}
force	newton	N	$kg\,m\,s^{-2}$
pressure	pascal	Pa	$N\,m^{-2}$, $kg\,m^{-1}\,s^{-2}$
energy	joule	J	$kg\,m^2\,s^{-2}$
power	watt	W	$J\,s^{-1}$, $kg\,m^2\,s^{-3}$
specific heat capacity	joule per kilogram kelvin	$J\,kg^{-1}\,K^{-1}$	$m^2\,s^{-2}\,K^{-1}$
thermal conductivity	watt per metre kelvin	$W\,m^{-1}\,K^{-1}$	$m\,kg\,s^{-3}\,K^{-1}$

Table B2 Selected physical constants and preferred values.

Quantity	Symbol	Value
speed of light in a vacuum	c	$3.00 \times 10^8\,m\,s^{-1}$
Planck constant	h	$6.63 \times 10^{-34}\,J\,s$
Boltzmann constant	k	$1.38 \times 10^{-23}\,J\,K^{-1}$
gravitational constant	G	$6.67 \times 10^{-11}\,N\,m^2\,kg^{-2}$
Stefan–Boltzmann constant	σ	$5.67 \times 10^{-8}\,W\,m^2\,K^{-4}$
Avogadro constant	N_A	$6.02 \times 10^{23}\,mol^{-1}$
molar gas constant	R	$8.31\,J\,K^{-1}\,mol^{-1}$
charge of electron	e	1.60×10^{-19} C (negative charge)
mass of proton	m_p	$1.67 \times 10^{-27}\,kg$
mass of electron	m_e	$9.11 \times 10^{-31}\,kg$
Astronomical quantities:		
mass of the Sun	$M_\odot$	$1.99 \times 10^{30}\,kg$
radius of the Sun	$R_\odot$	$6.96 \times 10^8\,m$
photospheric temperature of the Sun	$T_\odot$	5770 K
luminosity of the Sun	$L_\odot$	$3.84 \times 10^{26}\,W$
astronomical unit	AU	$1.50 \times 10^{11}\,m$

Table B3 Some useful conversions from alternative unit systems to SI units.

Quantity	Unit	SI equivalent
angle	1 degree	$(\pi/180)$ rad
pressure	1 bar	10^5 Pa
temperature	1 °C	1 K
energy	1 erg	10^{-7} J
	1 electron volt	1.60×10^{-19} J
	1 ton of TNT	4.18×10^9 J
length	1 foot	0.305 m
	1 mile	1.61×10^3 m
area	1 square inch	6.45×10^{-4} m^2
	1 square mile	2.59×10^6 m^2
mass	1 pound	0.454 kg
speed, velocity	1 mile per hour	0.447 m s^{-1}

Table B4 The Greek alphabet.

Name	Lower case	Upper case
Alpha	α	A
Beta (bee-ta)	β	B
Gamma	γ	Γ
Delta	δ	Δ
Epsilon	ε	E
Zeta (zee-ta)	ζ	Z
Eta (ee-ta)	η	H
Theta (thee-ta – 'th' as in theatre)	θ	Θ
Iota (eye-owe-ta)	ι	I
Kappa	κ	K
Lambda (lam-da)	λ	Λ
Mu (mew)	μ	M
Nu (new)	ν	N
Xi (cs-eye)	ξ	Ξ
Omicron	o	O
Pi (pie)	π	Π
Rho (roe)	ρ	P
Sigma	σ	Σ
Tau	τ	T
Upsilon	υ	Y
Phi (fie)	ϕ	Φ
Chi (kie)	χ	X
Psi (ps-eye)	ψ	Ψ
Omega (owe-me-ga)	ω	Ω

APPENDIX C THE ELEMENTS

Table C1 The elements and their abundances.

The relative atomic mass, A_r, is the average mass of the atoms of the element as it occurs on Earth. It is thus an average over all the isotopes of the element. The scale is fixed by giving the carbon isotope $^{12}_{6}C$ a relative atomic mass of 12.0. By convention, the Solar System abundance is normalized to 10^{12} atoms of hydrogen, whereas the CI chondrite abundance is normalized to 10^6 atoms of silicon. To directly compare chondrite abundance to Solar System abundance (by number), you would multiply chondrite abundance by 35.8.

Atomic number, Z	Name	Chemical symbol	Relative atomic mass, A_r	Solar System abundance		CI chondrite abundance by number
				by number	by mass	
1	hydrogen	H	1.01	1.0×10^{12}	1.0×10^{12}	2.79×10^{10}
2	helium	He	4.00	9.8×10^{10}	3.9×10^{11}	2.72×10^{9}
3	lithium	Li	6.94	2.0×10^{3}	1.4×10^{4}	57.1
4	beryllium	Be	9.01	26	2.4×10^{2}	0.73
5	boron	B	10.81	6.3×10^{2}	6.8×10^{3}	21.2
6	carbon	C	12.01	3.6×10^{8}	4.4×10^{9}	1.01×10^{7}
7	nitrogen	N	14.01	1.1×10^{8}	1.6×10^{9}	3.13×10^{6}
8	oxygen	O	16.00	8.5×10^{8}	1.4×10^{10}	2.38×10^{7}
9	fluorine	F	19.00	3.0×10^{4}	5.7×10^{5}	843
10	neon	Ne	20.18	1.2×10^{8}	2.5×10^{9}	3.44×10^{6}
11	sodium	Na	22.99	2.0×10^{6}	4.7×10^{7}	5.74×10^{4}
12	magnesium	Mg	24.31	3.8×10^{7}	9.2×10^{8}	1.074×10^{6}
13	aluminium	Al	26.98	3.0×10^{6}	8.1×10^{7}	8.49×10^{4}
14	silicon	Si	28.09	3.5×10^{7}	1.0×10^{9}	1.00×10^{6}
15	phosphorus	P	30.97	3.7×10^{5}	1.2×10^{7}	1.04×10^{4}
16	sulfur	S	32.07	1.9×10^{7}	6.0×10^{8}	5.15×10^{5}
17	chlorine	Cl	35.45	1.9×10^{5}	6.6×10^{6}	5240
18	argon	Ar	39.95	3.6×10^{6}	1.5×10^{8}	1.01×10^{5}
19	potassium	K	39.10	1.3×10^{5}	5.2×10^{6}	3770
20	calcium	Ca	40.08	2.2×10^{6}	8.8×10^{7}	6.11×10^{4}
21	scandium	Sc	44.96	1.2×10^{3}	5.5×10^{4}	34.2
22	titanium	Ti	47.88	8.5×10^{4}	4.1×10^{6}	2400
23	vanadium	V	50.94	1.0×10^{4}	5.3×10^{5}	293
24	chromium	Cr	52.00	4.8×10^{5}	2.5×10^{7}	1.35×10^{4}
25	manganese	Mn	54.94	3.4×10^{5}	1.9×10^{7}	9550
26	iron	Fe	55.85	3.2×10^{7}	1.8×10^{9}	9.00×10^{5}
27	cobalt	Co	58.93	8.1×10^{4}	4.8×10^{6}	2250
28	nickel	Ni	58.69	1.8×10^{6}	1.0×10^{8}	4.93×10^{4}
29	copper	Cu	63.55	1.9×10^{4}	1.2×10^{6}	522
30	zinc	Zn	65.39	4.5×10^{4}	2.9×10^{6}	1260
31	gallium	Ga	69.72	1.3×10^{3}	9.4×10^{4}	37.8
32	germanium	Ge	72.61	4.3×10^{3}	3.1×10^{5}	119

Atomic number, Z	Name	Chemical symbol	Relative atomic mass, A_r	Solar System abundance		CI chondrite abundance by number
				by number	by mass	
33	arsenic	As	74.92	2.3×10^2	1.8×10^4	6.56
34	selenium	Se	78.96	2.2×10^3	1.8×10^5	62.1
35	bromine	Br	79.90	4.3×10^2	3.4×10^4	11.8
36	krypton	Kr	83.80	1.7×10^3	1.4×10^5	45
37	rubidium	Rb	85.47	2.5×10^2	2.1×10^4	7.09
38	strontium	Sr	87.62	8.5×10^2	7.5×10^4	23.5
39	yttrium	Y	88.91	1.7×10^2	1.5×10^4	4.64
40	zirconium	Zr	91.22	4.1×10^2	3.7×10^4	11.4
41	niobium	Nb	92.91	25	2.3×10^3	0.698
42	molybdenum	Mo	95.94	91	8.7×10^3	2.55
43	technetium	Tc[a]	98.91	–[b]	–[b]	–[b]
44	ruthenium	Ru	101.07	66	6.8×10^3	1.86
45	rhodium	Rh	102.91	12	1.3×10^3	0.344
46	palladium	Pd	106.42	50	5.3×10^3	1.39
47	silver	Ag	107.87	17	1.9×10^3	0.486
48	cadmium	Cd	112.41	58	6.5×10^3	1.61
49	indium	In	114.82	6.6	7.6×10^2	0.184
50	tin	Sn	118.71	140	1.6×10^4	3.82
51	antimony	Sb	121.76	11	1.3×10^3	0.309
52	tellurium	Te	127.60	170	2.2×10^4	4.81
53	iodine	I	126.90	32	4.1×10^3	0.90
54	xenon	Xe	131.29	170	2.2×10^4	4.7
55	caesium	Cs	132.91	13	1.8×10^3	0.372
56	barium	Ba	137.33	160	2.2×10^4	4.49
57	lanthanum	La	138.91	16	2.2×10^3	0.4460
58	cerium	Ce	140.12	41	5.7×10^3	1.136
59	praseodymium	Pr	140.91	6.0	8.5×10^2	0.1669
60	neodymium	Nd	144.24	30	4.3×10^3	0.8279
61	promethium	Pm[a]	146.92	–[c]	–[c]	–[c]
62	samarium	Sm	150.36	9.3	1.4×10^3	0.2582
63	europium	Eu	151.96	3.5	5.3×10^2	0.0973
64	gadolinium	Gd	157.25	12	1.8×10^3	0.3300
65	terbium	Tb	158.93	2.1	3.4×10^2	0.0603
66	dysprosium	Dy	162.50	14	2.3×10^3	0.3942
67	holmium	Ho	164.93	3.2	5.2×10^2	0.0889
68	erbium	Er	167.26	8.9	1.5×10^3	0.2508
69	thulium	Tm	168.93	1.3	2.3×10^2	0.0378
70	ytterbium	Yb	170.04	8.9	1.5×10^3	0.2479
71	lutetium	Lu	174.97	1.3	2.3×10^2	0.0367

Atomic number, Z	Name	Chemical symbol	Relative atomic mass, A_r	Solar System abundance		CI chondrite abundance by number
				by number	by mass	
72	hafnium	Hf	178.49	5.3	9.6×10^2	0.154
73	tantalum	Ta	180.95	1.3	2.4×10^2	0.0207
74	tungsten	W	183.85	4.8	8.8×10^2	0.133
75	rhenium	Re	186.21	1.9	3.5×10^2	0.0517
76	osmium	Os	190.2	24	4.6×10^3	0.675
77	iridium	Ir	192.22	23	4.5×10^3	0.661
78	platinum	Pt	195.08	48	9.3×10^3	1.34
79	gold	Au	196.97	6.8	1.3×10^3	0.187
80	mercury	Hg	200.59	12	2.5×10^3	0.34
81	thallium	Tl	204.38	6.6	1.4×10^3	0.184
82	lead	Pb	207.2	110	2.3×10^4	3.15
83	bismuth	Bi	208.98	5.1	1.1×10^3	0.144
84	polonium	Po[a]	209.98	–[c]	–[c]	–[c]
85	astatine	At[a]	209.99	–[c]	–[c]	–[c]
86	radon	Rn[a]	222.02	–[c]	–[c]	–[c]
87	francium	Fr[a]	223.02	–[c]	–[c]	–[c]
88	radium	Ra[a]	226.03	–[c]	–[c]	–[c]
89	actinium	Ac[a]	227.03	–[c]	–[c]	–[c]
90	thorium	Th[a]	232.04	1.2	2.8×10^2	0.0335
91	protoactinium	Pa[a]	231.04	–[c]	–[c]	–[c]
92	uranium	U[a]	238.03	0.32	7.7×10^1	0.0090
93	neptunium	Np[a]	237.05	–[c]	–[c]	–[c]
94	plutonium	Pu[a]	239.05	–[c]	–[c]	–[c]
95	americium	Am[a]	241.06	–[c]	–[c]	–[c]
96	curium	Cm[a]	244.06	–[c]	–[c]	–[c]
97	berkelium	Bk[a]	249.08	–[c]	–[c]	–[c]
98	californium	Cf[a]	252.08	–[c]	–[c]	–[c]
99	einsteinium	Es[a]	252.08	–[c]	–[c]	–[c]
100	fermium	Fm[a]	257.10	–[c]	–[c]	–[c]
101	mendelevium	Md[a]	258.10	–[c]	–[c]	–[c]
102	nobelium	No[a]	259.10	–[c]	–[c]	–[c]
103	lawrencium	Lr[a]	262.11	–[c]	–[c]	–[c]

[a] No stable isotopes.

[b] Detected in spectra of some rare evolved stars.

[c] Too scarce to have been detected beyond the Earth (i.e. abundance value not well known).

GLOSSARY

absorption bands The result of the blending together of many close absorption spectral lines.

absorption spectral line A narrow wavelength range over which a spectrum is darker than at adjacent wavelengths.

acidophiles An extremophile that thrives in conditions of low pH, typically below 5.0.

albedo The fraction of the total light or other radiation falling on a non-luminous body, such as a planet, that is reflected from it.

alkaliphiles An extremophile that thrives in conditions of high pH, typically above 9.0.

amphiphiles Polar molecules capable of forming lamellar structures that could provide templates for natural synthesis of organic molecules.

anhydrobiosis A state of dormancy entered by some organisms in order to survive desiccation or other extreme stresses.

apolar Indicates the absence of electrical poles on a molecule.

asteroid Asteroids are small planetary bodies (typically a few kilometres in size), occupying a range of orbits between those of Mars and Jupiter.

asteroid belt Most asteroids are members of the asteroid belt, which lies between the orbits of Mars and Jupiter, from about 2.2–3.2 AU.

astrometry The branch of astronomy concerned with measuring precise positions of objects on the sky.

autotrophic/autotrophs Describing an organism that manufactures its own organic material from inorganic compounds, literally ‘self-feeding’. Autotrophs are primary producers, and include plants and photosynthesizing bacteria, and hydrothermal vent bacteria that use chemical energy rather than solar energy.

banded iron formations (BIFs) Finely banded, siliceous, iron-rich deposits normally found in rocks older than 600 Ma. They form thick strata often several hundred metres thick and persistent over 150 km or more.

bilayer A membrane consisting of a double layer of molecules.

bilayer vesicle A spherical membrane consisting of a double layer of molecules.

binary The encoding of numbers as a sequence of 0s and 1s.

binary systems A system of two stars close enough to be in orbit around each other.

bioload (sometimes termed bioburden) The number of viable organisms present on an item.

black hole Another type of remnant of a massive star (greater than about 8 solar masses) that is left after a supernova explosion. It is a point mass with infinite density, and is surrounded by a region from which not even light can escape.

brightness The energy carried by radiation crossing a unit area per unit time. The brightness of a source is also referred to as its flux.

brown dwarfs Hydrogen-rich objects in the approximate mass range 13–80 Jupiter masses. They are sufficiently massive for thermonuclear fusion to occur in their cores, but insufficiently massive to become main sequence stars.

carbohydrates Collective term for sugars and polysaccharides, which are chain-like molecules made up of carbon, hydrogen and oxygen.

carbonaceous chondrite A class of meteorites that contains 2–5% by mass of carbon in the form of organic compounds.

catalyst A substance that speeds up or facilitates a chemical reaction, but remains unchanged after the reaction. In living organisms, all enzymes are catalysts.

centre of mass The point about which two masses will execute their mutual orbit. (Note: this defines how centre of mass has been defined in Chapter 6.)

CETI Communication with Extraterrestrial Intelligence.

channel A narrow bandwidth in the radio spectrum. Modern radio receivers can monitor many channels simultaneously.

chaos In the internationally agreed nomenclature for features on planetary surfaces, chaos is a descriptor term to indicate a distinctive area of broken terrain. On Europa it is applied to regions where the icy shell appears to have become broken into separate slabs.

chemical equilibrium For a reversible reaction chemical equilibrium occurs when the rate of the forward reaction equals the rate of the back reaction, so that the concentrations of products and reactants reach steady-state values.

chemosynthesis Formation of organic compounds by reactions other than those of photosynthesis, and using an energy source *other* than solar radiation.

chirality The property of 'handedness' such that an object and its mirror image cannot be exactly superimposed one on the other (for example a left hand and a right hand). Applied to molecules, a molecule is said to be chiral if it has a structure such that a model of that molecule and a model of its mirror image cannot be superimposed.

circumstellar habitable zone The region surrounding a star throughout which the surface temperatures of any planets present might be conducive to the origin and development of life. The temperatures required are generally taken to be those necessary for liquid water to exist on a planet's surface.

clathrate A compound in which molecules of one component are physically trapped within the crystal structure of another.

coacervates A spherical aggregation of lipid molecules which is held together by hydrophobic forces.

column mass This parameter describes the mass of gas in a column of unit cross-sectional area (i.e. $1\ m^2$) extending from the surface of a planet vertically upwards to the very top of the atmosphere.

comets Planetary bodies often described as 'dirty snowballs', typically around 10 km in diameter, that follow some eccentric orbit around the Sun. They are aggregates of ice, interstellar dust and organic molecules. As a comet approaches the Sun, it heats up and some of the ice is vaporized. Both ice and dust then stream away from the Sun, and may produce a spectacular tail visible from Earth with the naked eye.

continuous habitable zone The region surrounding a star where the surface temperatures of any planets present have remained conducive to the origin and development of life throughout a star's life.

Copernican principle The principle stating that humankind has no special place in the Universe.

cryptoendoliths An extremophile that lives below the surface of a rock.

Darwinian evolution The theory that evolution occurs by the natural selection of individuals with characteristics that enable them to successfully reproduce and pass on their traits to their offspring.

DNA (deoxyribonucleic acid) A polymer in which the monomer is a composite molecule consisting of a phosphate group joined to a deoxyribose molecule, which in turn is joined to one of four different organic bases – adenine, guanine, cytosine or thymine. Molecules of DNA can be very large: some molecules of human DNA have relative molecular masses as high as 10^{12}.

Doppler effect The effect where the wavelength of a spectral line is shifted because radiation is being emitted from the body that is moving towards or away from the observer. A shift to longer, redder wavelengths indicates the source is moving away.

Doppler spectroscopy The method of measuring radial velocities from spectral lines that has been very successful at detecting exoplanets.

double helix The configuration of the two strands of polydeoxyribonucleotide wound around each other to give a molecule of DNA.

Dyson spheres An artificially constructed surface surrounding a star which derives its energy from the star that is designed to support very large populations.

ecliptic plane The plane of the Earth's orbit around the Sun. The orbits of the planets, apart from Pluto, and the Moon lie very near the ecliptic (as it is often termed).

Einstein radius The radius of the Einstein ring, either quoted as a physical distance or as an angular distance.

Einstein ring The circle of light that would be seen on the sky when a gravitational lens and source lie exactly on the observer's line of sight.

emission spectral line A narrow wavelength range over which a spectrum is brighter than at adjacent wavelengths.

endoliths An extremophile that lives on or inside rock or in the pores between mineral grains.

enzymes Catalysts found in living organisms that allow complex biochemical reactions to occur. Most enzymes are proteins.

equilibrium constant At equilibrium, the ratio of the concentrations of the reactants and products is a constant for a given reversible reaction and fixed temperature, called the equilibrium constant.

escape velocity The minimum velocity required to escape from the gravitational attraction of a body. At the surface of a planet, the escape velocity is $\sqrt{(2GM/R)}$ where M and R are its mass and radius respectively and G is the Universal gravitational constant. For the Earth, for example, the escape velocity is about 11.2 km s^{-1}.

exoplanets Planets that are not in the Solar System.

extremophiles Organisms (normally micro-organisms) that can grow under extreme conditions of, for example, heat, cold, high pressure, high salt concentrations or high or low pH.

fermentation Energy-yielding metabolic process by which sugar and starch molecules are broken down to carbon dioxide and ethanol in the absence of oxygen.

Fermi paradox The Fermi paradox is based on the supposition that intelligent life is common in the Universe. If this supposition is true then the paradox is: why have no extraterrestrials contacted us yet?

flux *see* brightness.

flux density *see* brightness.

forced eccentricity A description of the very slightly exaggerated elliptical shape of a satellite's orbit that results from repeated mutual gravitational attraction when adjacent satellites are in orbital resonance, and which leads to tidal heating.

galactic habitable zones Regions of the Galaxy that are considered to be more conducive to the origin and development of life.

Galilean satellites The four major satellites of Jupiter, named Io, Europa, Ganymede and Callisto, which were discovered by Galileo Galilei in 1610.

Gas Exchange experiment (GEX) One of the three main biological experiments carried aboard each Viking lander and designed to test for life under two different conditions. (1) Addition of water alone – it was assumed that organisms that had been dormant for a very long time under dry conditions on Mars would be revived and stimulated back into metabolic activity. (2) A rich organic broth was 'fed' to the sample as a further encouragement to induce metabolism.

genetic code The sequence of bases in the RNA (and DNA) that codes for the correct position of amino acids relative to each other.

globular star cluster A group of a few hundred thousand, very old stars that can be more than 10 Ga old.

gravitational lensing The effect whereby the presence of matter alters the path of rays of light, as if gravity was acting as a lens.

habitable zone The range of distances from a star in which the stellar radiation on an Earth-like planet would sustain water as a liquid on at least part of its surface.

halophiles An extremophile that thrives in extremely saline environments.

heteronuclear molecules Molecules in which there are at least two types of atom, e.g. HCl, CO_2, H_2O.

heterotrophic/heterotrophs Describing an organism that builds its tissues using organic compounds synthesized by other organisms, literally 'other-feeding'. Heterotrophs are also known as consumers or secondary producers, and include animals, fungi and some bacteria.

homonuclear diatomic molecules Molecules consisting of two identical atoms, e.g. O_2.

hot Jupiters A term given to massive planets (comparable with Jupiter in mass or greater) that are orbiting close to their star. These giant exoplanets are well within the ice-line of their star.

hydrazine A volatile highly reducing liquid (N_2H_2) used in rocket fuels.

hydrodynamic escape The process of loss of gas from an atmosphere as a result of a rapidly moving outward flow.

hydrothermal vents Sources of hot, mineral-rich waters located on fractures on deep-ocean submarine ridges. In some cases, precipitation of insoluble minerals takes place when the hydrothermal waters encounter the water of the ocean. This results in the phenomenon of 'black smokers'.

hyperthermophiles An extremophile that thrives in extremely high temperature environments, up to about 105 °C, with a few tolerating 113 °C, and generally fails to multiply below 90 °C.

ice Sometimes used simply to refer to frozen water, this can also mean other volatiles in a frozen state (either singly or in a mixture) such as methane, ammonia, carbon monoxide, carbon dioxide and nitrogen. Frozen water is sometimes distinguished (for clarity) as water-ice.

ice-line The boundary around a star beyond which water can condense. This can make available much more condensable material because water is abundant.

interferometer A device which brings different rays of light together so as to improve the resolution of several (usually radio) telescopes or to eliminate the image of a star next to its much fainter planet (see null).

interstellar medium The stellar matter that thinly fills the space between stars. It consists of gas (mainly hydrogen), with a trace of dust.

isomers Different molecules are isomers if they have the same molecular formula but differ in the arrangement of their atoms.

isotope fractionation The separation of the isotopes of an element during naturally occurring processes as a result of the mass differences between their nuclei.

kernels Bodies of the order of 10 Earth masses, mainly water, that can form beyond the ice-line and can capture gas from the circumstellar disc to form giant planets.

last common ancestor The earliest recognizable organisms from which all life evolved.

late heavy bombardment A period about 4 billion years ago during which planetary bodies suffered intense bombardment by fragments left over from the formation of the planets.

light-curve The graph of an astronomical object's brightness versus time.

limb darkening A phenomenon in which a planet or star appears darker at its edge or limb due to the presence of an atmosphere.

lipids Organic compounds related to fats, also providing an energy store.

luminosity The rate at which energy in the form of electromagnetic radiation leaves a star.

M dwarf A main sequence star with a mass less than about half that of the Sun. They are particularly abundant and very long-lived.

main sequence star The period during the evolution of a star's energy output and surface temperature where stars spend most of their lives. Most of the observed stars lie in the main sequence.

mesophiles Organisms that grow best at temperatures between 25 °C and 40 °C.

metallicity The proportion of a star (usually by mass) that consists of elements heavier than hydrogen and helium.

meteorites Extraterrestrial materials that fall to the surface of the Earth.

micell A spherical monolayer.

microbial Of or pertaining to micro-organisms such as bacteria.

micro-lensing Gravitational lensing where only the change in brightness due to the lens can be detected because the image cannot be resolved.

microspheres Microscopic spherules which form on the cooling of hot saturated solutions of protein-like substances.

moderation The process by which neutrons are slowed down by collisions of the neutrons with the nuclei of atoms.

molecules The smallest freely existing parts of a substance that retain the chemical identity of the substance. Molecules do not have to contain atoms of different elements.

monolayer A membrane consisting of a single layer of molecules.

monomer An individual organic unit which can be linked with similar units to form a polymer.

near-infrared (NIR) Wavelengths just beyond the red end of the visible spectrum, from 0.78 μm to about 2 μm.

neutron star One type of remnant of a massive star (greater than about 8 solar masses) that is left after a (Type II) supernova explosion. It is only a few tens of kilometres in diameter, but is very dense.

nitriles A class of organic compound containing the group CN.

noble gases A group of monatomic gaseous elements forming group 18 of the Periodic Table: helium (He), neon (Ne), argon (Ar), krypton (Kr), xenon (Xe), and radon (Rn).

nucleic acids There are two types of nucleic acid: DNA (deoxyribonucleic acid) and RNA (ribonucleic acid). DNA and RNA are polymers in which the monomer is called a nucleotide. DNA and some types of RNA are chemicals that carry the genetic information of cells.

nucleotides Monomers that form a nucleic acid. Each monomer consists of a phosphate group and a base attached to a ribose or deoxyribose molecule.

null A point where the trough of a waveform meets a peak resulting in complete cancellation of the wave. An interferometer is used to produce a null to eliminate a star's image next to its much fainter planet.

null result A result that arises from a scientific search where no evidence is found. Such a result can give us information, usually by placing limits on the occurrence of the sought after phenomenon.

observables Quantities that can either be directly measured or measured with the supposition of assumptions, or a particular model for what is being observed.

open star clusters A group of between a few hundred to a few thousand recently formed stars; typically tens of millions of years old.

orbital resonance A state in which one orbiting body (a planetary body about the Sun, or a satellite about a planet) has an orbit whose period is a simple ratio (e.g. 2:1, 3:2, 4:1, etc.) of another's.

OSETI Optical SETI – that is SETI performed in the visible part of the electromagnetic spectrum.

osmosis A process by which solvent molecules pass through a semi-permeable membrane from a less concentrated solution into a more concentrated one.

oxidation A chemical reaction that involves the loss of one or more electrons by an atom or molecule (always part of an oxidation–reduction reaction in which those electrons are gained by another molecule or atom). Previously, the term was more strictly applied to a reaction in which oxygen combines with another element or compound to form an oxide.

oxidation/oxidize/oxidizing A chemical reaction that involves a loss of one or more electrons by an atom or molecule (always part of an oxidation–reduction reaction in which those electrons are gained by another atom or molecule). Previously the term was more strictly applied to a reaction in which oxygen combines with another element or compound to form an oxide.

panspermia The arrival of living organisms on the surface of planets from space.

partial pressure The fractional contribution of a component to the total pressure of a gas. Can be obtained by multiplying total pressure by the volume ratio and has the same units as pressure.

periastron distance The closest distance of a planet to its star.

phase diagram A plot of temperature against pressure for a substance showing the relation between the solid, liquid and vapour states.

photochemical A range of chemical reactions which are a consequence of the absorption of photons of electromagnetic radiation.

photodissociation A particular photochemical process in which a molecule dissociates as a result of the absorption of a photon of electromagnetic radiation.

photolysis Breakdown of molecules into radicals or individual atoms as a result of exposure to short-wavelength (usually ultraviolet) radiation. Also called photodissociation.

photosynthesis The synthesis, by green plants (including algae) and some bacteria, of organic compounds using the energy of sunlight. Carbon dioxide is reduced to carbohydrate through addition of hydrogen atoms derived (in all photosynthetic organisms except certain bacteria) from water, and oxygen is produced as a consequence.

$$n\mathrm{CO_2} + n\mathrm{H_2O} + \text{energy} \longrightarrow (\mathrm{CH_2O})_n + n\mathrm{O_2}$$

phylogenetic tree A diagram which relates the evolutionary history of organisms.

planetary body A convenient term that can be used to describe planets, satellites and asteroids.

planetary embryos A hypothetical body of something like a hundredth to a tenth the mass of a planet, produced as the end product of runaway growth of planetesimals.

polar Indicates the presence of electrical poles on a molecule.

polymer Large molecules in which a group of individual organic units are repeated.

prebiotic Things such as organic molecules which formed on, or were introduced to, the Earth before life originated, are said to be prebiotic.

protein Large organic compounds made of chains of amino acids.

pulsar A particular kind of neutron star that is believed to rotate rapidly.

racemic Describes equal quantities of left- and right-handed objects.

radial velocity The velocity component measured along the line of sight, usually by exploiting the Doppler effect. Usually it is positive when the source is moving away from the observer.

radial velocity method See Doppler spectroscopy.

radicals/free radicals Uncharged atom or group of atoms (typically highly reactive) with one or more unpaired electron.

radiolysis Breakdown of molecules into radicals or individual atoms as a result of being hit by charged particles.

red giant star A star after its main sequence phase, with a mass less than about 8 solar masses, before it sheds material to leave a white dwarf. A red giant is much larger than it was at its main sequence phase.

redox pair A chemical reaction between two atoms or molecules, in which one atom/molecule is reduced and the other atom/molecule is oxidized.

reduced/reduction A chemical reaction in which atoms or molecules either lose oxygen or gain hydrogen or electrons. (*See also* oxidation)

reflectance spectrum A display of the amount of radiation reflected by a surface or a volume of substance over a range of wavelengths.

regolith Surface material on a planetary body that has been fragmented, mixed and widely distributed as a result of impact cratering.

resolution The ability of an optical instrument to distinguish fine detail. Usually measured as an angle or in true distance on a planetary surface, for example. The smaller the value, the better an instrument is able to recognize small details.

respiration/aerobic respiration The metabolic process through which biomass $(CH_2O)_n$ initially formed by photosynthesis is oxidized by atmospheric oxygen forming carbon dioxide and water, liberating useful energy.

$$(CH_2O)_n + nO_2 \longrightarrow nCO_2 + nH_2O + \text{energy}$$

RNA (ribonucleic acid) A polymer in which the monomer is a composite molecule consisting of a phosphate group joined to a ribose molecule, which in turn is joined to one of four different organic bases – adenine, cytosine, guanine and uracil. Molecules of RNA are much smaller than those of DNA, with relative molecular masses as low as 35 000.

Rubisco The acronym for the enzyme that mediates the first stages of carbon fixation in photosynthesis: ribulose bisphosphate carboxylase/oxygenase.

scan To monitor a radio signal as the frequency of the receiver is continuously increased (or decreased) over time.

Schwarzchild radius A quantity involved in defining the Einstein ring radius, equal to GM/c^2 for a body of mass M.

seeing A term used by astronomers to mean the distortion caused by the Earth's atmosphere.

selection effect An effect which means that a particular method is more likely to detect certain kinds of objects, for example Doppler Spectroscopy is most likely to detect massive planets that orbit close to their star.

SETI The Search for Extraterrestrial Intelligence.

SLiME An acronym for **S**ubsurface **Li**thoautotrophic **M**icrobial **E**cosystem, communities of bacteria that thrive in water-filled rock deep below the Earth's surface.

Snowball Earth The name given to the theory that the entire Earth was ice-covered for long periods 600–700 Ma ago.

solar nebula The hypothetical cloud of gas and dust within which the Sun and the other constituents of the Solar System formed.

spatial resolution The angle below which two objects cannot be seen as being separate in an image.

spectral resolution The minimum difference in wavelength for which features in a spectrum can be discriminated.

spectrum A display of the amount of radiation received from any source over a range of wavelengths.

stromatolites Laminated structures that commonly form mounds, built up over long periods of time by successive layers or mats of cyanobacteria that trapped sedimentary material.

supernova The explosion of a massive star (several solar masses at the main sequence phase) at the end of its life.

thermal emission Emission of radiation from a source that is determined by the temperature of the source.

thermal hysteresis The process by which substances produced by an organism in order to prevent freezing of its tissues or body fluids when subject to subzero environmental temperatures depress the freezing point of water to well below the melting point.

thermophiles An extremophile that thrives in environments where the temperature is high, typically up to 60 °C.

tholin Red-brownish substances made of complex organic compounds. They do not exist naturally on earth, because our present oxidizing atmosphere blocks their synthesis. However, tholins can be made in the laboratory by subjecting mixtures of methane, ammonia, and water vapour to simulated lightning discharges.

transits Events in which one astronomical body moves in front of another.

triple point The (unique) point on a phase diagram representing the pressure and temperature at which all three phases (solid, liquid and vapour) of a substance can exist in equilibrium with each other.

troposphere The lowest layer of a planetary atmosphere extending from the surface up to the region at which the temperature stops falling with increasing altitude.

Type I migration Migration of a planet or kernel due to gravitational interaction with a circumstellar disc, where no gap has opened in the disc. The rate of migration is much more rapid than in Type II migration, and is proportional to the mass of the planet/kernel, and to the mass of the disc. The disc migrates (inwards) less rapidly.

Type II migration Migration of a planet or kernel due to gravitational interaction with a circumstellar disc, where a gap has opened in the disc. The rate of migration is much less than in Type I migration, and the disc and planet/kernel migrate together.

ultraviolet circularly polarized light (UVCPL) Polarized light (light waves that have electromagnetic vibrations in only one direction) with wavelengths or frequencies between those of X-rays and visible light. There are three types of polarized light: plane-polarized, circularly polarized, and elliptical-polarized light – each depending on the net direction of the vibrations.

volatiles Elements or compounds that melt or boil at relatively low temperatures, or (equivalently) condense from a gas at a low temperature. Hydrogen, helium, carbon dioxide, and water are examples.

volume ratio (or mole fraction) The fraction by number of atoms or molecules present.

wavenumber The inverse of wavelength, i.e. 1/wavelength.

wet adiabatic lapse rate The lapse rate that prevails when a parcel of air reaches saturation.

white dwarf The remnant of a red giant after it has shed much of its mass. A white dwarf is about Earth-size though very dense, and starts its life very hot.

FURTHER READING

Beatty, J.K., Peterson, C.C. and Chaikin, A. (1999) *The New Solar System*, Cambridge University Press, Cambridge.

Bennett, J.O., Shostak, G.S., Jakosky, B.M. (2003) *Life in the Universe*, Addison Wesley, San Francisco, Calif.

Cockell, C.S. (2003) *Impossible Extinction: Natural Catastrophes and the Supremacy of the Microbial World*, Cambridge University Press, Cambridge.

Conway-Morris, S. (2003) *Life's Solution: Inevitable Humans in a Lonely Universe*, Cambridge University Press, Cambridge.

Gibson, E.K. *et al*. (1997) The Case for Relic Life on Mars, *Scientific American*, **277**, pp.58–65.

Green, S.F. and Jones, M.H. (2004) *An Introduction to the Sun and Stars*, Cambridge University Press, Cambridge.

Jones, B.W. (2004) *Life in the Solar System and Beyond*, Springer-Praxis, Heidelberg, Germany, and Chichester, UK.

Jones, M.H. and Lambourne, R.J.A. (2004) *An Introduction to Galaxies and Cosmology*, Cambridge University Press, Cambridge.

Knoll, A.H. (2003) *Life on a Young Planet: the first three billion years of evolution on Earth*, Princeton University Press, Princeton, New Jersey.

Lorenz, R. and Mitton, J. (2002) *Lifting Titan's Veil*, Cambridge University Press, Cambridge.

McBride, N. and Gilmour, I. (2004) *An Introduction to the Solar System*, Cambridge University Press, Cambridge.

Ward, P.D. and Brownlee, D. (2000) *Rare Earth: why complex life is uncommon in the Universe*, Copernicus, New York.

Ward, P.D. and Brownlee, D. (2003) *The life and death of planet Earth: how the new science of astrobiology charts the ultimate fate of our world*, Times Books, New York.

ACKNOWLEDGEMENTS

The production of this book involved a number of Open University staff, to whom we owe a considerable debt of thanks for their commitment and the high professional standards of their contributions. Jennie Neve Bellamy managed the production of the associated Open University course, ensured authors met deadlines, tracked down copyrights, and generally kept the project on track. Valerie Cliff styled the text for handover to Pamela Wardell and Peter Twomey who copy-edited it and steered the project through the production process. The graphic artwork was prepared by Sara Hack with considerable skill and the design and layout was undertaken in an exemplary fashion by Debbie Crouch. The index was prepared by Jane Henley. We are also grateful to Giles Clark (Open University) and Susan Francis (Cambridge University Press) for their support and help with co-publication.

In addition, we wish to thank the following people who commented on earlier versions of the text: David Hughes (University of Sheffield) and Alan Penny (Rutherford Appleton Laboratory), together with anonymous referees appointed by Cambridge University Press. We should also like to acknowledge the contribution made by the following members of The Open University in commenting on early drafts of the text: Philip A. Bland, Neil McBride, Elaine A. Moore, Mike Widdowson and Ian Wright. Many other individuals and organizations furnished and/or granted permission for us to use their diagrams or photographs and to them we also express our gratitude.

Grateful acknowledgement is made to the following sources for permission to reproduce material in this book:

Cover

Background image: NASA; *Thumbnail images:* (from the left) first and second images NASA; third image © 1999 Photo Disc Inc; fourth image ESA)

Figures

Figure 1.1 Robert Thom; *Figure 1.2* © The Natural History Museum, London; *Figures 1.3–1.5 and 1.7a* Reprinted from *Origins of life on the Earth and in the Cosmos,* Zubay, G., Copyright © 2000, with permission of Elsevier Science; *Figure 1.10* 'The Pulse of Life', Lowenstein, J. M & Zihlman, L. in Gribben, J. (ed.), *A Brief History of Science,* Weidenfeld and Nicholson Ltd; *Figure 1.11* © Dan Sudia; *Figure 1.13* Courtesy of I. D. J. Burdett; *Figure 1.14* From *Biogenesis: Theories of Life's Origins* by N. Lahav, copyright © 1999 by Oxford University Press, Inc. Used by permission of Oxford University Press, Inc; *Figure 1.15* © European Space Agency; *Figure 1.16* Adapted from de Muizon et al., 1986 in Pendleton, Y. J. and Tielens, A. G. G. M. (1997) *From Stardust to Planetesimals,* Astronomical Society of the Pacific; *Figures 1.17 and 1.18* © NASA; *Figure 1.26* © Anglo-Australian Observatory, photography by David Malin; *Figure 1.30a* Sourced from www.angelfire.com; *Figure 1.30b* Sourced from University of Hamburg website, www.biologie.uni-hamburg.de; *Figure 1.31* Dr. David Deamer, UC Santa Cruz; *Figure 1.36* D. Thomson/GeoScience Features; *Figure 1.38* Lahav, N. (1999), *Biogengesis,* courtesy of Noam Lahav.

Figures 2.1, 2.3, 2.4, 2.23, 2.25: © NASA; *Figure 2.2* © NASA George Curruthers; *Figure 2.9* © The Natural History Museum; *Figure 2.15* Hammersley Iron Pty Ltd; *Figure 2.16* Andrew A. Knoll; *Figure 2.17* Professor J. W. Schopf;

Figure 2.18 Commonwealth Palaeontological Collections of the Australian Geological Survey; *Figure 2.20* Schidlowski, M. 'A 3,800-million-year isotopic record of life from carbon in sedimentary rocks', *Nature,* **333**, p.316, Macmillan Magazines. Reprinted with permission from the author; *Figures 22a and b* Simon Conway Morris, University of Cambridge; *Figure 22c and d* Peter Crimes, Liverpool University.

Figure 3.1 Courtesy Yerkes Observatory, University of Chicago; *Figure 3.3a–c* Mary Evans Picture Library; *Figure 3.4* http://www.nasm.si.edu/ceps/etp/mars/marsimg/mars_lowell.jpg. Copyright © Smithsonian, National Air and Space Museum; *Figures 3.5, 3.10, 3.11a–c, 3.12, 3.13, 3.18* NASA; *Figure 3.9* NASA/JPL/Malin Space Science Systems/USGS Flagstaff; *Figure 3.21a* Copyright © Proszynski I S-ka SA 1999–2001. Wszystkie prawa zastrzezone; *Figure 3.21c* Monica Grady; *Figure 3.21b, d, and e* Douglas A. Kurtze, North Dakota State University, Department of Physics; *Figure 3.22a and b* Everett Gibson (NASA/JSC); *Figure 3.23* Photograph courtesy of Stephen Hyde.

Figures 4.1, 4.2 and 4.8 © Science Photo Library; *Figure 4.3* © National Portrait Gallery; *Figures 4.4–4.6, 4.7, 4.10a, 4.11–4.14, 4.16, 4.17, 4.19, 4.21–4.26, 4.28–4.29 and 4.33* © NASA; *Figure 4.10b* © Calvin J. Hamilton; *Figure 4.18* From *Satellites of the Outer Planets,* 2nd edition by David A. Rothery, copyright © 2000 by David A. Rothery. Used by permission of Oxford University Press, Inc; *Figure 4.20* USGS/Cascades Volcano Observatory; *Figure 4.30* © Rob Wood/Wood Ronsaville Harlin, Inc; *Figure 4.32* Image courtesy of Marc. W. Buie/Lowell Observatory.

Figure 5.1 © Royal Astronomical Society Library; *Figure 5.2* NASA; *Figure 5.11* Painting by Duragel, courtesy of the Observatoire de Paris; *Figure 5.13* ESA; *Figure 5.14* Coustenis, A. and Taylor, F. (1999) 'Titan: The Earth Like Moon', World Scientific Publishing Ltd; *Figure 5.17* Peter H. Smith and NASA.

Figure 6.1 © The Science Photo Library; *Figure 6.5* Steve Mandel/Galaxy Images; *Figure 6.8* ESA; *Figure 6.9a* NASA; *Figure 6.9b* Jim Ferreira; *Figure 6.10* Observatoire de Paris; *Figure 6.12* The Macho Project; *Figure 6.17* Korzennik, S. Harvard University. Smithsonian Centre for Astrophysics.

Figure 7.1 Geneva Observatory; *Figure 7.2a* Marcy & Butler; *Figure 7.3* Copyright © Lynette R. Cook; *Figure 7.7* Julian Baum/Take 27 Ltd; *Figure 7.8* Pawel Artymowicz; *Figure 7.9* Pawel Artymowicz; *Figure 7.14* European Southern Observatory (ESO).

Figure 8.1 Copyright © Take 27 Ltd; *Figures 8.2, 8.3, 8.13a–e* NASA; *Figure 8.15* Copyright © 2001 by Don Dixon. All Rights Reserved. http://www.cosmographica.com

Figure 9.1 © SyracusePhotographer.com; *Figure 9.3* 8444611 Photographed by David Malin. Copyright © UKATC/AAO, Royal Observatory, Edinburgh; *Figure 9.4* US Naval Observatory, Washington DC; *Figure 9.6* Copyright © Woodruff T. Sullivan III, University of Washington; *Figure 9.7* David Parker, 1997/Science Photo Library; *Figure 9.8* NAIC, Arecibo Observatory; *Figure 9.9* Courtesy of Scientific American from 'The search for extraterrestrial intelligence' by Carl Sagan and Frank Drake. Copyright © 1975 by Scientific American, all rights reserved.

Every effort has been made to contact copyright owners. If any have been inadvertently overlooked, the publishers will be pleased to make the necessary arrangements at the first opportunity.

FIGURE REFERENCES

Abe, Y., Ohtani, E., Okuchi, T., Righter, K. and Drake, M. (2000) Water in the early Earth, in Canup, R.M. and Righter, K. (eds) *Origin of the Earth and Moon*, University of Arizona Press.

Benner, S.A., Ellington, A.D. and Tauer, A. (1989) Modern metabolism as a palimpsest of the RNA world, *Proceedings National Academy of Sciences*, USA, **86**, pp.7054–58.

Carroll, S.B. (2001) Chance and necessity: the evolution of morphological complexity and diversity, *Nature*, **409**, pp.1102–09.

Coustenis, A. and Taylor, F.W. (1999) *Titan: the Earth-like Moon*, World Scientific, Singapore.

de Muizon, M., Gabelle, T.R., d'Hendecourte, L. and Baas, F. (1986) in Pendleton, Y.J. and Tielens, A.G.G.M. (1997) *From Stardust to Planetesimals*, Astronomical Society of the Pacific, San Fransisco, Calif.

Kasting, J.F., Whitmore, D.P., Reynolds, R.T. (1993) Habitable Zones around Main Sequence Stars, *Icarus*, **101**, p.108.

Lahav, N. (1999) *Biogenesis: theories of life's origin*, Oxford University Press, Oxford.

Lowenstein, J.M. and Zihlman, A. (1998) The Pulse of Life, in Gribbin, J. (ed.) *A Brief History of Science*, Weidenfeld and Nicholson Ltd.

Melosh, H.J. (1989) *Impact Cratering, a Geological Process*, Oxford University Press, Oxford.

Porcelli, D. and Pepin, R.O. (2000) Rare gas constraints on early Earth history, in Canup, R.M. and Righter, K. (eds) *Origin of the Earth and Moon*, University of Arizona Press.

Rothery, D.A. (2000) *Satellites of the Outer Planets*, Oxford University Press, Oxford.

Rothschild, L.J. and Mancinelli, R.L. (2001) Life in extreme environments, *Nature*, **409**, pp.1092–101.

Sagan, C. and Drake, F. (1975) The search for extraterrestrial intelligence, *Scientific American*, **232**, pp.80–89.

Schidlowski, M., Hayes, J.M. and Kaplan, I.R. (1983) Isotopic inferences of ancient biochemistries: carbon, sulfur, hydrogen, and nitrogen, in Schopf, J.W. (ed.) (1983) *Earth's Earliest Biosphere: Its Origin and Evolution*, Princeton University Press, Princeton, New Jersey.

Schopf, J.W. (ed.) (1983) *Earth's Earliest Biosphere: Its Origin and Evolution*, Princeton University Press, Princeton, New Jersey.

Wallace, J.M. (1977) *Atmospheric Science*, Academic Press, San Diego.

Zolotor, M.Y. and Shock, E.L. (2001) Composition and stability of salts on the surface of Europa and their oceanic origin, *J Geophys Res*, **106**, p.32815–27.

Zubay, G.L. (2000) *Origins of Life on the Earth and in the Cosmos*, Academic Press, San Diego.

INDEX

Entries and page numbers in **bold type** refer to key words which are printed in **bold** in the text. Italics indicate items mainly, or wholly, in a figure or table.